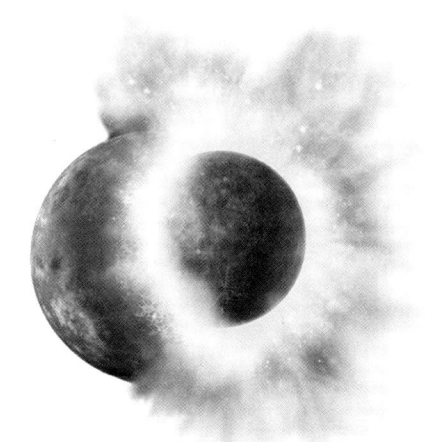

On the
Origin of Planets
By Means of Natural Simple Processes

Montante Family Library
D'Youville College

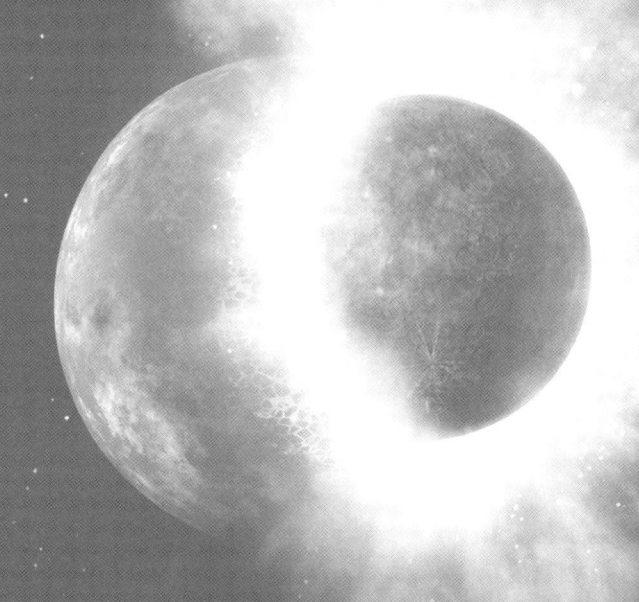

On the Origin of Planets

By Means of Natural Simple Processes

Michael M. Woolfson
University of York, UK

Imperial College Press

Published by

Imperial College Press
57 Shelton Street
Covent Garden
London WC2H 9HE

Distributed by

World Scientific Publishing Co. Pte. Ltd.
5 Toh Tuck Link, Singapore 596224
USA office: 27 Warren Street, Suite 401-402, Hackensack, NJ 07601
UK office: 57 Shelton Street, Covent Garden, London WC2H 9HE

British Library Cataloguing-in-Publication Data
A catalogue record for this book is available from the British Library.

ON THE ORIGIN OF PLANETS
By Means of Natural Simple Processes

Copyright © 2011 by Imperial College Press

All rights reserved. This book, or parts thereof, may not be reproduced in any form or by any means, electronic or mechanical, including photocopying, recording or any information storage and retrieval system now known or to be invented, without written permission from the Publisher.

For photocopying of material in this volume, please pay a copying fee through the Copyright Clearance Center, Inc., 222 Rosewood Drive, Danvers, MA 01923, USA. In this case permission to photocopy is not required from the publisher.

ISBN-13 978-1-84816-598-4
ISBN-10 1-84816-598-6
ISBN-13 978-1-84816-599-1 (pbk)
ISBN-10 1-84816-599-4 (pbk)

Typeset by Stallion Press
Email: enquiries@stallionpress.com

Printed in Singapore by Mainland Press Pte Ltd.

Preface

I began to write this book in the year 2009, the 200th anniversary of the birth of Charles Darwin and the sesquicentenary of the publication of *The Origin of Species*. The BBC has commemorated this double anniversary with two splendid series of television programmes, one highlighting the experimental side of Darwin's work and the other the impact made by his theory of natural selection on both the scientific and religious establishments of his time and on political philosophies until quite recently. Come to think of it, the impact can be felt even today. There are still many who believe that the world and all it contains was made some 6,000 years ago by an act of divine creation and that, in particular, mankind was created in its present form. Such fundamentalist views are quite common in the United States but perhaps less so in most European countries.

Stimulated by the BBC programmes, I looked again at my copy of *The Origin of Species*, the sixth edition published in 1888. Included on the title page is the line 'THIRTY-THIRD THOUSAND', the total number of copies printed by John Murray, the publisher, up to that date. This is quite a modest print run for a work of such importance and I find it surprising that so few copies of the book could have made such a great impact. Probably the newspaper cartoons depicting Darwin's head on the bodies of apes made more of an impression on the public than the book itself, a lengthy scholarly presentation that few laymen would have regarded as attractive reading. Browsing through it, I am struck by the detail in the presentation of the arguments in favour of natural selection. Darwin was worried about the consequences of presenting his theory, fearing adverse reactions from both the clergy and scientists. To defend himself against potential criticism he laid out his ideas in meticulous detail, building his case step by step in a logical progression. Even so, it took the intervention of Alfred Russel Wallace, a young English anthropologist living in the East Indies at the time, to push Darwin into publication. In 1858 Darwin received from Wallace a draft paper which described ideas on evolution and natural selection very similar to his own. To his credit, Darwin helped in the publication of this paper but covered his own proprietary rights on the theory by arranging for a presentation of both his and Wallace's ideas to the Linnean Society of London, the world's leading society devoted to natural history, in July 1858. One is left wondering when, and if, *The Origin of Species* would have appeared without Wallace's influence on events.

A striking aspect of Darwin's style is that it is written in the first person, so the reader feels as if the narrative is related to him personally — at least this is the impression I have. The kinship I feel with Darwin is a particular one. Much of my scientific work has been in a field where the word 'origin' has relevance; in my case it is the origin of planets. Darwin took 20 years to develop his ideas while I have taken 47 years to develop my own. Darwin's ideas on evolution were themselves evolving as he uncovered more and more evidence to support the basic concept of natural selection. My own ideas have evolved as observational evidence has revealed more and more facts about the presence of planets in our galaxy and the locations in which stars, and presumably planets, are formed. Let it not be thought that I am claiming equal importance for the topic of planet formation as that of natural selection. The Universe is full of material and finding out how this is assembled by physical processes to form planets is mundane and humdrum compared with how new species of both animate and inanimate life evolved from simple forms through to creatures as complex as ourselves. However, there are problems even more fundamental than those addressed by Darwin; natural selection describes how new types of life develop from previous forms but does not deal with the origin of life in the simplest forms of single-celled bacteria and archaea. The origin of life and the origin of the Universe — these are the really important, fundamental problems to which, as yet, we have no solutions.

Humdrum or not, the problem of how planets originated is regarded to be of sufficient importance for space agencies worldwide to assign considerable resources to support projects that make observations that might help to solve this problem. Even if it is not one of the great fundamental problems of our age, looking for the mechanisms by which planets may have formed is of interest and has exercised, and still exercises, the minds of many scientists, especially during the last 200 years.

Looking at Darwin's great opus, I am inspired to take it as a model and, to the best of my ability, to emulate his thoroughness in describing my own ideas in *The Origin of Planets*. There must, of course, be important differences between the general form of presentation of *The Origin of Species* and the one I shall be using. A treatise on natural history can be cast in everyday language and *The Origin of Species* deals with the problems of the general reader by providing a glossary of scientific terms at the end of the book. By contrast, astronomy, dealt with in a formal way, must be expressed in terms of physics and mathematics — foreign languages to the majority of non-specialist readers. Many of the physical relationships that describe astrophysical processes are most clearly expressed in the form of equations and would be impossibly difficult to describe in words, except in a most handwaving way. Another difference is that, whereas Darwin made do with a single simple figure illustrating an evolutionary tree, modern printing technology allows for the prolific use of pictures and diagrams to assist the reader. I shall try to walk the tightrope of giving a proper formal presentation of the science while also expressing ideas in words that should convey the essence of the science to a non-specialist. In place of Darwin's glossary is a series of appendices. Some of these are designed to help the non-specialist reader to understand the physical concepts and terminology; others will derive physical relationships and will be of more interest to the specialist reader.

I shall begin this book in much the style that Darwin did in *The Origin of Species*, with an historical review of the important theories and ideas that have gone before. Thereafter I shall revert to a more modern format, although I shall try to maintain the detailed and rigorous reasoning that Darwin employed. Will I succeed in spanning the divide between the professional scientist and the non-specialist? We shall see.

Contents

Preface   v

An Historical Sketch of the Progress of Opinion on the Origin of Planets   xix

1. Observations of Stars   1
   - 1.1 Locations of Stars   1
   - 1.2 Stellar Material   4
   - 1.3 Determining the Distances of Stars   5
     - 1.3.1 The distances of nearby stars   5
     - 1.3.2 Distance measurements using variable stars   7
   - 1.4 The Temperature of Stars   9
   - 1.5 Stellar Radii   12
   - 1.6 Estimating Stellar Masses   13
   - 1.7 The Physical Properties of Main-Sequence Stars   14
   - 1.8 Stellar Spin Rates   15
   - 1.9 Summary   15

2. Producing Protostars — Embryonic Stars   19
   - 2.1 Star-Forming Regions   19
   - 2.2 The Formation of Dense Cool Clouds   21
   - 2.3 Maser Emission from Star-Forming Regions   26
   - 2.4 The Process of Protostar Formation   27
   - 2.5 The Formation of Binary Systems   28

	2.6	Modelling the Collapse of a Cloud	32
	2.7	The Spin of Stars	33
	2.8	Summary	36
3.	The Life and Death of a Star		39
	3.1	The Journey to the Main Sequence	39
	3.2	Energy Generation in Main-Sequence Stars . .	42
	3.3	Leaving the Main Sequence for Low- and Moderate-Mass Stars	46
	3.4	The Evolution of Higher-Mass Stars	50
	3.5	Summary	53
4.	The Evolution of a Galactic Cluster		55
	4.1	Embedded Clusters	55
	4.2	The Formation of Massive Stars	58
	4.3	The Embedded Cluster Environment and Binary Star Frequencies	60
	4.4	The Progress of Star Formation in a Galactic Cluster	61
	4.5	Summary	62
5.	Exoplanets — Planets Around Other Stars		65
	5.1	Planets Orbiting Neutron Stars	65
	5.2	The Characteristics of Orbits	68
	5.3	Planets Around Main-Sequence Stars; Doppler-Shift Detection	69
	5.4	The Direct Imaging of Exoplanets	77
	5.5	Exoplanets and the Solar System	82
	5.6	Summary	83
6.	The Formation of Planets: The Capture Theory		85
	6.1	The Interaction of a Star with a Protostar . . .	86
	6.2	The Interaction of a Star with a High-Density Region	90
	6.3	Summary	93

7.	Orbital Evolution	95
	7.1 The Nature of the Disk	96
	7.2 The Force on a Planet Due to the Medium	99
	7.2.1 Viscosity-based resistance	99
	7.2.2 Mass-based resistance	100
	7.3 Modelling the Medium and Details of the Calculation Method	101
	7.4 Calculations of Orbital Decay and Round-off	105
	7.5 Orbits of High Eccentricity	107
	7.6 The Range of Semi-Major Axes	113
	7.7 Simple Ratios of Orbital Periods	113
	7.8 Stellar Spin Axes	114
	7.9 Summary	117
8.	The Frequency of Planetary Systems	119
	8.1 Observations and Observational Constraints	119
	8.2 Initial Formation Statistics	122
	8.3 The Disruption of Planetary Systems	126
	8.4 Summary	129
9.	Satellite Formation	131
	9.1 Angular Momentum Considerations	132
	9.2 The Form of the Disk	135
	9.3 The Settling of Dust	137
	9.4 The Formation of Satellitesimals	141
	9.5 Satellite Formation	144
	9.6 Comments	147
	9.7 Summary	148
10.	Features of the Solar System	151
	10.1 The Planets	151
	10.1.1 The terrestrial planets	152
	10.1.2 The major planets	154
	10.1.3 Tilts of planetary spin axes	155

	10.2	Satellites	157
		10.2.1 The satellites of Jupiter	158
		10.2.2 The satellites of Saturn	162
		10.2.3 The satellites of Uranus	167
		10.2.4 The satellites of Neptune	169
		10.2.5 Other satellites	171
	10.3	Dwarf Planets and the Kuiper Belt	172
	10.4	Asteroids	177
		10.4.1 Types of asteroids and their orbits . . .	177
		10.4.2 The composition of asteroids	180
	10.5	Comets	181
	10.6	Summary	183
11.	**Interactions Between Planets**		**185**
	11.1	The Precession of Planetary Orbits	185
	11.2	Close Interactions of Planets and the Tilts of Spin Axes	186
	11.3	The Problem of the Terrestrial Planets	190
	11.4	Deuterium and the Major Planets	191
	11.5	Earth and Venus	196
	11.6	Summary	201
12.	**The Moon**		**203**
	12.1	The Earth–Moon Relationship	203
	12.2	Satellites of the Colliding Planets	206
	12.3	Features of the Moon	208
	12.4	The Hemispherical Asymmetry of the Moon . .	212
	12.5	The Evolution of the Moon's Orbit	215
	12.6	Summary	217
13.	**Mars and Mercury**		**219**
	13.1	Larger Solid Bodies of the Solar System	219
	13.2	Mars as a Satellite	221
		13.2.1 The hemispherical asymmetry of Mars	221

		13.2.2	Mars — now and in the past	223
		13.2.3	The Martian spin axis and hemispherical asymmetry	226
	13.3	Mercury as a Satellite		228
	13.4	The Orbits, Spins and Tilts of Mercury and Mars		232
	13.5	Summary		233
14.	Neptune, Triton and Pluto			235
	14.1	The Neptune–Pluto Relationship		235
	14.2	The Strange Satellites of Neptune		235
	14.3	The Neptune–Triton–Pluto Relationship Explained		237
	14.4	Summary		239
15.	Dwarf Planets, Asteroids, Comets and the Kuiper Belt			241
	15.1	Dwarf Planets		241
		15.1.1	Ceres	241
		15.1.2	The outer dwarf planets	242
	15.2	Asteroids and Comets		245
		15.2.1	Asteroids	245
		15.2.2	Comets and the Kuiper Belt	248
		15.2.3	Long-period comets and the Oort cloud	249
		15.2.4	The survival of the Oort cloud	251
	15.3	Summary		252
16.	Meteorites: Their Physical and Chemical Properties			255
	16.1	The Broad Classes of Meteorites		256
	16.2	The Physical and Chemical Characteristics of Meteorites		258
		16.2.1	Stony meteorites	258
		16.2.2	Iron meteorites	262
		16.2.3	Stony-iron meteorites	263

	16.3	Interpreting the Physical Properties and Appearance of Meteorites	264
	16.4	Summary	266

17. Isotopic Anomalies in Meteorites — 269

	17.1	Isotopes and Anomalies	269
	17.2	The Planetary Collision and Nuclear Reactions	271
	17.3	Explanations of the Anomalies	275
	17.4	Individual Isotopic Anomalies and How They Are Produced	275
		17.4.1 The oxygen anomaly	276
		17.4.2 The magnesium anomaly	279
		17.4.3 Neon in meteorites	282
		17.4.4 Anomalies associated with silicon carbide	284
		17.4.4.1 Silicon in silicon carbide	284
		17.4.4.2 Carbon and nitrogen in silicon carbide	286
		17.4.4.3 Neon in silicon carbide	288
	17.5	General Remarks Concerning Isotopic Anomalies	291
	17.6	Summary	291

18. Overview and Conclusions — 295

	18.1	What Constitutes a Good Theory?	295
	18.2	Protostars and Stars	298
	18.3	Creating the Conditions for the Capture-Theory Process	300
	18.4	The Capture-Theory Process	301
	18.5	The Frequency of Planetary Systems	302
	18.6	Satellite Formation	303
	18.7	The Tilts of Spin Axes of the Planets and Stars	303
	18.8	A Planetary Collision — Earth and Venus . . .	304

18.9	The Moon, Mars and Mercury	305
18.10	Neptune, Triton and Pluto	307
18.11	Small Bodies of the Solar System	307
18.12	The Characteristics of Meteorites	308
18.13	Conclusions	309

Appendix A: Angular Momentum   313

Appendix B: Equipotential Surfaces of a Tidally Distorted Star   316

Appendix C: The Instability of a Gaseous Filament   318

Appendix D: The Jeans Critical Mass   320

Appendix E: The Lynden-Bell and Pringle Mechanism   322

Appendix F: Grains in Molecular Clouds   324

Appendix G: The Structure of a Spiral Galaxy   328

Appendix H: The Centre of Mass and the Orbits of Binary Stars   330

Appendix I: The Doppler Effect   333

Appendix J: Atomic Energy Levels and Stellar Spectra   335

Appendix K: Stellar Masses from Observations of Binary Systems   338

Appendix L: Smoothed-Particle Hydrodynamics   341

Appendix M: Free-Fall Collapse   345

Appendix N: Fragmentation and Binary Characteristics	348
Appendix O: Spin Slowing Due to a Stellar wind	351
Appendix P: The Virial Theorem and Kelvin–Helmholtz Contraction	353
Appendix Q: The Lifetime of Stars on the Main Sequence	356
Appendix R: The Eddington Accretion Mechanism	358
Appendix S: The Mass and Orbit of an Exoplanet	360
Appendix T: Radiation Pressure and the Poynting–Robertson Effect	361
Appendix U: Active Stars and Their Effect on a Stellar Disk	364
Appendix V: The Structure and Decay of a Stellar Disk	369
Appendix W: The Formation of Exoplanets	372
Appendix X: Disrupting a Planetary System	376
Appendix Y: From Dust to Satellitesimals	383
Appendix Z: From Satellitesimals to Satellites	387
Appendix AA: The Tidal Heating of Io	391
Appendix AB: The Trojan Asteroids	395
Appendix AC: Orbital Precession	399

Appendix AD: The Temperature Generated by Colliding Planets	401
Appendix AE: Heating by Deuterium-Based Reactions	414
Appendix AF: The Thermal Evolution of the Moon	416
Appendix AG: The Abrasion of a Hemisphere of the Moon	420
Appendix AH: The Rounding-off of a Highly Eccentric Satellite Orbit	422
Appendix AI: Continental Drift on Mars	426
Appendix AJ: The Oort Cloud and Perturbing Stars	429
Appendix AK: Planetary Perturbation of New Comets	433
Appendix AL: Reactions and Decays	435
Appendix AM: Cooling and Grain Formation	440
Index	445

An Historical Sketch of the Progress of Opinion on the Origin of Planets

Ptolemy and Copernicus

It is only since 1995 that the presence of planets around stars other than the Sun has been detected. I shall have more to say about this later. Early work on planet formation was concerned just with the Solar System since it was not known whether or not planets existed around other stars. It was suspected that other planets might exist and, indeed, this was a proposition put forward by Giordano Bruno (1548–1600), an Italian Dominican monk and philosopher. He supported the Copernican theory that the Earth, along with all the other planets in the Solar System, orbited the Sun and not, as the previously well-established Ptolemaic system proposed, that the Earth was the centre of the Universe, orbited by every heavenly body. In Bruno's view, the Sun was just a star, like other stars, and these other stars, like the Sun, had planets accompanying them. He might have got away with that view — at the time the Roman Catholic Church seemed to be quite relaxed about the Copernican model of the Solar System. Copernicus (1473–1543), sensing that his ideas might provoke some reaction from the Church, had taken the precaution of dedicating the work describing his theory, *De Revolutionibus Orbium Coelestium* (On the Revolutions of the Heavenly Spheres), to Pope Paul III to deflect any possible criticism. However, Bruno went further and claimed that these planets around stars contained other races of men, and that claim challenged the biblical view of the central role of mankind, created in the image of God. This was only one, and possibly not a major one, of Bruno's many blasphemies, which included the rejection of the divinity of Jesus. Bruno was brought before the Inquisition and was ordered to renounce his heretical views, but, despite being given many opportunities to do so, he refused to comply. As a result of his obduracy, he was condemned to death and burnt at the stake. It seems likely that this event, together with others, gradually contributed to a view by the Church that the Copernican model was a source of heresy and in 1616 *De Revolutionibus* was added to the *Index Librorum Prohibitorum*, the list of books that Catholics were forbidden to read, where it remained until 1758.

The condemnation of the Copernican model of the Solar System had repercussions for one of the greatest scientists of the 17th century, Galileo Galilei (1564–1642). In 1608 Galileo made a telescope, an instrument that had recently been invented by a spectacle-maker, Hans Lippershey, in Holland. With his telescope, Galileo made a number of astronomical observations, studying mountains on the Moon and the rings of Saturn (although he did not recognize them for what they were), and he discovered the four large satellites of Jupiter, now collectively referred to as the *Galilean satellites* in his honour. However, his most significant observations were of the planet Venus. Sometimes Venus is seen as a complete disk, but distant and small in size, and at other times in a crescent phase, but nearer and hence much larger in extent. This was not something that the Ptolemaic theory would allow since, for that theory, Venus would always be between Earth and the Sun and close to the line joining those two bodies. In that arrangement Venus would always be seen as a crescent. By contrast, the observations were entirely consistent with the Copernican model. When Venus is at its greatest distance from Earth, with the Sun between the two planets, then Venus appears as a small disk. When Venus and Earth are on the same side of the Sun, then Venus is between Earth and the Sun and is seen as a large crescent. Galileo was in a quandary. He was a devout man, despite fathering a number of illegitimate children, and he had no wish to offend the Church. But he was also a scientist who believed in the conclusions that followed from his observations. What could he do? He had an idea: he would write a book. The book, published in 1632, was entitled *Dialogue Concerning the Two Chief World Systems* (Fig. I), and was in the form of a discussion between two men, Simplicio, who supported the Ptolemaic model, and Salviati, who championed the Copernican model. Also present was a third individual, Sagredo, who asked questions and acted as a catalyst to maintain the discussion. To an intelligent reader there could be no doubt who got the better of the debate — and, for all their fundamentalist fervour, the Inquisitors who read the book were very intelligent readers! They were not taken in by this pretence of presenting a disinterested neutral discussion and Galileo was arrested and arraigned before the Inquisition. He was given a simple choice: recant or suffer the consequences. Galileo was a sensible and

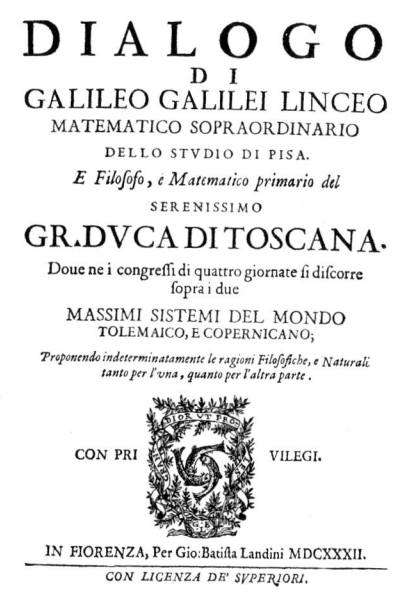

Fig. I. The frontispiece and title page of *Dialogue Concerning the Two Chief World Systems*.

practical man — he was no Bruno — and he gave way to brute force and renounced his heretical views. It was as well for him that the Inquisitors could hear his voice but could not read his mind! Despite his concession to the demands made on him, Galileo did not get away scot-free and spent the rest of his life under house arrest. Fortunately, truth prevailed in the end and, through the work of outstanding scientists such as Johannes Kepler (1571–1630) and Isaac Newton (1642–1727), the Copernican heliocentric model eventually became firmly established. Theories of the origin of the Solar System that were produced from the 18th century onwards were theories that attempted to explain the presence of planets orbiting the Sun.

I shall now briefly describe a selection of theories from the last 200 years or so. There have been many theories presented in that period and to describe them all would add confusion rather than clarity to an understanding of what the main difficulties are in producing a viable theory. Just four theories will be described that, taken together, establish the main problems and illustrate the range of physical mechanisms that have been proposed to solve those problems.

Laplace and the Nebular Model

Early attempts by eminent scientists — for example, René Descartes (1596–1650) and Immanuel Kant (1724–1804) — to explain the formation of planets did not have sound scientific foundations and had little influence on subsequent work. The first theory with serious scientific credentials was put forward in 1796 by the eminent French mathematician and astronomer, Pierre-Simon Laplace (1749–1827; Fig. II), in his book *Exposition du Système du Monde*. The starting point of Laplace's model was a gaseous nebula, a giant, slowly spinning sphere of gas and dust that was collapsing under the influence of self-gravitational forces. The development of the sphere is shown in Fig. III. As the material of the sphere fell inwards towards the spin axis, the rate of spin had to increase to conserve *angular momentum* (Appendix A). At first the increase in angular speed led to the gas taking the shape of an oblate spheroid, like a flattened football, then it became ever flatter and smaller, eventually taking

Fig. II. Pierre-Simon Laplace.

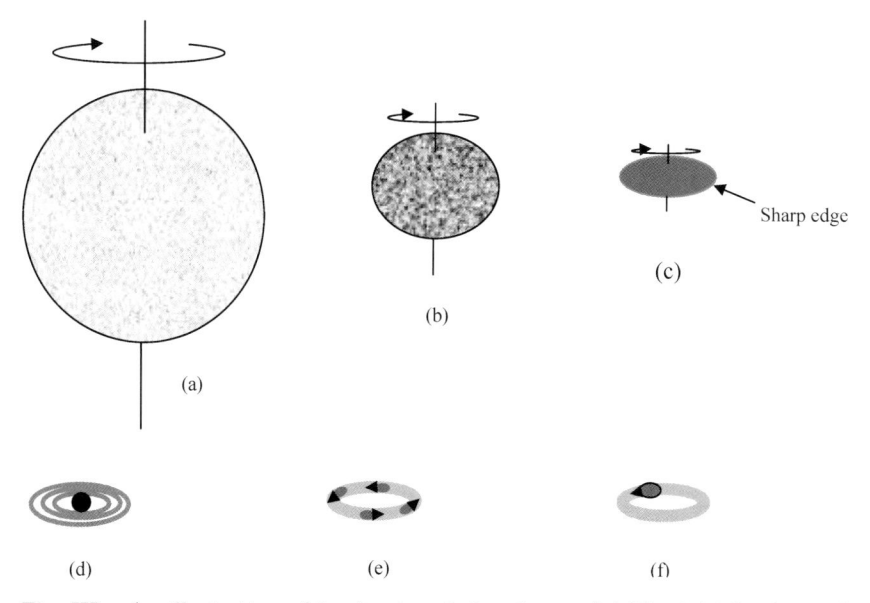

Fig. III. An illustration of Laplace's nebular theory. (a) The initial sphere of gas and dust (b) Forming an oblate spheroid (c) A lenticular shape with a sharp edge (d) The formation of gas rings (e) Condensations forming within a ring (f) Coalescence to form a planetary blob.

a lenticular shape (like a lens) with a sharp edge. At this stage the material at the edge was in free orbit around the central mass and further collapse of the gas left this material in orbit. More and more material was left behind in the equatorial plane of the spinning mass and Laplace postulated that, under the influence of gravitational forces, this would have formed a series of concentric gaseous rings. As the central material collapsed to form the Sun, each gaseous ring became the source of a planet. Gravitational forces caused the material of each ring to form a series of condensations and, because these condensations orbited the central mass at slightly different rates, all the condensations in one ring eventually coalesced to form a planet.

This theory was widely accepted by the astronomical community. Its attraction was that, by spontaneous processes, the Sun and the planets could form together from a very plausible starting point. It explained why all the planets orbited in the same sense and why the orbits were coplanar, at least to a first approximation. However, towards the middle of the 19th century, serious doubts were being expressed about the nebular theory. The main problem, which could be expressed in a number of different ways, concerned the physical quantity angular momentum. The Sun is by far the dominant member of the Solar System, comprising 99.86% of its total mass. However, it spins slowly at a rate that varies with latitude, with a spin period of 25 days at the equator and increasing to 33 days near the poles. Although the masses of the planets are much smaller than that of the Sun, and their orbital angular speeds are lower than its spin angular speed, the radii of their orbits are so large that they contribute by far the major part of the angular momentum of the Solar System — about 99.5% of the total. That was the problem. It was difficult to envisage how Laplace's nebula could evolve in such a way that 0.14% of the mass ended up with 99.5% of the angular momentum.

Another way of expressing this problem is to consider what the total angular momentum of the whole nebula would have been if, at the time it had the radius of Neptune's orbit, its period of spin equalled that of Neptune's orbital period — which it would have needed to be in order to release a ring in the right place to produce Neptune, the outermost planet. Whatever extreme assumptions are made about the distribution of mass in the nebula, this condition would give the Solar System an angular momentum hundreds of times its present one. A third and different way of expressing the problem is to imagine the material of the Solar System with its present angular momentum turned into a spherical nebula with the radius of Neptune's orbit. The angular speed at the boundary would then be far less than the orbital speed, so no ring would be left behind and hence Neptune would not form.

There were various attempts to rescue the theory. Another French astronomer, Édouard Roche (1820–1883), who made many important contributions to celestial mechanics, proposed extreme distributions of material in the nebula to explain the way that mass and angular momentum were partitioned. However, this meant there was insufficient mass at the outside of the system to explain the outer planets. By the end of the 19th century, it was clear that Laplace's nebular theory was not the answer; some other theory was needed to explain the origin of the planets.

The Chamberlin–Moulton Theory

In the mid-19th century, astronomical structures called spiral nebulae (Fig. IVa) were observed. We now know that these are galaxies containing billions of stars but they were first observed at low resolution and could be interpreted as fuzzy gaseous clouds surrounding a central body that could have been a star. Two American scientists, the geologist Thomas Chamberlin (1843–1928) and the astronomer Forest Moulton (1872–1952), linked the appearance of spiral nebulae with the phenomenon of solar prominences, large eruptions of matter from the solar surface (Fig. IVb) that eventually loop back to rejoin the Sun.

The idea put forward by Chamberlin and Moulton was that, at a time when the Sun was producing particularly large prominences, the tidal force from a passing star pulled prominence material outwards and then, through the action of its gravitational force, gave it

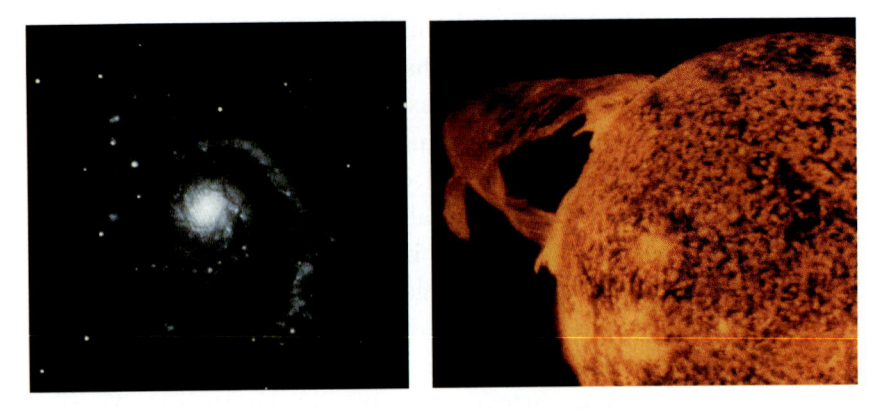

Fig. IV. (a) A spiral nebula. (b) A solar prominence (SOHO, NASA, ESA).

sufficient energy and angular momentum to be left in orbit. This material would have been orbiting more slowly at larger distances so the streams of matter pulled out of the Sun would have wound up to give the appearance of the arms of a spiral nebula.

The streams of matter pulled from the Sun were assumed to have been in the form of bunches of matter, due to the loss of solar material in a spasmodic series of pulses. Within each bunch, material would quickly cool, forming solid particles and liquid droplets that would eventually coalesce to form the individual planets.

The theory was of a rather handwaving kind with virtually no detailed analysis to back it up. It had a large number of weaknesses which its detractors described in great detail, an obvious one being that close interactions between stars would be far too rare to explain the numbers of spiral nebulae being observed. Once the true nature of spiral nebulae was established, the whole basis of the theory was lost. Its scientific credentials were always weak but I mention it here because it introduced an important new concept into the topic of theories of planet formation, the idea of tidal forces.

The Jeans Tidal Theory

If the Chamberlin–Moulton theory was bereft of sound analysis, the next one I describe is based on a solid bedrock of theory — theory that has had, and still has, a wide range of astrophysical applications.

Fig. V.   James Hopwood Jeans.

The originator of all this, and of a new model of planet formation, first formulated in 1916, was the British astronomer James Jeans (1877–1946; Fig. V). The Jeans model of planet formation was loosely based on that of Chamberlin and Moulton, except that solar prominences were not required to assist in the removal of solar material. Material was removed through the direct action of the tidal force due to a passing massive star on the surface material of the Sun. The material came off the Sun in the form of a filament that was pulled by the gravitational attraction of the passing star into a solar orbit. The filament became unstable and broke up into a series of blobs, rather like beads on a string, and each blob then collapsed under self-gravitational forces to form a planet. The sequence of events is illustrated in Fig. VI.

There are three distinct steps in this process of planet formation:

(i) the tidal distortion of the Sun and the formation of an escaping filament,
(ii) the break up of the filament into a string of blobs,
(iii) the collapse of a blob to form a planet.

Jeans developed a comprehensive theory to deal with each of these steps.

Step (i)

When stars approach each other, each becomes distorted due to the tidal effect of the other. Jeans found the form of this distortion

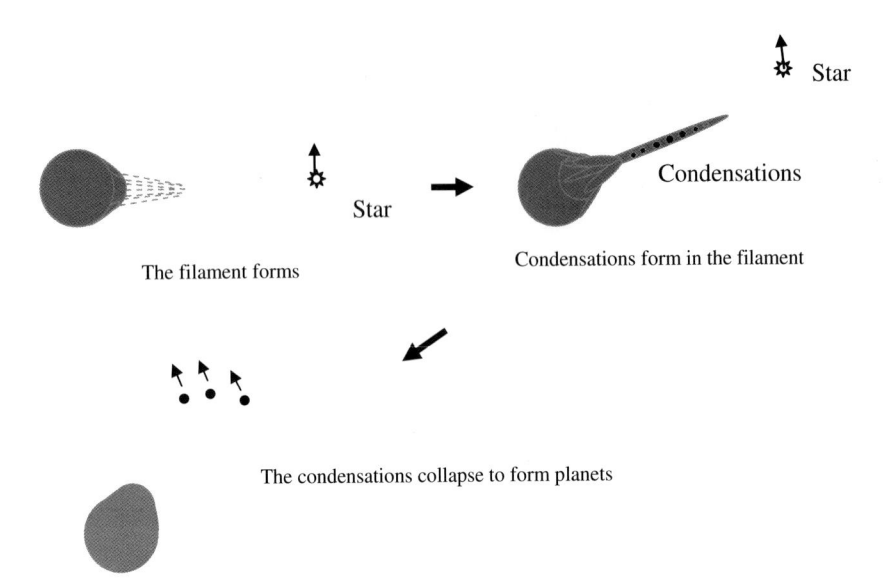

Fig. VI. Jeans' tidal theory.

[Illustration labels: Star; The filament forms; Star; Condensations; Condensations form in the filament; The condensations collapse to form planets]

under the assumption that the mass of a star is concentrated at its centre and that its volume stays constant when it is distorted. Both conditions are closely true for stars. They are very centrally condensed, which is to say that most of their mass is close to the centre. Also, not too far from the surface, the pressures are very high and gases under high pressure are very incompressible, meaning that they behave rather like liquids and their volume remains nearly constant. Figure VII shows how the profile of a solar-mass star changes as another star of two solar masses approaches ever closer.

These contours are projections of equipotential surfaces (Appendix B), and represent a necessary condition for the star to be in a stable configuration under the prevailing tidal force. To give an obvious, everyday example of an equipotential surface, we take the flat, horizontal surface of a lake under terrestrial gravitational forces. If the surface were other than horizontal and flat, the water would flow until it became so. For the distorted stars shown in Fig. VII, if they had any other shape, the material would flow until the star took up the indicated shape. It can be noted that the final profile, corresponding to a distance of 2.99 solar radii between the star

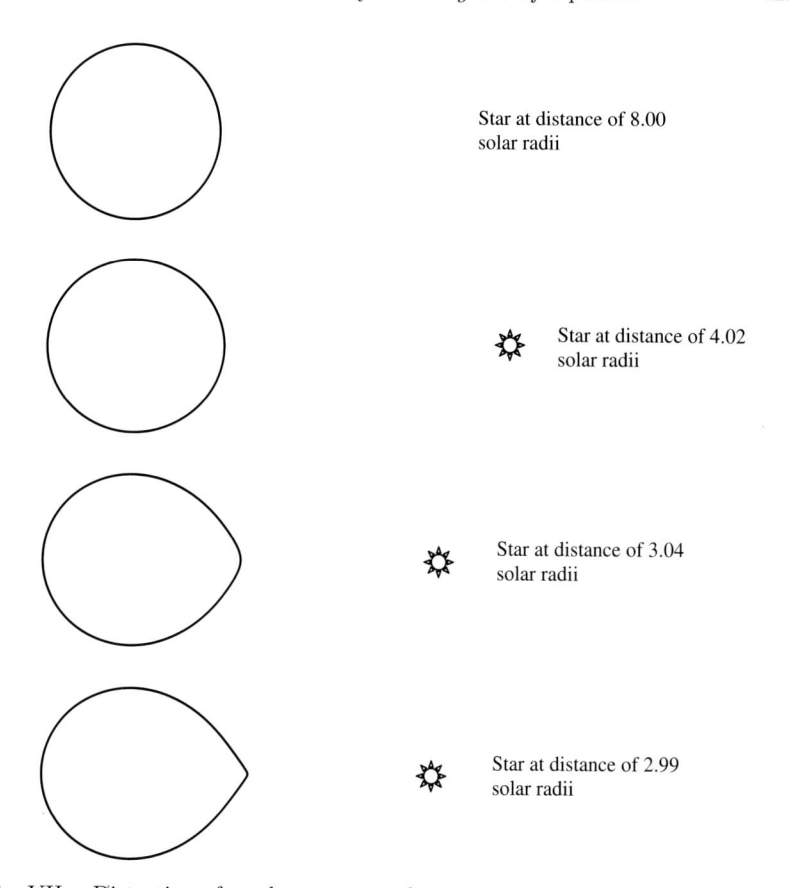

Fig. VII. Distortion of a solar mass star by one of two solar masses for different separations of the star centres.

centres, forms a point directed towards the tide-raising star. This is a critical contour; if the distance between the stars is any less, then there is no closed equipotential surface enclosing a sufficient volume to contain the material of the star. Once the tide-raising star approaches closer than this critical distance, the solar-mass star ejects material in the form of a filament from the region of the point.

Step (ii)

By a purely analytical approach, Jeans showed that a uniform gaseous filament would be unstable and would tend to break up into a string of higher density regions. The basis of this can be understood

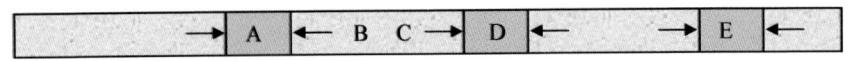

Fig. VIII. The formation of a string of high-density regions due to a density enhancement at A.

by a consideration of how an originally uniform gas stream would react to an enhancement of density at some position. Figure VIII shows a gas stream with an enhancement of density at region A. Gas on either side of A, with an imbalance of the forces acting on it, is attracted towards A and this effect is strongest for the gas closest to A. This creates a deficit of mass in a region such as B and hence gas at C will tend to move away from B, creating a higher density at D where gas is not moving away so quickly from B. Now the high-density region at D acts like that at A and creates another higher density region at E, and so on throughout the stream on both sides of A.

Jeans' analysis showed that the distance between the higher density regions was proportional to the speed of sound in the gas and inversely proportional to the square root of the density of the gas (Appendix C). The analysis assumes an initially uniform stream of gas of uniform density, a condition that nature is unlikely to provide in practice. I shall show a very convincing illustration of this instability phenomenon, produced as part of a numerical simulation, in Chapter 6.

Step (iii)

The next part of the tidal theory requires the high-density regions produced in the filament to collapse spontaneously to form planets. To find the conditions under which this would happen, Jeans considered the behaviour of a spherical mass of gas of uniform density and temperature. While, once again, this is a rather idealized model, it does indicate the general behaviour pattern of more realistically shaped clouds of gas. There are two different kinds of force acting. The first is an inwardly acting self-gravitational force exerted by the sphere of gas on itself. For a fixed density of gas, this just depends on the radius of the sphere and hence on its total mass. The second is the outwardly acting force due to the pressure of the gas, which is just like the pressure of the gas in a balloon that forces outwards against the balloon's surface. This pressure depends on the product of density and temperature (on an absolute scale[1]) of the gas. For a small sphere, the outward force would prevail and the gas would dissipate. As the radius, and hence mass, of the sphere increased so the ratio of the inward force to outward force would increase until, at a certain mass, the forces were equal. Now the gas is at a critical condition where it will neither collapse nor expand; this mass, which depends on the density, temperature and nature of the gas, is known as the *Jeans critical mass*. Any mass greater than the Jeans critical mass would start to collapse. For a particular gas the Jeans critical mass is proportional to $\sqrt{T^3/\rho}$ where T is the absolute temperature of the gas and ρ its density and the constant of proportionality depends on the composition of the gas (Appendix D). Figure IX shows the variation of the Jeans critical mass for various temperatures and densities for a mixture of hydrogen and helium typical of the atmosphere of a planet such as Jupiter. If a condensation in the filament had a mass exceeding the Jeans critical mass for the prevailing density and temperature, then it would collapse. For example, a gaseous sphere of mass 10^{30} kg and density 10^{-10} kg m^{-3}, represented by the point X in Fig. IX, would collapse at a temperature of 100 K but dissipate at a temperature of 1,000 K.

The Jeans model of planet formation, underpinned by some very elegant theory, was widely accepted by the astronomical community and others, until 1935. Prior to that, eminent British geophysicist Harold Jeffreys (1891–1989), who had originally been a strong supporter of the tidal theory, spoke out against it on the grounds that the passing star would have to be very massive and the probability that such a star would approach closely enough to the Sun, even given a long period of time, would be infinitesimally small. It was not a strong argument. No other planetary systems were known at

---

[1]The absolute, or Kelvin, temperature scale is one in which the lowest theoretical temperature is 0 K and the increments of temperature are the same as those of the Celsius scale. On the Celsius scale the absolute zero of temperature is −273 C.

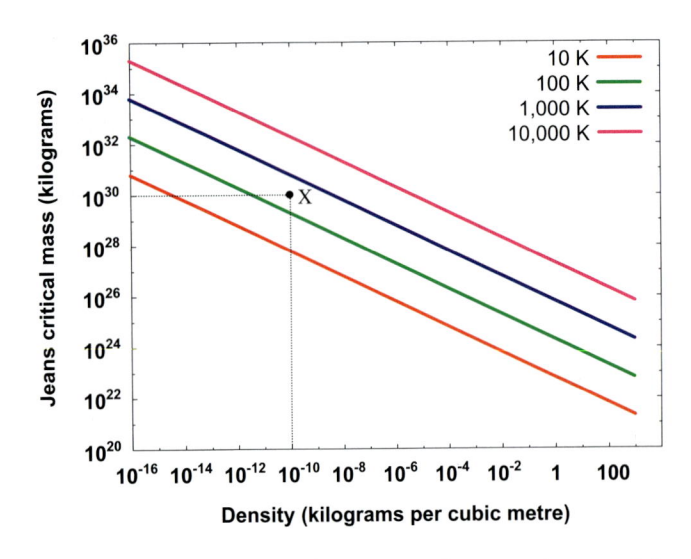

Fig. IX. The Jeans critical mass for a range of temperatures and densities.

that time and the Solar System might have been the only planetary system in the galaxy — or even the Universe. There is a principle often quoted by scientists, called the *anthropic principle*, that asserts that if a particular event *must* have happened in order to explain the existence of mankind on Earth, then, no matter how unlikely, that event did happen.

A much more telling argument was made against the tidal theory in 1935 by the American astrophysicist Henry Norris Russell (1877–1957). He argued that if material was pulled from the Sun, then, even allowing for the effect of the gravitational pull of the star, some part of the final orbit of the ejected material would return it close to the Sun. This placed an upper limit on the intrinsic angular momentum (angular momentum per unit mass) of the ejected material and this would be very much less than would be required to explain the orbit of Mercury, the innermost planet, let alone those much further out. This was a strong argument and showed that, like Laplace's nebular theory, the tidal theory had an angular momentum problem, albeit of a different kind.

Another problem with the tidal theory was put forward in 1939 by the American astrophysicist Lyman Spitzer (1914–1997), rather curiously using theory that Jeans himself had provided. Spitzer showed that if a quantity of material sufficient to produce Jupiter (2×10^{27} kg) were pulled out of the Sun, then it would have about the mean density of the Sun (1.4×10^3 kg m^{-3}) and a temperature of about 10^6 K. Such a mass of material would be far below the Jeans critical mass for that density and temperature and hence it would not collapse to form a planet but would violently dissipate.

Jeans recognized that his theory, in the way that he had presented it, was untenable and so, honest scientist that he was, he made the statement, 'The theory is beset with difficulties and in some respects appears to be definitely unsatisfactory.' Some years later, long after belief in the tidal theory had been abandoned, another difficulty was found, relating to the light elements lithium, beryllium and boron. These light elements are missing from the Sun because at temperatures of the solar interior they are removed by nuclear reactions. However, these elements are found on Earth, and presumably on other planets, implying that the substance of the planets could not have come from the Sun in its present state.

At the end of this episode in the development of theories of planetary origin, yet another model was found wanting, but Jeans had developed a body of theory that was both sound and useful and that has found applications in many other astronomical contexts.

The New Solar Nebula Theory

From the 1940s to the 1960s, several ideas were put forward to explain the origin of the Solar System but none of them carried much conviction and were quickly put to rest by the criticism of the astronomical community. This was a very active period in the study of meteorites — solid bodies either of silicate, iron or a mixture of the two materials — that fall to Earth. They are a valuable source of extraterrestrial material, obtained at little cost, and it is generally accepted that they contain information about the early stages of solar-system formation. From the mineral compositions of silicate meteorites and their physical appearance, a pattern was discerned of what came to be known as *condensation sequences*. When minerals are heated to such a high temperature that they are vaporized, the chemical framework structures of which they are comprised break up into smaller, stable units which are the basic units of the minerals themselves. Thus the mineral olivine, one form of which is Mg_2SiO_4,[2] breaks up by

$$Mg_2SiO_4 \Rightarrow MgO + MgO + SiO_2.$$

Similarly, two units of the mineral albite will break up by

$$2(NaAlSi_3O_8) \Rightarrow Na_2O + Al_2O_3 + 6SiO_2$$

where Na is the symbol for sodium and Al that for aluminium. Thus a vaporized assortment of minerals will provide an alphabet soup of large numbers of these stable sub-units, consisting mostly of metallic oxides and silica (silicon dioxide), and when the vapour cools the sub-units will assemble themselves into minerals that take no account of the original composition. The first minerals to condense will be those with the highest condensation temperatures — the most refractory materials — and as the vapour cools, minerals with lower condensation temperatures will be able to form. The evidence from meteorites suggested that some of them had condensed from a cooling gaseous state and the idea quickly took hold that this gaseous state was that of a high-temperature nebula which at one time had surrounded the Sun. By the late 1960s, Laplace's idea had been revived and large numbers of researchers began to develop what came to be known as the Solar Nebula Theory (SNT).

The first idea of how planets would form from the nebula involved the concept of Jeans critical mass. Just as a filament of gas would be intrinsically unstable, so the same is true for a large volume of gas. It will tend to break up into a series of condensations, each of which would be of about a Jeans critical mass. However, it did not take long to find out that this model of planetary formation was untenable under the conditions of the SNT. If the nebula were to be of about a solar mass and to extend to a distance of the order of the extent of

---

[2]This mineral consists of a combination of two atoms of magnesium (Mg), one atom of silicon (Si) and four atoms of oxygen (O).

the Solar System, then the density of the nebula would certainly be less that 10^{-8} kg m^{-3} and, from Fig. IX, to produce a condensation with the mass of Jupiter, 2×10^{27} kg, would require a temperature not much greater than 10 K. This conclusion is incompatible with the idea of a hot nebula. In 1976 the Canadian astronomer A.W.G. Cameron wrote:

> 'At no time, anywhere in the solar nebula, anywhere outwards from the orbit of Mercury, is the temperature in the unperturbed solar nebula ever high enough to evaporate completely the solid materials contained in interstellar grains.'

This statement cut away the whole *raison d'être* for the revival of the solar nebula concept but, by this time, other ideas had been advanced about the way that planets could form from nebula material, now assumed to be cold.

The passage of time since Laplace put forward his theory has not removed the need to resolve the angular momentum problem. Somehow angular momentum had to be transferred from inner to outer material so that a slowly spinning Sun could form. One early idea, put forward by the British astronomers Donald Lynden-Bell and Jim E. Pringle in 1974,[3] considered the effect of energy dissipation in the rotating nebula. This could come about in a number of ways. For example, the nebula would not rotate like a solid object with all its parts having the same angular speed. The angular spin rate would fall with increasing distance from the centre of the nebula, as is the case for individual planets, and the drag between neighbouring material moving at different speeds would then transform mechanical energy into heat energy. What Lynden-Bell and Pringle showed was that if the nebula lost kinetic energy while satisfying the principle of the conservation of angular momentum, then inner material would have to move inwards while outer material moved outwards, tantamount to the outward transfer of angular momentum (Appendix E). However, this did not solve the basic

---

[3]Lynden-Bell, D. and Pringle, J.E., 1974, *Mon. Not. R. Astron. Soc.*, **168**, 603.

angular momentum problem. Material at any distance from the centre would be moving in free orbit around the central mass and the Sun produced in this way would be spinning at a speed just below that at which it would fly apart. The Sun is spinning at less than one two-hundredth of that rate.

Another idea,[4] first suggested by the British astrophysicist Fred Hoyle (1915–2001), involves the formation of a gap between inner material and outer material of the nebula with material bordering the gap at such a high temperature that the gaseous material was ionized — that is, it contained negatively charged electrons and positively charged ions. If a magnetic field were established between the inner and outer parts then the magnetic field lines would be frozen into the conducting material at the edge of the gap. These field lines would behave like elastic strings connecting the inner and outer material that were firmly attached to it at both ends (Fig. X). The collapse of central material would tend to stretch the strings that would resist stretching by pulling on outer material so causing it to spin faster and inhibiting the more rapid spin of inner material. This would transfer angular momentum outwards as required. There are several problems with this process — for example, the magnetic

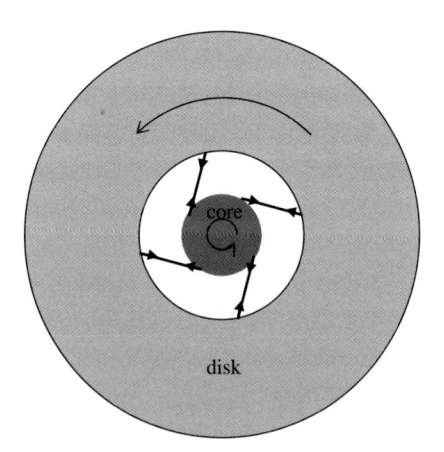

Fig. X. Hoyle's model of flux lines linking a collapsing core to a disk.

---

[4]Hoyle, F., 1960, *Q. J. R. Astron. Soc.*, **1**, 28.

field required would be very large, more than 1,000 times that of the Sun, and it is dubious that temperatures in an early nebula would be high enough to give the required ionization of disk material.

It is probably fair to say that the angular momentum problem has not yet been convincingly solved but that it is not impossible that a solution may eventually be found.

Granting that the nebula evolves into a central slowly spinning Sun and a surrounding flattish disk, there remains the question of the mechanism for forming planets. The first idea was called *top-down* because it assumed that the first bodies produced were the largest ones and smaller bodies were formed by their disruption. We have seen that this could not occur in a hot nebula and it turns out that having a cool nebula offers different, insurmountable difficulties. On the one hand, if the original nebula were cool and very massive, the density would be high enough to get spontaneous condensations of planetary mass but there would be so many of them that the problem of disposing most of them would be insoluble. It would require a tremendous input of energy from somewhere. On the other hand, if the total mass of the nebula were not much more than the combined present mass of the Sun and planets, the density would be too low to produce planets by spontaneous condensation. Top-down did not seem to offer a solution.

The second process, on which a large amount of development work has been done, is referred to as *bottom-up*, implying that small bodies are first produced and then gradually assembled into larger bodies. As envisaged, the formation of a major planet is a four-stage process:

(1) Solid grains in the nebula disk fall towards the mean plane of the disk to form a dust carpet.
(2) The dust carpet is intrinsically unstable and breaks up into a set of solid condensations, called *planetesimals*, a term first used by Chamberlin and Moulton in relation to intermediate bodies formed in their theory.
(3) Planetesimals collect together to form larger solid bodies. In the inner regions of the Solar System, these directly form the

terrestrial planets. In the outer part of the system, they become the solid cores of the major planets.
(4) The major planets capture nebula gas that eventually forms the bulk of their mass.

Before commenting on these stages, it should be recorded that in the mid-1980s the credibility of the SNT received an enormous fillip due to the detection of dusty disks around very young stars, setting the right scenario for the planet-forming process to operate. The disks were detected by examining the variation of light intensity with wavelength coming from the star. Bodies at a particular temperature have a characteristic emission of light, peaking at a wavelength that decreases for higher stellar temperature, and with a progressive fall-off on both sides of the peak. For many very young stars, it was found that there was a slight bump in the long wavelength tail of the distribution of light. This could be explained by the presence of a low-temperature source in the vicinity of the star. However, low-temperature sources are very poor radiators of energy per unit area so the fact that the bumps could be seen implies that the low-temperature sources had a very large area. The obvious, and probably the only, feasible explanation is that it was due to the existence of an extensive disk accompanying the star. The masses of the disks were estimated to be in the range of one-hundredth to one-tenth of the mass of the Sun. However, there was a sting in the tail of this discovery. Towards the end of the 20th century, it was discovered that the lifetimes of these circumstellar disks were in the range of 1 to 10 million years, mostly at the lower end of that range. This meant that the four stages of planetary formation must be accomplished within a few million years — a new constraint for the theoreticians.

Stage 1

The solid grains that constitute about 1% of the gaseous material of our galaxy are very tiny, usually quoted as less than 1 micron[5] in dimension. That being so, they are heavily buffeted by collisions

---

[5]1 micron (μm) is 10^{-6} metre.

An Historical Sketch of the Progress of Opinion xxxix

due to the thermal motion of the molecules of the gas in which they are situated and hence will not settle very quickly. The timescales for settling would be several million years and so possibly longer than the observed lifetime of disks. To resolve this problem, in 1989 the American planetary scientist Stuart Weidenschilling and his colleagues suggested that the dust would stick together to form larger aggregates that would settle more quickly.[6] In 2000 a Space Shuttle experiment called CODAG (COsmic Dust AGgregation) investigated this possibility and found that dust did indeed stick together. However, rather than forming a compact aggregation, the grains formed long wispy strings that would not settle much more quickly than the original dust grains. Later it was suggested, but without supporting evidence, that if the wispy strings became long enough, they would tangle up into a ball-like form and so settle more quickly.

There is a strong possibility that the estimated long settling times, predicated on sub-micron dust particles, may, in fact, be much shorter because there are much larger grains in the disk even without the aggregation of smaller particles. The estimated size of particles in the interstellar medium (ISM), the gas and dust that occupies the space between stars, is based on the absorption of light by the dust, which depends mainly on the very smallest particles. Particle sizes much larger than one micron could not be detected by their absorption. However, an experiment carried out in 2001 using the Ulysses spacecraft, launched in 1990 to study the Sun, measured dust properties in the local ISM and found radii in the range of 0.01–5 microns.[7] Although larger particles are ineffective in terms of light absorption, it is possible that they provide the majority of the total mass of the dust of the ISM, and presumably of the dust in the circumstellar disks (Appendix F). The larger a dust particle, the more quickly it settles — the time of settling is inversely

---

[6]Weidenschilling, S.J., Donn, B. and Meakin, P., 1989, *The Formation and Evolution of Planetary Systems*, eds. H.A. Weaver and L. Danley (Cambridge: Cambridge University Press).
[7]Mann, I. and Kimura, H., 2001, *Space Sci. Rev.*, **97**, 89.

proportional to the diameter, according to the theory developed by Weidenschilling.

On the whole, the evidence supports the idea that a dust carpet could form in most disks, except perhaps in disks of the shortest duration.

Stage 2

If a dust carpet formed, it would have a mass between one-tenth that of Jupiter and that of Jupiter, and would be very substantial with a surface density between 3 and 30 kg m^{-2}. In 1973 the American planetary scientists Peter Goldreich and William Ward showed that such a carpet would be unstable — like a two-dimensional version of the filament considered by Jeans — and would form a number of substantial solid bodies.[8] The size of these bodies, the planetesimals, would depend on the local density of the dust carpet and the distance from the Sun and it was estimated that they would vary in size from being a few hundred metres across to tens of kilometres. This stage of the process of forming planets has a good theoretical foundation and presents no timescale problems.

Stage 3

The seminal theoretical work on the aggregation of planetesimals to form larger bodies was due to the Russian earth scientist, Victor Safronov (1917–1999),[9] in relation to a theory of planetary formation proposed in 1944 by another Russian, Otto Schmidt. In the Safronov theory, the planetesimals were first produced in circular orbits but over the course of time, due to close approaches with each other, they were perturbed and their motions became more chaotic. Sometimes the perturbed planetesimals collided and this tended to reduce the differences of velocity and hence decreased the amount of chaos in the system. Eventually the processes of increasing and reducing chaos came into balance and at that stage it was estimated that the relative

---

[8]Goldreich, P. and Ward, W., 1973, *Astrophys. J.*, **183**, 1051.
[9]Safronov, V.S., 1972, *Evolution of the Protoplanetary Cloud and Formation of the Earth and Planets* (Jerusalem: Israel Program for Scientific Translations).

speeds of planetesimals would have been similar to the escape speed[10] from the largest planetesimal in the system. A small planetesimal striking the largest one and losing some of its energy of motion, for example, in the form of heat or through the work done in fragmenting material, would be accreted. In this way, in different regions of the disk, the largest local planetesimal became dominant and tended to grow until it eventually became a terrestrial planet or the core of a major planet.

When Safronov's theory was applied to the conditions given by the SNT, it was found that the time for forming the Earth was a few million years, for Jupiter about 250 million years and for Neptune about 10 billion years — twice the age of the Solar System. At the time that these results were first found, it was clear that Neptune presented a problem but Jupiter was not seen as such because the constraint imposed by disk lifetimes was not known then. The time of formation by Safronov's mechanism is proportional to the period of a circular orbit in the position of planet formation and inversely proportional to the local averaged surface density of solid material, which will fall off with distance from the Sun. The period of Neptune's orbit is 165 years and if the density of material in that region of the disk were one-fiftieth of that where the Earth forms, then, given a 1 million year formation time for Earth, the formation time of a Neptune core would be $10^6 \times 165 \times 50 = 8.25 \times 10^9$ years, similar to the figure of 10^{10} years previously given.

Attempts have been made to rescue the theory of planetesimal aggregation by considering factors that might speed up the process but these have not been very convincing. Even postulating unsupportable extreme conditions in the solar nebula, the problem of the outer planets cannot be solved. An attempt to deal with this problem is the development of theory on what is termed *migration*, whereby planets may be produced in the inner parts of the system, where times of formation are acceptable, and then moved

---

[10]The escape speed from an astronomical body is the lowest speed at which an object, projected away from the body, would not return to it. The escape speed from Earth is about $11\,\mathrm{km\,s^{-1}}$.

outwards to where they now are. The processes that have been suggested usually involve the most massive planet Jupiter which, by a small movement inwards, could provide enough energy and angular momentum to move a smaller planet outwards a long way. However, the mechanisms that have been suggested do not solve the problem. For example, one mechanism involves the action of Jupiter on the gaseous material of the nebula, producing nebula gas that moves towards and away from the planet in the form of spiral waves. These are illustrated in Fig. 7.6. The outward-moving wave contains material with more angular momentum per unit mass than the material into which it moves, and that moving inwards contains less. The waves share their angular momentum with the material on which they impinge so the net effect is that material outside the orbit gains angular momentum while that inside loses angular momentum. SNT theorists have suggested that the outward-moving spiral waves could push outer planets and so cause them to migrate outwards. Since a planet is such a small target, the proportion of Jupiter's energy of motion and angular momentum that could be transferred in this way is extremely small. Even if the solar-system problem could be solved for the outer planets Uranus and Neptune, which is extremely doubtful, there has been recent direct imaging of planets around other stars that are much further from those stars than Neptune is from the Sun (Section 5.4). The presence of planets at many tens, or even hundreds, of astronomical units[11] from their parent stars presents an insuperable problem for the standard SNT.

There is a migration process that *is* well founded, whereby a planet moving in the gaseous nebula would experience retarding forces and gradually spiral inwards. This is the kind of process that leads to orbital decay of an Earth satellite in the upper reaches of the atmosphere, so that it eventually plunges to Earth. It has been shown that a growing terrestrial embryo would spiral in towards the

---

[11] The mean distance of the Earth from the Sun is one astronomical unit (au) or 1.496×10^8 km.

Sun quite rapidly and would plunge into the Sun before it was fully formed. Not only does migration not solve the problem of locating the outer planets, it also prevents the formation of the terrestrial planets.

Stage 4

If a core could be produced in the presence of a considerable amount of nebula gas, then the capture of the gas presents no theoretical difficulties. The time of the accumulation of gas by Jupiter would be tens of thousands of years and hence impose no timescale problems.

I would like to add a final footnote regarding the status of the SNT, relating to an observation made as this book was being completed. In August 2009 a team of British astronomers observed a planet that was orbiting a star in the opposite sense to the direction of the star's spin — something completely impossible for the SNT. *Ad hoc* ideas are being put forward to explain this observation, mainly involving a gravitational interaction between the observed planet and another planet, which caused its direction of motion to be reversed.

It is interesting to compare Jeans' gracious acceptance that his theory had been critically weakened and the rather dogged reluctance of modern scientists to do likewise for their theories when adverse evidence is discovered. I think this relates to the different environments in which science is carried out then and now. In the second decade of the 20th century, Jeans sat alone in his room at St John's College, Cambridge, teasing out the problems of the tidal theory; the time spent was his own and the main expenditure was on paper and pencils. Nowadays, a theory like the SNT involves large numbers of scientists working in teams in many locations all over the world, with perhaps tens of millions of dollars, euros or pounds acquired from research agencies being expended on salaries and equipment. This kind of situation exists now in many areas of science and the heavy investments in human and financial resources make the abandonment of any major development in science, be it theoretical or experimental, much more difficult.

General Conclusions

A consideration of the four theories described here shows that two difficulties of a general kind would need to be dealt with by any successful theory of planet formation. The first is that of solving the angular momentum problem which all the theories exhibit in different ways. For the Solar System, this means that the mechanism must produce both a slowly spinning Sun and planets in more or less circular orbits out to the distance of Neptune. There are similar considerations for other planetary systems. Most stars similar to the Sun spin slowly and some of their planets have orbital radii much greater than that of Neptune. The second problem is that of producing the planets. The SNT has a perfectly good theoretical mechanism for producing planets but it cannot happen within the timescale that observations allow. The Jeans method of producing planets, by the collapse of blobs in an unstable filament would give planets very quickly, in a few thousand years, if only the blobs exceeded the Jeans critical mass. This requires a suitable combination of density and temperature — in particular a low temperature since the Jeans critical mass is highly temperature-dependent.

The theories that have commanded the most support and have had the greatest input of theory are the Jeans tidal model and the SNT. An important distinction between these two theories is that the SNT is a *monistic theory* that describes the formation of the Sun and the planets as part of a single process — the evolution of an initially diffuse nebula. If the nebula were not acted on by outside forces, its mass and angular momentum could not have been too much greater than that of the present Solar System, unless a mechanism for losing considerable amounts of mass and angular momentum can be found. It is possible that the Sun could have lost some of its original angular momentum after it formed; a process by which this could happen will be described in Chapter 2. However, this would not change the basic requirement that the initial mass and angular momentum could not have been much greater than that which characterizes the Solar System at present, and that this must somehow be partitioned so that a tiny part of the mass has most of the angular momentum.

By contrast, the tidal theory is a *dualistic theory* that assumes that different mechanisms produced the Sun and the planets. A preexisting, slowly rotating Sun is taken as the starting point and the whole emphasis of the theory is concerned with extracting some of its mass, concentrating it into a set of planets and then, through the gravitational attraction of the passing star, giving those planets the angular momentum necessary to explain their orbits. This goal was not realized but at least the problem of the slow rotation of the Sun was sidestepped. I suppose it could be argued that it is an observational fact that slowly rotating Sun-like stars exist, so this can be a legitimate starting point for a theory of solar-system formation. Why does the Sun rotate slowly? Let the stellar theorists worry about that — it isn't a problem that planetary scientists should worry about. That is not a view to which I subscribe. I take the purist view that the Solar System consists of the planets *and the Sun*, and that the problems of the formation and properties of *all* these bodies should be within the remit of the planetary scientist. For this reason, I shall begin by considering how stars are formed and how it is that they acquire their properties.

In approaching the task of describing a proposed model of the formation of planets, with particular reference to the Solar System, I shall adopt the Baconian procedure of first establishing the observational background of the topic under consideration and then relating the theory to it. Following this model for the topic of star formation, the first thing to do is to review what is known about stars, taking into account the most recent observations.

Chapter 1
Observations of Stars

The formation of stars and the formation of planets are intricately connected in monistic theories. The theory that I shall describe later, which forms the main topic of this book, is essentially a dualistic theory but one that is still dependent on the processes of star formation. However, before I give an account of how stars form, I shall first describe some of the properties of stars and how they are determined.

Looking up at the night sky, we see points of light that are individual stars and using telescopes, many more can be seen than with the naked eye. If we examine individual stars with optical instruments we find that the vast majority are indeed *point light sources* of no discernable size. Yet, for all that, by the use of optical instruments and a knowledge of physics, we can learn a great deal about these stars — their distances, their temperatures, how luminous they are, their masses, their radii and how fast they spin on their axes. How we achieve this miracle from just the evidence provided by points of light, I shall now explain.

1.1. Locations of Stars

In considering where stars are located, I shall limit myself to a very tiny part of the contents of the Universe, the Milky Way galaxy of which the Solar System is a member. It may seem a little strange to use the adjective 'tiny' in relation to the collection of about 10^{11} stars that constitutes the Milky Way, but so it is, for there are of the order of 10^{11} galaxies in the Universe, according to the latest estimates. The other restriction I shall observe is mainly to concentrate on stars that, like the Sun, are generating energy by nuclear reactions that convert hydrogen to helium in their interiors. Stars in this state are said to be on the *main sequence*. The Sun has been a main-sequence star for about 5 billion years and will remain so for another 5 billion.

The stars in our galaxy are in a variety of environments. First, there are stars like the Sun that move through the galaxy in splendid isolation without any stellar companions. By contrast, many stars occur in large associations, *stellar clusters*, of which there are two main types. The first is the *galactic* or *open cluster* that usually consists of between 100 and 1,000 stars (Fig. 1.1) and is between 6 and 60 light years (ly)[1] across. The alternative names describe the structure of such clusters and where they are situated. They are 'open' in the sense that the stars are sufficiently separated for them to be seen individually and they are 'galactic' in the sense that they only occur in the galactic plane (Appendix G). The second type, the *globular cluster*, is much larger, containing as many as several tens of thousands to a million stars (Fig. 1.2) and with diameters of 60 to 200 ly. The name 'globular' describes the shape of these clusters. The

Fig. 1.1 The Pleiades open cluster (John Lanoue).

---

[1] A light year (ly) is the distance that light will travel in one year. 1 ly = 9.46×10^{12} km.

Fig. 1.2 The globular cluster M80 (NASA/ESA).

number density of stars is so high in the central core of the cluster that individual stars cannot be easily resolved there, but stars in the outer regions can be seen individually. Globular clusters occur in all parts of the galaxy — the galactic plane, the nucleus and the halo (Appendix G).

Both within and outside clusters, many stars exist in pairs, known as *binary systems*, in which the two stars orbit around their centre of mass (Appendix H). For a small proportion of binary systems, the stars are sufficiently far apart for them to be seen individually; these are *visual binaries*. Greatly outnumbering visual binary systems are *spectroscopic binaries* in which the stars are so close together that they cannot be individually resolved, even with the most powerful telescopes. The fact that they consist of a pair of stars is determined by looking at the light that comes from them. As the two stars move around their orbits, the variation in their velocities relative to the Earth can be detected by Doppler shifts (Appendix I) in the wavelengths of spectral lines (Fig. 1.3) in the light emitted. There are about as many binary systems in the galaxy as there are individual stars, meaning that a minority of stars are isolated stars like the Sun. There are also some rare multiple-star close systems containing more than two, but very few, stars.

Fig. 1.3   Fraunhofer lines in the solar spectrum.

1.2. Stellar Material

The Sun is a typical main-sequence star, although its mass is somewhat above the average for such stars. The light emitted from the Sun comes from a very thin layer, called the *photosphere*, which defines its visible surface. Light that comes from material below the photosphere is absorbed by material above it before it can escape and so does not add to the emitted light. The material above the photosphere is so diffuse that, although it is at a very high temperature, it emits very little light. However, the material above the photosphere does have an important effect on the emitted light; it contains different kinds of atoms in different states of ionization[2] and these absorb some of the light coming from the photosphere. The absorbed light is at many well-defined wavelengths, characteristic of particular atoms and their states of ionization, and produces in the solar spectrum a set of dark lines, known as Fraunhofer lines (Fig. 1.3). These absorption lines also occur in the spectra of the light from distant stars, and from the presence and strength of these lines it is possible to determine which elements are present in the star and to what extent.

The stars in the two types of cluster are distinguished by their chemical compositions. The vast majority of material in the Universe is either hydrogen or helium that together account for 98% of the total by mass. These are the elements, together with tiny amounts of lithium and beryllium, two other light elements, which were produced in the *Big Bang*, the cataclysmic event that is thought to have produced time, space and all matter.[3] The extent to which a star

---

[2]Ionized atoms are those that have lost one or more of their electrons.
[3]A description of the Big Bang can be found in the author's book, *Material, Matter and Particles: A Brief History*, 2009 (London: Imperial College Press).

contains heavier elements — those from carbon with atomic number six and heavier — is described as the *metallicity* of its material, although most of the heavier elements are not actually metals. Stars in open clusters, similar in type to the Sun, have high metallicity with typically 1–2% of their mass in the form of heavier elements. These open-cluster stars, and field stars of similar composition, are called *Population I stars*. Stars in globular clusters have much lower metallicities with typically 0.01% by mass of heavier elements. Such stars are called *Population II stars*. It is theorized that there may be *Population III stars* consisting of pure Big Bang material with no heavier element content whatsoever, but such stars have never been observed.

1.3. Determining the Distances of Stars

Starting with finding the distances of some nearer stars by a technique similar to that used in military rangefinders, astronomers have found ways of finding the distances of stars out to the limit at which individual stars can be seen. I shall now describe two techniques that enable us to determine the distances of stars within our galaxy — my self-imposed limited region of interest.

1.3.1. *The distances of nearby stars*

The principle of measuring the distance of nearer stars is easily demonstrated. If you observe a vertical finger held at arm's length, first with one eye and then with the other, it will be seen to move relative to distant objects. In the same way, if a nearby star is observed at times six months apart, during which period the Earth has moved from one point of its orbit to the diametrically opposite point, then it will appear to have moved relative to those stars very much further away. This situation is illustrated in Fig. 1.4.

The best arrangement for making the observations is when the line joining the viewing positions is perpendicular to the direction of the star. The figure is completely out of scale and the angle α is very

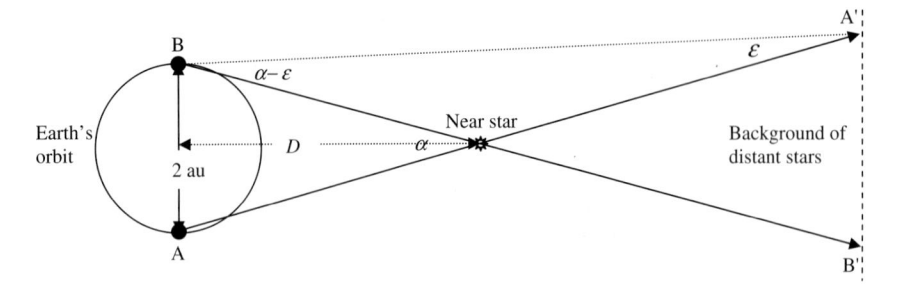

Fig. 1.4 Finding the distance of a nearby star by the parallax method.

small, of the order of one second of arc.[4] For the calculation of the star's distance we need to express the angle in radians (Appendix A) and then the distance of the star is given by

$$D = \frac{2}{\alpha} au. \tag{1.1}$$

The question then arises of how α is measured since we cannot travel to the star and measure the angle between the viewing positions A and B. The answer is that we measure the angle from either of the viewing points between the positions A′ and B′ on the distant star field. As shown in Fig. 1.4, that angle is actually $\alpha - \varepsilon$ but since the background is so distant, ε is extremely small — very much smaller than α.

By this method, called the *parallax method*, the distances of stars can be estimated out to about 100 ly by ground observations and up to about 600 ly by measurements from Hipparcos, a satellite launched by the European Space Agency in 1989. The distances of one million of the nearest stars have been measured by the parallax method. These measurements provide the launching pad for the measurement of stars at ever increasing distances.

You may have noticed that there was an assumption made in the above account, which is that the star did not move during the six-month interval between the observations. If it did move, Eq. (1.1) would not be valid. What is done in practice is to make

---

[4] 60 seconds of arc equals one minute of arc and 60 minutes of arc equals one degree.

three measurements — from positions A, B and then A again. This not only gives the distance of the star but also its transverse component of velocity, the component at right angles to the direction of view.

There is a unit of distance, commonly used by astronomers, that is derived from the parallax method. The unit, the *parsec* (pc), is the distance at which 1 au subtends an angle of one second of arc and is equivalent to 3.26 light years. In Fig. 1.4, if the star were at a distance of 1 pc, then the angle α would be two seconds of arc. If it were at a distance of 2 pc, then α would be one second of arc. It will be seen that measuring the angle α directly gives the distance in parsecs — hence its attraction.

1.3.2. *Distance measurements using variable stars*

In the 18th century, in the city of York, there lived a very interesting English astronomer, John Goodricke (1764–1786; Fig. 1.5(a)). He was a deaf mute at a time when society looked down on people with such impediments but he came from a distinguished family and so was able to overcome this disadvantage. His first education was at a special school in Edinburgh run by Thomas Braidwood, who had developed effective techniques for teaching deaf pupils. Later, Goodricke was able to attend a normal school, Warrington Academy,

Fig. 1.5 Personalities connected with Cepheid variables: (a) John Goodricke (b) Henrietta Leavitt.

where he acquired a deep interest in astronomy. When he returned to his home, the Treasurer's House immediately next to York Minster, he started a programme of stellar observations. He specialized in observing variable stars, those that vary in intensity, and one such star was δ-Cephei. It is believed that exposure while observing this star lead to his untimely death at the age of 21.

This star, δ-Cephei, is a prototype of a whole class of stars, *Cepheid variables*, which vary in brightness in a periodic way. The observed brightness of a star depends on its distance; car headlights at a close distance are dazzling but may be seen quite comfortably if far away. The intrinsic brightness of a star, a measure independent of its distance, is its *luminosity*, the rate at which it emits electromagnetic energy. The luminosity of the Sun is approximately 3.8×10^{26} W.[5]

The observed brightness of a star depends on the energy per unit area per unit time received from the star by a detector (the eye, for example) on Earth. This decreases as the square of the distance to the star, so the observed brightness of a star of known distance enables its luminosity to be determined. Conversely, if the luminosity of the star were known, then its observed brightness would enable its distance to be estimated.

In 1908 a Harvard astronomer, Henrietta Leavitt (1868–1921; Fig. 1.5(b)), from a study of Cepheid variables within the range of parallax distance measurement, found that there was a relationship between the maximum luminosity of a Cepheid variable and its period (Fig. 1.6). Some Cepheid variables are extremely bright, with luminosities up to 30,000 times that of the Sun, so they can be seen at great distances. They occur, and can be seen, in the outer regions of some distant galaxies. From their periods, their maximum luminosities are known, as shown in Fig. 1.6. Then from their observed maximum observed brightness, their distances can be found. In this way the distances of galaxies can be estimated out to

---

[5] The joule (symbol J) is the Standard International Unit of energy and the watt (symbol W) is $1\,\mathrm{J\,s^{-1}}$.

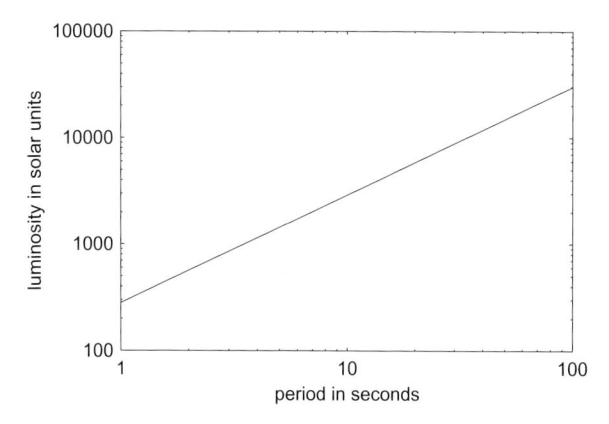

Fig. 1.6 Relationship between the luminosity of a Cepheid variable and its period.

80 million light years and also the distances of clusters of stars within our own galaxy if they happen to contain a Cepheid variable.

Astronomers have many more techniques available to extend estimates of distances to the limits of the observable Universe but we shall explore these no further — remember, I am only concerned with a tiny part of the Universe, our Milky Way galaxy.

1.4. The Temperature of Stars

In domestic situations we normally measure temperature with a thermometer in the form of a device that takes up the temperature of the object of interest and reveals it in some visual way — for example, by the expansion of mercury up a fine glass tube. Another way, which is used in metal foundries, is to measure the radiation coming from the hot metal and to use this to determine temperature. This, essentially, is how the surface temperatures of stars can be found. To understand how this can be done, we first need to know something about electromagnetic radiation, the form of energy that includes visible light, radio waves and x-rays. A representation of the full electromagnetic spectrum is shown in Fig. 1.7.

At the long wavelength end, there are radio waves of the type that are used to broadcast radio and television signals. Next, we

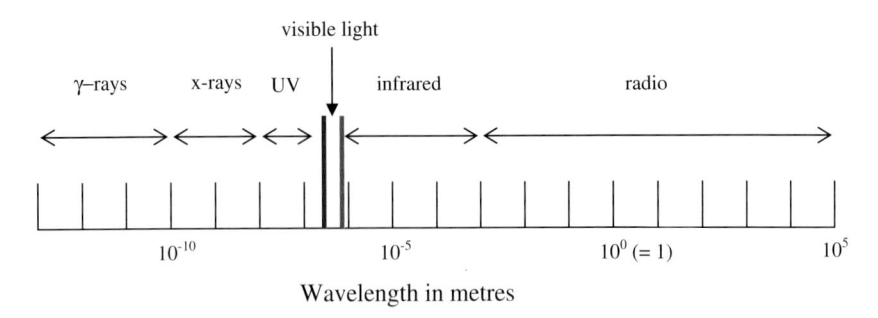

Fig. 1.7 The electromagnetic spectrum.

find the infrared region of the spectrum; wavelengths beyond the visible range but detectable by their heating effects. The visible part of the spectrum is a small region with wavelengths between 4×10^{-7} m (violet) and 7×10^{-7} m (red). It is surprising that we use so little of the spectrum in vision — infants can see a small distance into the ultraviolet region, as can some insects. After the ultraviolet region come x-rays, which are used in medical diagnosis and also in various ways by scientists as a tool for exploring the nature of matter. Finally we come to γ-rays, very high-energy electromagnetic radiation that is a component of the cosmic rays which permeate the whole Universe and also comes from some compact high-energy astronomical sources. Astronomers exploit the whole of this range of electromagnetic radiation in exploring the heavens, using devices that range from γ-ray telescopes to radio telescopes.

Figure 1.8 shows the intensity distribution against the wavelength of the radiation coming from sources with the same areas of emission at three different temperatures. Two properties of the curves shown in the figure are evident. The first is that the higher the temperature, the shorter the wavelength corresponding to the maximum. Theory shows that the product of the absolute temperature and peak wavelength is a constant (Wien's law). The second is that the total energy emitted per unit time, which is proportional to the area contained under the curve, increases with temperature; theory shows that it is proportional to T^4 where T is the absolute temperature. I have already referred to this relationship in a non-quantitative way in describing how disks were detected around young stars. It is only

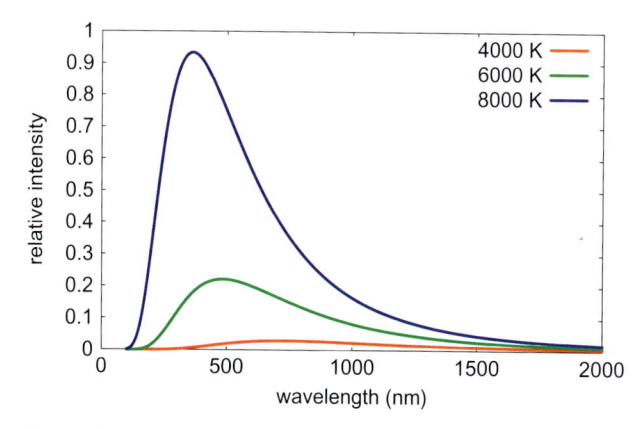

Fig. 1.8 The radiation intensity curves from sources at 4,000 K, 6,000 K and 8,000 K (1 nm = 10^{-9} m).

possible to detect the radiation from a low temperature source when it has a large area.

In principle, by looking at the overall intensity distribution of the light coming from a star, or finding the wavelength of the maximum of the distribution, it should be possible to determine the temperature of the star. This is difficult in practice and only very approximate temperatures can be found in this way. Astronomers have found a much more accurate and subtler way of determining temperature based on the absorption lines shown in Fig. 1.3. The intensities of these lines are temperature-dependent (Appendix J) and the different lines vary in intensity with temperature in different ways. Figure 1.9 illustrates the variation of the intensities of various absorption lines for different temperatures.

If the spectra of many stars are available, they can be arranged in sequence such that each varies only slightly from its immediate neighbours. In the figure, three spectral lines produced by hydrogen — Hα, Hβ and Hγ — are shown and the gradual change from one spectrum to the next is evident. According to which spectral lines are most prominent, stars are assigned to *spectral classes* O, B, A, F, G, K or M; each class has ten subdivisions indicated as A0, A1, A2 ··· A9, for example. Most astronomers remember the sequence of spectral-class letters by the mnemonic 'Oh Be A Fine Girl — Kiss Me'. The spectral classifications for the illustrated spectra are

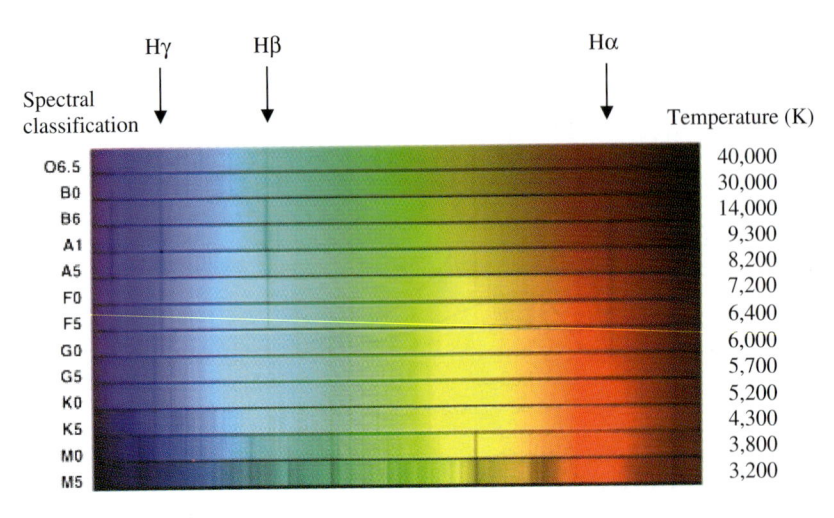

Fig. 1.9 The absorption spectra of stars at different temperatures.

shown on the left-hand margin in Fig. 1.9. By accurately comparing the relative intensities of a number of different absorption lines, it is possible to estimate stellar temperature with an accuracy of 10 K in the most favourable situations.

1.5. Stellar Radii

For about a dozen red giant stars, it is possible to produce an image of their disk and so estimate radii directly and, for a small number of other nearby stars, radii can be found by an optical technique known as stellar interferometry. However, the majority of stars are simple point sources of light and so estimates of their radii must be made by indirect means.

The areas under the curves in Fig. 1.8 are proportional to the energy emitted per unit area per unit time, a quantity that we have already indicated as proportional to T^4. The luminosity of a star is the total energy per unit time emitted by its surface, which has area $4\pi R^2$, where R is the radius of the star. Putting these two statements together, we may express the luminosity as

$$L = \sigma 4\pi R^2 T^4 \qquad (1.2)$$

where σ is Stefan's constant, equal to $5.67 \times 10^{-8}\,\text{W}\,\text{m}^{-2}\,\text{K}^{-4}$. From this, if both L and T have been estimated from astronomical observations, the radius of the star may be found from

$$R = \sqrt{\frac{L}{4\pi\sigma T^4}}. \tag{1.3}$$

For the Sun, a G2 star with a surface temperature of 5,770 K and a luminosity of 3.8×10^{26} W, this formula gives a radius of 6.94×10^5 km which compares well with the accepted value of 6.96×10^5 km.

1.6. Estimating Stellar Masses

If all stars existed as single isolated stars, it would be virtually impossible to know anything about their masses, except for that of the Sun. The mass of a star manifests itself through its gravitational influence and the characteristics of the planetary orbits enable the mass of the Sun to be found. The existence of binary-star systems, in which the actions of gravitational forces also manifest themselves through the motions of the two stars, detected by Doppler shifts of spectral lines, enables the masses of the individual stars to be estimated (Appendix K).

Very massive stars are comparatively rare and stars are ever more common with decreasing mass. To put this into a quantitative form, the mass frequency function, $f(M)$, is used to describe the relative numbers of stars per unit mass range: the proportion of stars with mass between M and $M + dM$, for small dM, is $f(M)dM$. From observation it is found that $f(M)$ is proportional to $M^{-\mu}$ where μ is in the range 2.3 to 2.6 for stars between one-tenth and ten times a solar mass. To give some feel of what this means for stars with masses in this given range, just one star in 30 has the mass of the Sun or more and one star in 100 has twice the mass of the Sun or more. Figure 1.10 shows, for $\mu = 2.4$, the proportion of main-sequence stars in the range from 0.1 to 10 solar masses that have a mass less than that indicated by the x-coordinate. This shows that very few stars have mass greater than a solar mass.

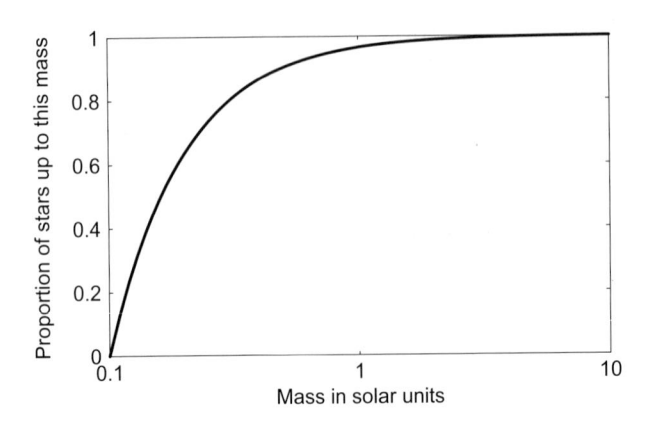

Fig. 1.10 The proportion of stars with mass less than the x-coordinate mass.

1.7. The Physical Properties of Main-Sequence Stars

The only property of a star that can always be measured is its temperature, as judged by its spectral class. If it is an individual, isolated star outside the range of parallax measurements, then neither its luminosity nor its mass can be directly estimated.

Fortunately, the masses, luminosities and radii of main-sequence stars are related to their spectral class so the general physical properties may be inferred for any main-sequence star for which the spectral class has been determined. Table 1.1 gives the properties of main-sequence stars for a sample of spectral classes.

Table 1.1 Properties of main sequence stars related to spectral class.

Spectral class	Temperature (K)	Mass (solar units)	Radius (solar units)	Luminosity (solar units)
O5	45,000	60	12	790,000
B5	15,000	5.9	3.9	830
A5	8,200	2.0	1.7	14
F5	6,400	1.3	1.3	3.2
G5	5,700	0.92	0.92	0.79
K5	4,300	0.67	0.72	0.15
M5	3,200	0.21	0.27	0.011

This enables the distances of main-sequence field stars to be estimated well beyond the parallax range. The spectral class gives the luminosity and the measured brightness gives the distance of the star.

1.8. Stellar Spin Rates

I have explained why an explanation of the slow spin of the Sun has proved to be a difficulty for monistic theories of solar-system origin, so it is clearly interesting to know whether or not the Sun is typical in this respect. In general, due to their spins, when stars are observed, different parts of the star are moving either towards the Earth or away from it, relative to the motion of the star's centre. If the star's spin axis were perpendicular to the line of sight, then material at one edge of the equator would be moving with the greatest speed away from the observer while that from the opposite edge would be moving with the greatest speed towards the observer. Light from different parts of the star would have different Doppler shifts and the effect of this would be a broadening of the observed spectral lines. From the extent of the broadening, the equatorial speed can be estimated. In general, the tilt of the stellar axis relative to the observer is unknown but from observations of many stars of similar mass, and with the assumption that their axes are randomly tilted with respect to the observer, an average equatorial speed for that mass can be found. The average equatorial speed for stars of different masses is shown in Fig. 1.11. It is clear from the figure that stars of a solar mass or less typically have low equatorial speeds while more massive stars have higher ones. The equatorial speed for the Sun, $2\,\mathrm{km\,s^{-1}}$, is on the low side when compared with that indicated for a solar mass in Fig. 1.11, but is within the normal range of variation for such stars.

1.9. Summary

In this chapter I have described some of the important features of main-sequence stars. They can occur within open clusters, outside a cluster, as an individual star or as a member of a binary pair.

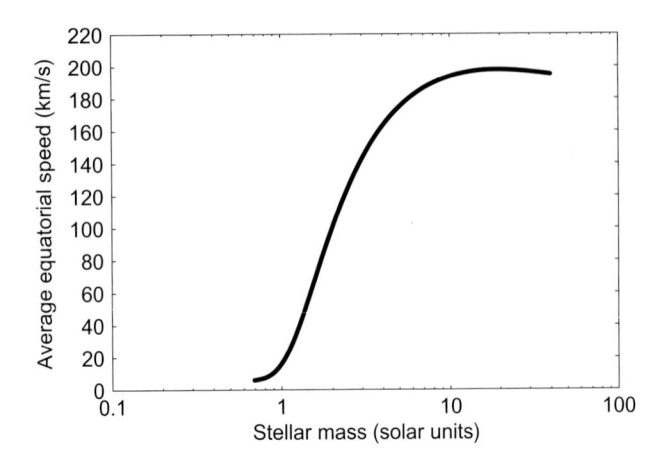

Fig. 1.11 The relationship between mean equatorial speed and stellar mass.

There are more stars existing within binary pairs than there are as individual stars and the majority of binaries are of the close spectroscopic type.

The Sun, and stars in open clusters, are Population I stars with 1–2% of their mass in the form of elements of atomic number equal to, or higher than, that of carbon. Stars in globular clusters are Population II stars with typically only 0.01% of their mass in the form of heavier elements.

The distances of about one million stars closest to the Sun have been determined by parallax measurements. By the use of Cepheid variables, distance measurements can reach out to all parts of the Milky Way galaxy and beyond.

Stellar masses can be deduced from observations of binary-star systems. The frequency of stellar masses decreases with increasing mass; only about 3% of main sequence stars have greater than a solar mass.

A combination of the apparent brightness of a star and its distance gives an estimate of its luminosity, while its spectral class gives an estimate of its temperature. From its temperature and luminosity, the radius of a star can then be found. For main-sequence stars, physical properties are linked to spectral class and thus these properties can be inferred for isolated individual stars outside the parallax range for distance measurement.

Stellar equatorial speeds generally increase with stellar mass, except for a slight fall-off for the most massive stars. In terms of spin rate, the Sun is typical of stars of the same or lower mass.

Chapter 2
Producing Protostars — Embryonic Stars

2.1. Star-Forming Regions

The interstellar medium (ISM), material that occupies most of the apparently empty space of the Milky Way galaxy, is tenuous to an amazing degree. Most of it is gas, a mixture of hydrogen and helium, which are the original main products of the Big Bang. It varies in density from place to place but typically contains one atom per cubic centimetre or 10^6 atoms m^{-3}. To put this in perspective, the air you breathe, a mixture mainly of nitrogen and oxygen, contains 6×10^{25} atoms m^{-3}. For another comparison, we take the ultra-high vacuum systems used by scientists to investigate materials when they want to avoid contamination of the material surfaces by adhering gas molecules. Within a very good ultra-high vacuum system, there are 10^{12} atoms or molecules m^{-3} — a million times the number density of the ISM.

A minor, but very important, component of the ISM is dust which accounts for 1–2% of its total mass. This dust is of three main kinds — stone, iron and ices composed of solidified volatile materials such as water, carbon dioxide and ammonia. The volatiles sometimes form coatings on the stony and iron dust. The dust is generally of sub-micron size, *i.e.* less than 10^{-6} m in average dimension (but see Appendix F) and there is about one such dust particle in a cube of side 100 m of the ISM. Although the dust is very widely dispersed in the ISM, its presence can be detected because it produces a slight reddening of the light coming from distant objects in the cosmos; the light from the blue end of the spectrum suffers greater scattering by the dust in the intervening space between object and observer.

The temperature in the ISM varies with location but, typically, is about 7,000 K. This temperature is a measure of the speed of the atoms within the ISM and on average a hydrogen atom moves with a speed of 13 km s^{-1}. However, despite this high temperature, the energy content per unit volume of the ISM is very low, about 1.4×10^{-13} J m^{-3}. The aftermath of the Big Bang left radiation traversing the Universe, corresponding to an overall temperature that falls with time as the Universe expands. The present temperature is about 2.7 K, giving an energy density of 4×10^{-14} J m^{-3}, somewhat smaller than, but similar to, that due to the temperature of the ISM.

It is not possible for a star to form directly from ISM material; the Jeans critical mass is about 10^{38} kg, equivalent to 50 million solar-mass stars. To produce stars, it is necessary to have material that is much denser and colder. Such conditions occur in dense cool clouds that can be detected as dark opaque regions by telescopic observation (Fig. 2.1). Within a dense cool cloud (DCC), the density may be as high as 10^{-18} kg m^{-3}, 1,000 times greater than that of the ISM, and the temperature in the range of 10–50 K. The Jeans critical mass for

Fig. 2.1 The Horsehead Nebula (ESO).

such material is of the order of a few hundred solar masses, typical of that of an open cluster. In these regions, new stars are forming and they can be detected either by their visible light emissions or, at an early stage of their development when they are cooler, by the emission of infrared radiation.

A critical first stage in the formation of stars is the conversion of ISM material, the source of everything in the galaxy, to the state corresponding to a DCC, a process that I shall now describe.

2.2. The Formation of Dense Cool Clouds

Material in the galaxy is exposed to constant sources of heating. One of these is starlight, the radiation coming from stars. This has a variable effect, depending on how close the material is to radiating stars. Of much greater uniformity throughout the galaxy, and similar in magnitude, is heating by *cosmic radiation*. The term cosmic 'radiation' is somewhat of a misnomer; the great majority of the energy contained in the 'radiation' is due to very high-energy particles of every kind of atom, mostly hydrogen, with atomic numbers all the way up to uranium. Only about 0.1% of the energy is in the form of radiation — very high-energy electromagnetic radiation, γ-rays. Cosmic rays impinge on galactic material wherever it is and in whatever form it exists; some of it is absorbed and will act as a heating agent. Unless there is a compensating cooling process, the material will increase in temperature without limit — something that does not happen.

One type of cooling is due to radiation by dust particles that act just like the radiator of a car or a central heating system. The hotter the dust particles, the more they radiate and, under a constant rate of heating, they will acquire a temperature such that the heating and cooling rates are in balance. The theory of dust cooling was given in 1966 by the Japanese astrophysicist Chushiro Hayashi (b. 1920).[1] However, even more effective is cooling due to the presence of atoms, ions and molecules in the ISM. I shall explain the

---

[1] Hayashi, C., 1966, *Ann. Rev. Astron. Astrophys.*, **4**, 171.

general principle of how such cooling occurs by considering the effect of carbon atoms. Within galactic material, there are free electrons — electrons unattached to atoms — due to the ionization of atoms by the action of cosmic rays and collisions of atoms with each other. In a medium in which there are many different kinds of particle interacting with each other, a rule called the *equipartition theorem* operates. This states that the average kinetic energy of a particle, $\frac{1}{2}mv^2$ where m is its mass and v its speed, is the same for all particles. Because of their low mass, free electrons move quickly and hence make many collisions with atoms and ions. In Fig. 2.2(a), a carbon atom is shown in schematic form with six protons and six neutrons in its nucleus (black circle) and with a full complement of six electrons (small white circles) making it into a neutral atom. Figure 2.2(b) shows the effect of the collision of a free electron with one of the carbon electrons, which is pushed into an allowed state of higher energy (Appendix J). To provide the increased energy of the carbon electron, the free electron loses energy. Atomic systems tend spontaneously to adopt a state of lowest energy, the most stable state, so the electron that was pushed into a higher energy state falls back again to its original state; the released energy appears in the form of a photon, a packet of electromagnetic radiation. If the gaseous medium in which this occurs is very diffuse, the photon leaves the region at the speed of light without being absorbed. The net effect of this process is that the free electron has lost some of its energy. The temperature of the medium is a measure of the average

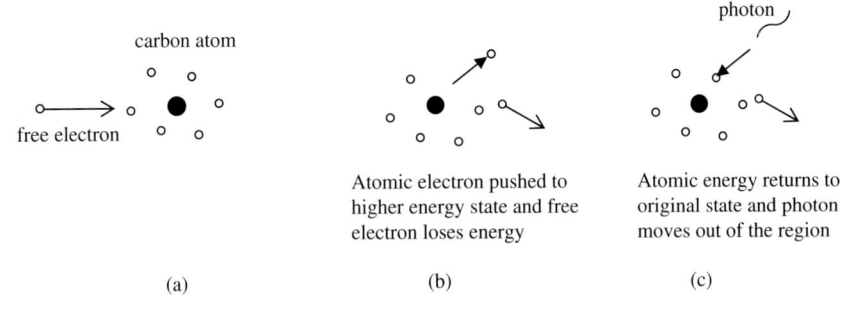

Fig. 2.2 The cooling of a gaseous medium by free-electron collision with a carbon atom.

kinetic energy of the particles within it; and if the free electrons, which are part of the medium, lose some of their kinetic energy, the medium cools. Similar processes can work with atomic ions and also with molecules, which can be pushed into configurations of higher energy by collisions with free electrons, atoms, ions or other molecules.

One of the strongest coolants is singly ionized carbon (C^+), carbon with one of the atomic electrons removed. The dependence of the carbon ion cooling rate on the density and temperature of the medium, with the expected concentration of carbon atoms, and for many similar cooling processes, was first deduced by the British astrophysicist Michael Seaton (1923–2007).[2] For C^+ it is of the form

$$\frac{dQ}{dt} = 1.04 \times 10^{14} \rho T^{-1/2} \exp\left(-\frac{92}{T}\right) W\,kg^{-1} \quad (2.1)$$

in which dQ/dt is the rate of loss of thermal energy per unit mass per unit time, ρ is the density of the medium and T is the absolute temperature. The coefficient before ρ in (2.1) depends on the assumed composition of the ISM. Other effective coolants are ionic silicon and iron; there are several important atomic coolants, especially oxygen which is very effective, and molecular coolants, such as molecular hydrogen (H_2).

The total cooling of the ISM, and of the material in DCCs, comes from a combination of radiation from dust and atomic, ionic and molecular cooling. When the gaseous medium, wherever it is situated, is in thermal equilibrium, then the total cooling rate must equal the total heating rate from a combination of starlight and cosmic rays. This heating rate is about $10^{-5}\,W\,kg^{-1}$ and is not heavily dependent on the state of the material as long as it is fairly diffuse. For any particular density, there will be a temperature at which the medium will be in thermal equilibrium. The equation that links the pressure of a gas to its density and temperature is

$$p = \frac{\rho k T}{\mu} \quad (2.2)$$

---

[2]Seaton, M., 1955, *Ann. Astrophys.*, **18**, 188.

in which p is the pressure, k is the Boltzmann constant[3] and μ the mean mass of the atoms or molecules comprising the gas. This relationship means that one can specify combinations of *density and pressure* of the medium that give particular cooling rates. These combinations are plotted in Fig. 2.3 for the cooling rate 10^{-5} W kg^{-1}, which is taken to be that which just balances the heating by cosmic rays and starlight. The sinuous nature of the curve plays an important role in the behaviour of the ISM.

Points A, B and C on the horizontal line in the figure give three states of the material for which both the pressure and the cooling rates are identical. Point A corresponds to a temperature of 8,150 K and a density of approximately 10^{-21} kg m^{-3}, appropriate to the ISM. Point C corresponds to a temperature of 10 K and a density of about 10^{-18} K, appropriate to a DCC. This shows that a DCC could exist in pressure equilibrium with the ISM while at the same time both the ISM and DCC could be in thermal equilibrium with the external heating sources. Point B is another equilibrium state but it is unstable; any slight variation of the conditions would precipitate a move either towards state A or state C. For medium material in

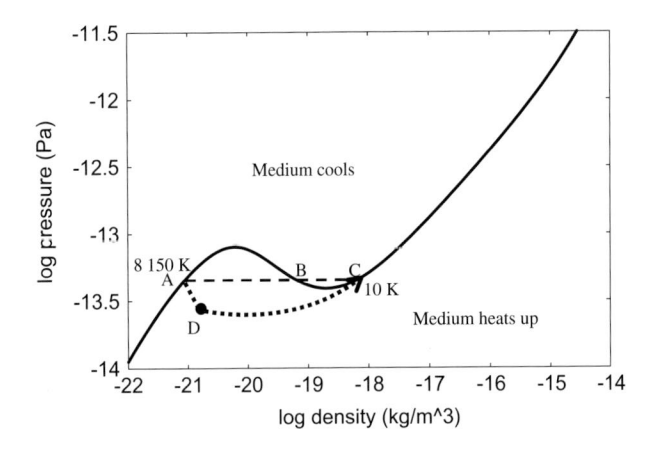

Fig. 2.3 The density–pressure curve giving a fixed cooling rate.

---

[3]The Boltzmann constant has dimensions energy per unit temperature rise and is 1.38×10^{-23} J K^{-1}.

a state above the curve, the pressure, and hence the temperature, is too high for thermal equilibrium for the given density and so the medium will cool. Conversely, for medium material in a state below the curve, the medium will heat up.

To understand how a DCC might form from the ISM, we must anticipate a topic that will be dealt with later, the occurrence of a *supernova*, the explosive event that follows the main sequence stage of the existence of some stars. When a supernova occurs, shock waves travel through the ISM, compressing it as it passes through. At the same time, the supernova, through nuclear reactions, produces large quantities of material in the form of heavy atoms that condense into dust, which is injected into the region through which the shock wave passes. The compression of the ISM increases the density and this, together with the injection of dust, increases the cooling rate within the affected region. Cooling is quite a rapid process and within a short time the pressure falls and the state of the material is changed from A to D. Now, two influences come into play. The first is the pressure of the ISM, which is greater than that of the affected material and so compresses it further and increases its density. The second influence is that the state of the affected material is below the line so that it will heat up. The net effect of the two influences is to move the state from point D to C, which creates a DCC embedded in the ISM.

The description given above is somewhat idealized. A computer-based simulation carried out by Yann Golanski and myself in 2001,[4] using a numerical technique known as smoothed-particle hydrodynamics (SPH; Appendix L), gave the evolutionary path from A to C, staying quite close to the thermal equilibrium line and sometimes crossing it — but the final result is substantially the same.

The process of forming a DCC does not happen quickly; the SPH calculation above gave a time of 27 million years to produce the DCC. The same calculation showed that after the DCC reached point C, it continued to collapse. The reason is that for the conditions at position C, the Jeans critical mass was about 90 solar masses and

---

[4]Golanski, Y. and Woolfson, M.M., 2001, *Mon. Not. R. Astron. Soc.*, **324**, 573.

the DCC region had a larger mass than that. The form of collapse of a mass appreciably greater than a Jeans critical mass approximates to what is known as *free-fall collapse* (Appendix M), which assumes that only gravity is acting without the braking effect of gas pressure. For a real DCC, the collapse will increasingly be slowed down by gas pressure, especially in the later stages when the cloud becomes more opaque and most or all of the heat energy being generated by the collapse is retained. Nevertheless, in a period known as the *free-fall time* (Appendix M), which is about 2 million years for the cloud we are considering, the cloud will have collapsed to something like one-tenth of its original dimensions.

2.3. Maser Emission from Star-Forming Regions

In the late 1960s radio telescopes were picking up very intense radio-wave emissions coming from compact sources within star-forming regions. These radio waves had the property of being at well-defined frequencies, which is not the usual pattern for most astronomical radio sources. The Sun, for example, sends out radio waves over a wide and continuous range of frequencies. The sources had the characteristics that, in the visible part of the electromagnetic spectrum, are associated with lasers. A laser (Light Amplification by Stimulated Emission of Radiation) is a physical system that amplifies light of a sharply defined frequency, corresponding to an atomic electron transition between one allowed state to another (Appendix J), and radiates it in a specific direction. Somehow, by a mechanism that is not understood, this is happening in star-forming regions for systems that operate at radio frequencies. The phenomenon is known as *maser emission* where the 'm' in 'maser' stands for 'microwave'.

The energy transitions that give these radio frequencies come from molecules such as water (H_2O), carbon monoxide (CO) and carbon dioxide (CO_2), or from chemical radicals — unstable entities that cannot exist for any length of time on Earth because they combine with something else to form a stable molecule, but that have a long lifetime in space — such as hydroxyl (OH) and carbon hydride

(CH). These energy transitions are between different vibration states, or rotation states, of the molecules or free radicals. Thus the hydroxyl radical can give several characteristic frequencies — 1,720 MHz,[5] 1,667 MHz, 1,665 MHz and 1,612 MHz — corresponding to wavelengths from 0.174 m to 0.186 m.

The importance of these maser emissions is that they give information about motions of the medium within which the stars are forming. It is found from Doppler shifts in the frequencies of the maser radiation that different parts of the medium are moving at different speeds, the characteristic of a region in a turbulent condition. Another measurement that can be made is of the sizes of the emitting regions. A general conclusion is that in star-forming regions, within which there are turbulent motions with characteristic speeds of up to $20\,\mathrm{km\,s^{-1}}$, there are discrete sources of diameter in the range of 10^8–10^9 km within a region of overall size of order 10^{12} km. The fact that there is turbulence within star-forming regions has a bearing on how stars form.

2.4. The Process of Protostar Formation

The starting point we have for the formation of stars is a collapsing DCC with a mass of a few hundred solar masses, density of $10^{-18}\,\mathrm{kg\,m^{-3}}$, temperature in the range of 10–50 K and with turbulent motions with average speeds of, say, $10\,\mathrm{km\,s^{-1}}$. As the cloud collapses, it releases gravitational energy that can be transformed into either heat energy or an increase in turbulent energy. In the early stages of the collapse the cloud will be transparent to radiation so that the heat energy will be radiated away and the temperature will remain approximately constant. Since the mean density of the cloud is increasing without change of temperature, this means that the Jeans critical mass of the cloud material is decreasing. If the initial temperature were 10 K and stayed constant, then by the time the density reached $10^{-16}\,\mathrm{kg\,m^{-3}}$ the Jeans critical mass would have fallen to 1 solar mass.

---

[5] 1 hertz (Hz) is one vibration per second. 1 MHz = 10^6 Hz.

The increase in turbulent energy due to the cloud collapse would be offset by an increased rate of collisions of streams of material within the cloud. This would generate heat, which would radiate away, and hence reduce the turbulent energy. However, it is these very collisions that assist in the process of star formation, for the effect of two streams of material colliding head-on is to produce a region of increased density. The effect of compressing the gas is to heat it and its subsequent behaviour is both to re-expand and also to cool by the processes described in Section 2.2. Cooling is a much faster process than re-expansion, so before the region has greatly expanded it has returned to approximately its original temperature; a compressed, heated gas that was previously below a Jeans critical mass has cooled and may now be above the critical mass, so that further collapse may then begin. Even if this happens, it is not inevitable that the collapsing mass will form a star. Free-fall collapse begins extremely slowly (Appendix M) and if the region is buffeted by further turbulent streams of material before it has substantially collapsed, then the material may be stirred up and reabsorbed into the general body of the cloud.

2.5. The Formation of Binary Systems

Given that an isolated body of gas, a *protostar*, has formed with above the Jeans critical mass, say of mass 1.5 solar masses (3×10^{30} kg), radius 3×10^{14} m (2,000 au), with material of mean molecular mass 4×10^{-27} kg and temperature 25 K, the next question to consider is how it would evolve. This critically depends on the angular momentum of the protostar. Let us say that the initial angular speed is 5×10^{-14} radians per second; this is just 50 times the angular speed with which the galaxy spins and is probably a reasonable rate. The density of the cloud, assumed to be spherical, would be 2.65×10^{-14} kg m^{-3} and the speed at the equator of the spinning sphere would be

$$\text{radius} \times \text{angular speed} = 15 \,\text{m s}^{-1}. \qquad (2.3)$$

If the cloud collapsed to stellar dimensions, the radius would have to shrink to about 10^8 m, and hence, to conserve angular momentum, the angular speed would increase by a factor of $(3\times 10^6)^2 = 9\times 10^{12}$, that is, to 0.45 radian per second. The equatorial speed would then be $4.5\times 10^4\,\mathrm{km\,s^{-1}}$, an absurd value since at that speed the star would fly apart.

The answer to the problem of what would happen as the protostar developed was, once again, given by that rich source of astrophysical theory, James Jeans. He considered the problem of the behaviour of collapsing spinning spheres and distinguished two extreme cases. The first is the case of a sphere of incompressible material, essentially a liquid sphere. Since liquid is incompressible, such a sphere cannot collapse but the effect of collapse while keeping the same density everywhere can be simulated by considering the effect of increasing the spin rate of the body. As it spins more rapidly, it takes on an increasingly distorted shape and finally breaks into two parts. The progression of shapes is shown in Fig. 2.4.

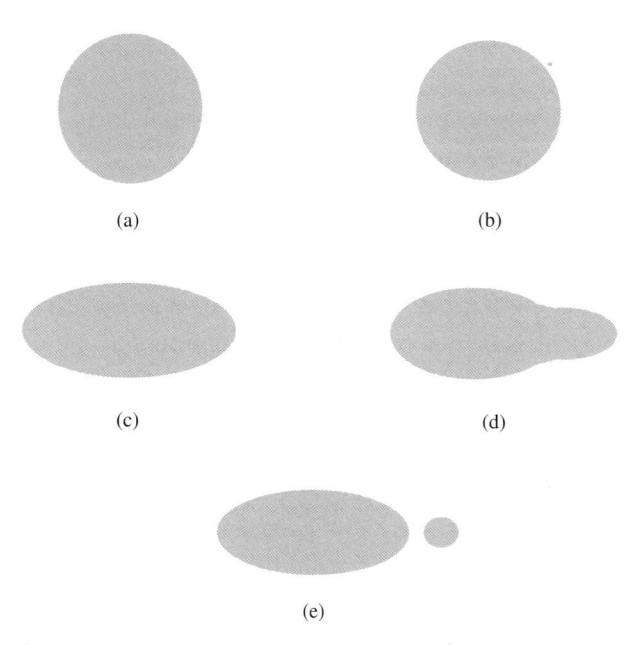

Fig. 2.4   The stages in the simulated collapse of a spinning liquid sphere. (a) The original sphere (b) A Maclaurin spheroid (c) A Jacobi ellipsoid (d) A pear-shaped configuration (e) Fission into two bodies.

The first departure from an almost spherical shape is shown in Fig. 2.4(b) and is a *Maclaurin spheroid*, a shape like a flattened football. It is a special form of ellipsoid,[6] an *oblate spheroid*, where the semi-major axis along the spin axis is shorter than the other two, which are equal. The flattening of the ellipsoid continues until the ratio of the three semi-major axes becomes 1.72:1.72:1.00, after which the shape becomes a *Jacobi ellipsoid*, shown in Fig. 2.4(c). This is where the three semi-major axes are all different and they gradually change as the body spins faster. When the ratio of the three semi-major axes becomes 1.89:0.81:0.65, the shape changes to that shown in Fig. 2.4(d), a pear-shape. As the increase of spin rate continues, the neck between the large and small ends of the 'pear' becomes more distinct until the small end breaks away and one is left with two bodies orbiting around their centre of mass.

The second extreme case is the Roche model for a body where all the mass is concentrated at the centre, surrounded by an 'atmosphere' comprising all the volume but a negligible amount of mass. In this case the form of evolution is just like that shown in Fig. III in relation to Laplace's nebular model. The body first takes the form of an oblate spheroid, then a lenticular form with a sharp edge and finally, with further collapse, material is left behind in the equatorial plane.

Initially, a collapsing protostar has the characteristics of neither of the extreme cases. It is certainly compressible; a characteristic of the second case, but it would be of more or less uniform density, a characteristic of the first case. At the beginning it would undergo something like a free-fall collapse since pressure forces would be low and heat energy would be radiated out of the system, thus limiting the build-up of pressure. Eventually, during the collapse, it would have become opaque and the pressure in it would increase more and more rapidly as it collapsed. The compressibility of a gas decreases as its pressure increases and it would begin to take on some of the

---

[6] An ellipsoid is a solid shape, the three principal projections of which are ellipses with semi major axes a_x, a_y and a_z. If all three semi-major axes are different, it is a general ellipsoid. If two are equal, it is a spheroid, and if all three are equal, it is a sphere.

characteristics of a liquid body, while still continuing to collapse, albeit at a reduced rate due to the increasing pressure.

The next stage in the development would be the adoption of a form like a Jacobi ellipsoid, followed by fission into two unequal parts. Gravity would form these into two roughly spherical blobs, distorted by their mutual gravitational tidal effects. They would form a binary system, with separation probably of order 1–10 au, the material of which was in rigid rotation around the centre of mass (Fig. 2.5(a)). The individual protostar blobs would have a spin rate equal to the orbital rate of the binary system — so that it would be spinning like a rigid body. However, the individual protostars would continue to collapse and, to conserve angular momentum, would have a tendency to spin faster. Now an interesting effect occurs that actually prevents them from spinning faster and, instead, causes the distance between the two protostars to increase. This is illustrated in Fig. 2.5(b). For each of the protostars, the tendency for them to spin faster due to collapse will cause the tidal bulges to point ahead of the line joining the centres of the two protostars. It can be seen from the figure that the gravitational effect of each star on the tidal bulge of the other is to exert a force with a component tending to retard its spin. The angular momentum taken from the spinning protostars has to go somewhere since angular momentum must be conserved, and it goes

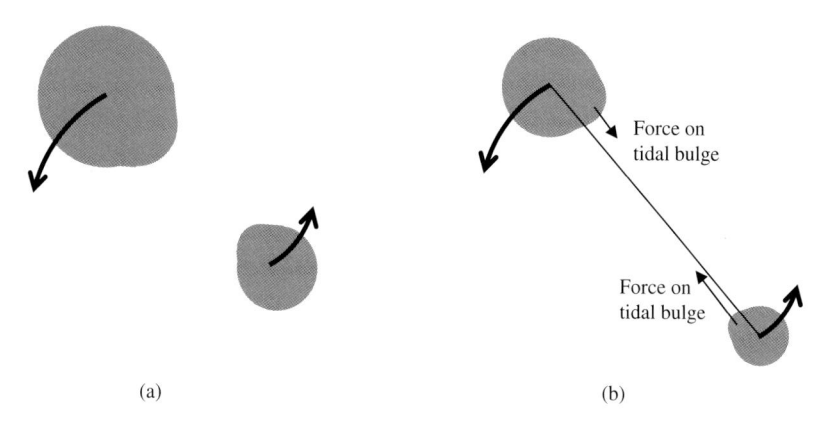

Fig. 2.5 (a) The initial binary star configuration. (b) Forces with components pulling backwards on tidal bulges leading to the binary components moving apart.

into pushing the two stars apart. As the protostars move further apart, their mutual tidal effects become weaker until eventually the distance between the protostars stabilizes. After they become tidally decoupled, angular momentum is conserved within each of the collapsing spinning protostars and their spin rates increase.

For the protostar with mass 3×10^{30} kg, radius 3×10^{14} m and spin angular speed 5×10^{-14} radians per second, we assume a final binary system with components of masses 2×10^{30} kg and 10^{30} kg. Eventually the binary system has a star separation of about 2.19 au and a period of 2.64 years (Appendix N). If a star then shrinks by a factor of 10 after tidal decoupling, then the final spin period of the star will be about 9.6 days (Appendix N).

The characteristics of the final binary system will depend on the conditions in the initial protostar, and will be very sensitive to its initial angular speed, so that there will be a wide variety of binary star separation distances and stellar spin rates.

2.6. Modelling the Collapse of a Cloud

In 2002 Matthew Bate, Ian Bonnell and Volker Bromm carried out a very comprehensive modelling of the formation of binary and multiple-star systems using SPH (Appendix L).[7] One objective of their calculation was to see how many close binaries, with star separations less than 10 au, would form. They argued that close binaries in which the separation of the components was less than about 10 au could not form directly by the process described in the previous section because heating effects, due to increased opacity of the gas, would prevent the collapse to the extent that would give such separations. They started their calculation with a turbulent cloud with mass equal to 50 solar masses, density 1.53×10^{-17} kg m^{-3} and temperature 10 K. When they stopped their massive calculation they had formed 23 stars and a number of sub-stellar masses known as *brown dwarfs*.[8] Several other smaller bodies were still accreting

---

[7]Bate, M., Bonnell, I. and Bromm, V., 2002, *Mon. Not. R. Astr. Soc.*, **332**, L65.
[8]A brown dwarf is a body intermediate in mass between a planet and a star. It is massive enough to generate a temperature that will initiate nuclear reactions involving

matter and would probably have ended up either as stars or brown dwarfs. Seven close binary systems were formed with separations less than 10 au but only one of these was completely separate from the remainder of the system; the remaining six were still gravitationally bound to other stars. The conditions of the calculation were designed to prevent close binary systems from being formed directly, in accordance with their initial assumption, but they came about by the dynamical interaction of initially wider binary systems with cloud material. Accretion of matter by the stars of a formed binary tends to drive them closer together as does the interaction of a binary system with a disk surrounding it. If a system of three interacting stars forms, then the action of a surrounding disk can drive the stars closer together and give an unstable system in which one star is ejected, leaving behind a close binary system. In the conditions under which a stellar cluster forms, stars are constantly interacting with each other and all types of outcome can result, from unattached stars, binary systems and other kinds of multiple systems.

2.7. The Spin of Stars

After a star forms, there are many influences that can affect the rate at which it spins. In 1979, in a scientific paper on star formation that was flawed because it did not consider the production of binary systems,[9] I showed that the accretion of material by a star after it formed steadily increased its angular momentum. The theory explained well, and quantitatively, the relationship of equatorial speed to mass as shown in Fig. 1.11. It is based on the assumption that material is accreted in discrete lumps coming in from random directions. Each lump striking the surface obliquely adds angular momentum to the star but since the lumps come in randomly from all directions, their effects tend to cancel each other. However, if there are n lumps, the ith lump of which provides an angular momentum component with magnitude dJ_i, then the expected square of the

---

deuterium, an isotope of hydrogen, but no other nuclear reactions. Brown dwarf masses are between about 13 and 70 Jupiter masses.
[9]Woolfson, M.M., 1979, *Phil. Trans. R. Soc. Lond.*, **A291**, 219.

magnitude of the final angular momentum is given by

$$J^2 = \sum_{i=1}^{n} (dJ_i)^2. \qquad (2.4)$$

Thus as the mass increases by accretion, so does the expected magnitude of the angular momentum.

For smaller mass stars, those that either do not accrete or accrete very little, it may be necessary to lose angular momentum to explain what is observed. Something that well-developed stars like the Sun exhibit is the phenomenon of a stellar wind, a constant loss of material in the form of charged particles — predominantly protons and electrons — moving with speeds of several hundreds of kilometres per second. In the case of the solar wind, the rate of mass loss amounts to nearly 2 million tonnes[10] per second, which may seem a great rate of loss, but maintained over a period equal to the present age of the Sun, it corresponds to a loss of just 0.015% of its initial mass. However, very early in their existence, when they approach the main sequence, sun-like stars go through a T-Tauri stage where they lose mass at a much faster rate, typically some 10^{-8} of a solar mass per year, and they may lose up to 10% of their total mass while in this state. These T-Tauri stars are also sources of strong magnetic fields and as the star rotates, so does the magnetic field. The path of a charged particle in a magnetic field is in the form of a helix around a magnetic field line[11] (Fig. 2.6). As the charged particle moves away from the star, so the magnetic field becomes weaker and the radius of the helix increases. Eventually the field becomes so weak that the charged particle breaks free and is no longer constrained by the field.

Helical path of changed particle

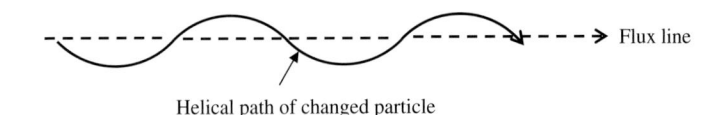

Fig. 2.6 The path of a charged particle linked to a magnetic flux line.

---

[10] 1 tonne = 1,000 kg
[11] A magnetic field line at any point shows the direction that would be indicated by a compass needle.

The important thing about this mechanism is that the lost material moves away from the star with a *constant angular velocity* because it is linked to a magnetic field line that rotates with the star. This means that, since the lost material is increasing its distance from the spin axis while retaining its angular velocity, it gains angular momentum; to compensate for this, so as to conserve angular momentum, the star spins at a reduced angular velocity.

The angular momentum of a spherically symmetric body of mass M and radius R, spinning with angular velocity ω about a diameter, is of the form

$$J = \alpha M R^2 \omega \text{ (Appendix N)} \tag{2.5}$$

where the *moment of inertia factor*, α, is 0.4 for a uniform sphere but less for astronomical bodies that have their mass concentrated towards the centre. For the Sun, α is 0.055. We now consider the loss of a mass m of the star from its equator with the radius of the star substantially unchanged by the mass loss. If this mass is coupled to a field line out to a distance βR, assuming that β and the rate of mass loss are constant, it is possible to calculate the ratio of the final angular velocity to the initial after any time (Appendix O). Figure 2.7 plots this ratio as a function of time for a mass loss equal to 10^{-8} of the initial mass of the star per year, and with $\beta = 3$.

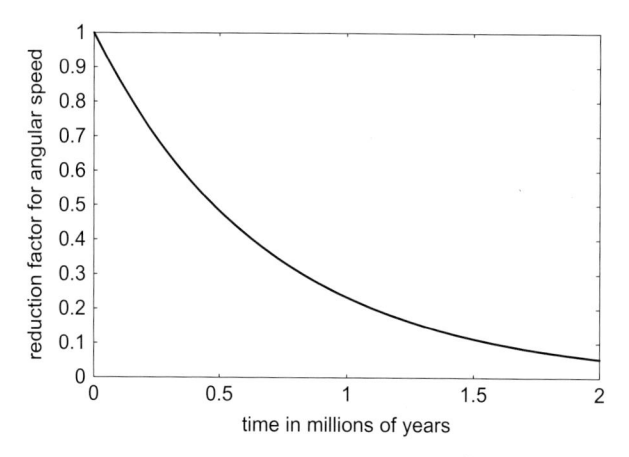

Fig. 2.7 The spin slowdown factor for mass loss 10^{-8} of initial mass per year and $\beta = 3$.

From Fig. 2.7 it is clear that during the T-Tauri stage of a star's existence, considerable slowdown of the spin can occur — perhaps up to a factor of ten. However, any mass gain and corresponding angular momentum gain after the T-Tauri stage will largely be retained.

2.8. Summary

The basic material from which stars are formed is the interstellar medium (ISM), consisting mainly of hydrogen and helium, typically with density 10^{-21} kg m^{-3} and temperature about 7,000 K. A minor, but important, component is dust — stone, iron or volatile-material sub-micron particles that account for 1–2% of the mass of the ISM.

The effects of a supernova, compressing a region of the ISM and injecting coolant materials into it, can precipitate the formation of a dense cool cloud (DCC) within which the density may be about 10^{-18} kg m^{-3} with temperature in the range of 10–50 K. Such a cloud can be in pressure equilibrium with the ISM and also in thermal equilibrium, with dust radiation, atomic, ionic and molecular cooling processes just balancing heating due to starlight and cosmic radiation.

If the mass of the DCC exceeds the Jeans critical mass, then it will collapse and turbulent motions, which can be detected by maser emissions, result in collisions between streams of gas. The compressed gas cools rapidly and the outcome can be a stellar mass of material exceeding the Jeans critical mass. If this is not disturbed by further turbulent collisions before it appreciably collapses, then it will become the starting point for a process of star formation.

As the material collapses it will spin more rapidly and distort first into the form of an oblate spheroid, then an ellipsoidal form, and next a pear-shaped form that eventually will fragment into two distinct protostars forming a binary pair. Tidal coupling of the stars will cause the stars to move apart but, once they become decoupled, the collapse of the individual stars will cause them to spin more rapidly. Detailed modelling of the evolution of a cluster of stars using smoothed particle hydrodynamics shows the formation of close binary systems, some with stellar separation of less than 10 au, individual stars and multiple star systems.

The accretion of matter by a star after it forms increases its angular momentum and this process explains the relationship of the mean equatorial speed and the mass of main-sequence stars. Stars can also lose angular momentum due to the coupling of stellar-wind particles and magnetic field lines that rotate rigidly with the spinning star.

Chapter 3

The Life and Death of a Star

In Chapter 2 I described what might be thought of as the birth of a star, a protostar, which is a big diffuse ball of gas and dust collapsing under the influence of gravitational forces. In the living world, birth is the creation of a small entity that becomes larger as it develops to maturity and then, finally, decays and dies. By contrast, for a star the path from infancy to maturity, and then to what might be thought of as its death, involves fluctuations of size with the final state being a very small object by stellar standards; it is a path that involves many stages. I shall now relate to you what theorists have deduced to be the life history of stars.

3.1. The Journey to the Main Sequence

Reference has already been made to the main-sequence stage of a star's existence. During this part of its life, a star is generating energy by nuclear processes that convert hydrogen into helium in its core, and externally the star seems to change very little during that period. The Sun is a middle-aged star, having spent nearly five billion years on the main sequence and with another five billion years still to pass before it reaches the notable change of behaviour that betokens the end of its main-sequence lifetime. The Japanese astrophysicist Chushiro Hayashi has described the journey from birth to the beginning of the main sequence. His description is best followed on a chart known as the *Hertzsprung–Russell* ($H-R$) *diagram* in which the state of an astronomical object is defined by its temperature, plotted on the x-axis, and its luminosity, plotted on

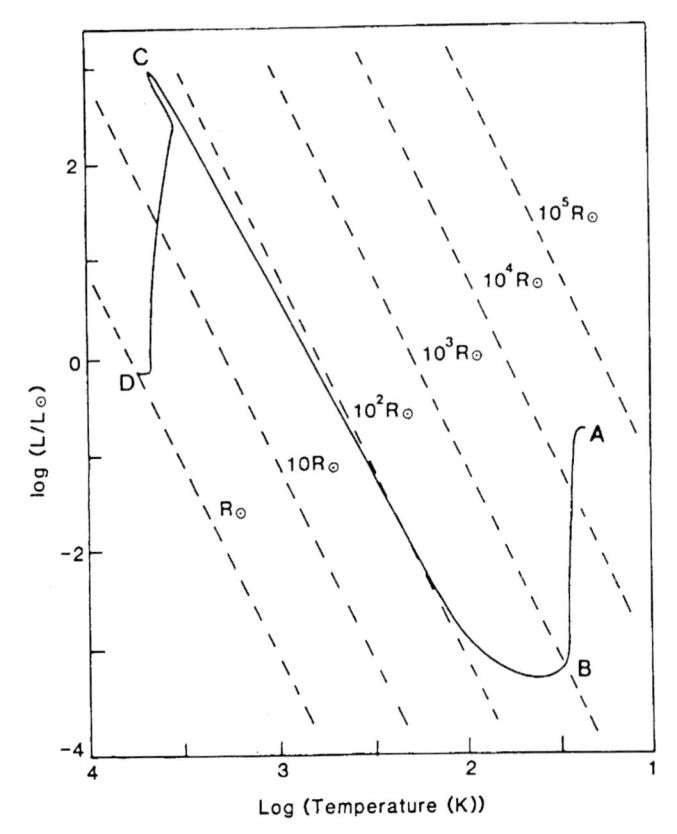

Fig. 3.1 The Hayashi plot for a star of one solar mass. $R_\odot$ is the radius of the Sun.

the y-axis. Because of Eq. (1.2), which links luminosity to radius and temperature, a point on the diagram also indicates the radius of the object. The *Hayashi plot*, showing the journey from a protostar to the main sequence is given in Fig. 3.1, with dashed lines representing the radius in units of the present solar radius.

Because a wide range of temperature and luminosity has to be covered in the plot, they are represented on a logarithmic scale so that each increment of 1 on either axis represents a factor of 10. This also has the advantage that states with the same radius fall on a straight line in the diagram. The initial state, represented by point A, corresponds to a temperature of 25 K, radius of 2×10^{13} m (about

30,000 times that of the present Sun), density of $6 \times 10^{-11}\,\mathrm{kg\,m^{-3}}$ and luminosity of about one-fifth that of the present Sun. It would not be luminous in the sense that it could be seen since the radiation it emits would mostly be in the infrared part of the spectrum. It might seem strange that a body so cool should be so luminous but this is due to its huge radius and, hence, radiating surface area. The first part of the journey, from A to B, involves a large reduction in radius but a comparatively small change in temperature. This is because throughout the period, the protostar is transparent to radiation so that the heat energy generated by compression of the material is radiated away. At point B the opacity[1] of the star has increased to a level such that much of the generated heat energy is being retained, so that the temperature, and hence the pressure, increases and the rate of collapse slows down. The collapse continues to point C at which a bounce occurs as the protostar moves through an equilibrium position and then returns to it again. The duration of this bounce, from Hayashi's calculation, is about 100 days; in 1936 an American astrophysicist, G.H. Herbig, observed a temporary increase in the luminosity of a young star, FU Orionis, by a factor of over 200 during a period of less than one year, which could have been a manifestation of the bounce predicted in the Hayashi plot. At point C the radius of the star is approximately 50 solar radii, corresponding to a density of $0.01\,\mathrm{kg\,m^{-3}}$, with a temperature of $4{,}000\,\mathrm{K}$ and luminosity 1,000 times that of the Sun.

At the end of the bounce, the star is in a temporary state of equilibrium — an equilibrium that cannot be permanent since the surface of the star is radiating energy. Rather paradoxically, the radiation of heat by the star makes it *hotter*. The reason for this is that the star is contracting and the energy released by the collapse both enables the star to radiate and also to increase in temperature (Appendix P). This stage of the process, which takes the star from C to D, takes about 50 million years for a solar-mass star and is called the *Kelvin–Helmholtz contraction*. Stars in this part of their life are

---

[1] The opacity of stellar material describes its propensity to absorb radiation passing through it.

usually referred to as *Young Stellar Objects* — YSOs. During this stage the YSO is in a state of equilibrium that is constantly changing. The surface temperature is slowly rising, eventually reaching about 5,800 K but, more importantly, the internal temperature is also increasing. When the temperature at the centre reaches a few million degrees, the first nuclear reactions begin to take place, but they do not generate enough energy substantially to affect the general progress of the Kelvin–Helmholtz contraction. However, eventually, at point D, the central temperature reaches 15 million degrees and then the energy generated by nuclear reactions just balances that being radiated from the surface. The contraction comes to a halt and the star has embarked on the main-sequence stage of its development.

The Kelvin–Helmholtz path of a star to the main sequence depends on its mass and Fig. 3.2 shows the paths for stars of a solar mass and greater. The path for a solar-mass star in this figure is not quite the same as that shown in Fig. 3.1 since it was calculated later with a better stellar model, but the essence of the development is the same. The times taken in the final contraction to the main sequence are mass-dependent as are the lifetimes on the main sequence. Table 3.1 shows both these times for a variety of stellar masses. The main-sequence lifetime depends on how long it takes to exhaust the hydrogen fuel for nuclear reactions in the centre of the star. The way in which this depends on the mass of the star is explained in Appendix Q.

3.2. Energy Generation in Main-Sequence Stars

As the star follows the Kelvin–Helmholtz path, the centre steadily increases in temperature. When the temperature reaches 2–3 million degrees, the first nuclear reactions take place. Before I explain what these are, it is first necessary to establish some of the conventions for describing the structure of nuclei. The basic structure of an atom is described in Appendix J. The electrons are the important part of the atom when it comes to chemical reactions or absorbing or emitting photons in, or close to, the visual range, but nuclear reactions, as the name implies, are concerned only with changes in

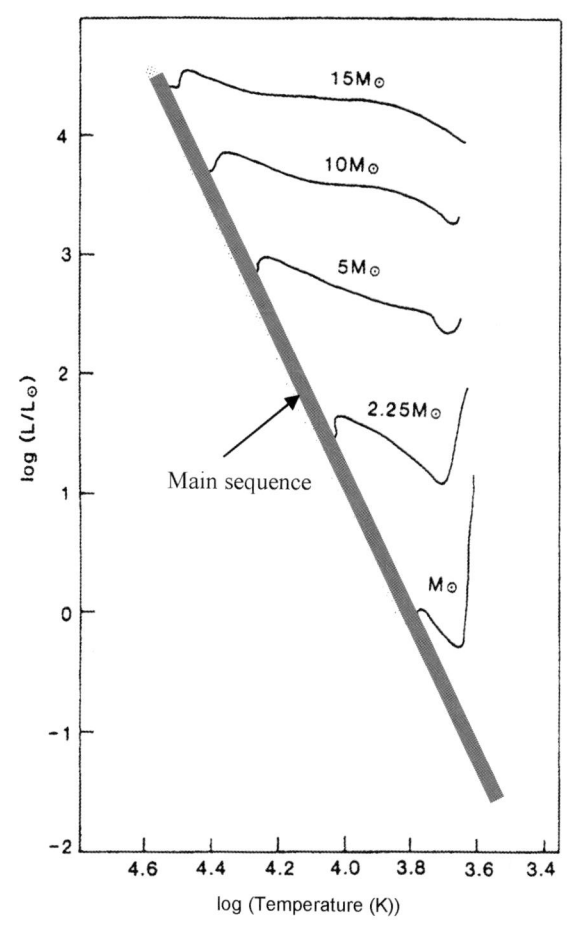

Fig. 3.2 Kelvin–Helmholtz paths for various stellar masses. $M_\odot$ is the solar mass.

nuclear structure. A carbon atom, in its most common form, contains six protons and six neutrons in its nucleus and can be represented as $^{12}_{6}$C. C is the chemical name for 'carbon'. The pre-subscript 6, gives the total charge of the nucleus, in units that are the magnitude of the electronic charge; for a nucleus, this charge equals the number of protons in the nucleus. The pre-superscript 12, gives the total number of protons + neutrons in the nucleus; since these particles have approximately equal mass, this number is an indication of the mass of the nucleus. There is some redundancy in this description

Table 3.1 Kelvin–Helmholtz and main-sequence lifetimes for stars of various masses.

Stellar mass (solar units)	Kelvin–Helmholtz lifetime (10^6 years)	Main-sequence lifetime (10^6 years)
15	0.062	10
9	0.15	25
5	0.58	100
3	2.5	350
2.25	5.9	900
1.5	18	2,000
1.25	29	4,000
1	50	10,000
0.5	150	50,000

since a carbon atom *must* contain six protons; often the carbon atom would just be represented by ^{12}C, with the unnecessary 6 removed. You may have noticed that I described the carbon atom as in its 'most common form' because there is another stable *isotope* of carbon that is represented by $^{13}_{6}$C. Carbon-13 forms just over 1% of all terrestrial carbon — for example, that which exists in trees or in the human body. Because it contains six protons, it is *carbon* and in chemical reactions will behave just like carbon-12. The extra neutron in carbon-13 makes no difference chemically.

The first reactions that take place in the centre of the developing star are those involving an isotope of hydrogen, one so important that it is given its own individual name, *deuterium*, and its own chemical symbol, D. Hydrogen is represented by $^{1}_{1}$H, showing that its nucleus is just a single proton, and deuterium is represented by $^{2}_{1}$D, showing that deuterium contains an additional neutron in its nucleus. The first reaction can be described by

$$^{2}_{1}D + ^{2}_{1}D \rightarrow ^{3}_{1}T + ^{1}_{1}H. \tag{3.1}$$

The left-hand side of (3.1) indicates that the nuclear reaction involves two deuterium nuclei. The right-hand side gives the products of the reaction. The first product is another isotope of hydrogen, *tritium*, which also has its own name and symbol, T. Tritium, which has a proton and two neutrons in its nucleus, is radioactive, which means that it is intrinsically unstable and spontaneously decays to give other products. Tritium has a half-life of 12.3 years, meaning that at the end of a period of 12.3 years, one half of the tritium atoms present at the beginning will have decayed. The second product on the right-hand side of Eq. (3.1) is a hydrogen nucleus or, in other words, a proton.

Only about 2 parts in 10^5 of the hydrogen in a newly formed star is deuterium, so the heating due to the energy released by (3.1) is rather small by stellar standards. However, once the central temperature in the star rises to 15 million K, a new nuclear reaction occurs that involves hydrogen, of which there is a plentiful supply. This reaction takes place in several stages. The first is

$$^1_1\text{H} + ^1_1\text{H} \rightarrow ^2_1\text{D} + ^0_1\text{e} + \nu. \tag{3.2a}$$

This reaction between two hydrogen nuclei (protons) gives a deuterium nucleus and a particle that has a positive charge but no nucleons — the kind of particles that constitute the nucleus — either protons or neutrons. This particle is a *positron*, which is exactly like an electron except that it has a positive charge. The other particle, indicated by ν, is a neutrino, a strange particle with no charge and with so little mass that at one time it was thought that it might have no mass. It is an important particle in the world of nuclear physics and actually there are many different kinds of neutrino indicated by different symbols. We need not concern ourselves greatly with it except to note its presence. The second stage is

$$^2_1\text{D} + ^1_1\text{H} \rightarrow ^3_2\text{He} + \gamma. \tag{3.2b}$$

The left-hand side shows that the reaction involves a deuterium nucleus and a hydrogen nucleus. The first product on the right-hand side is a stable isotope of helium, helium-3. This is a very tiny part of terrestrial helium that is mostly helium-4. The second product on the right-hand side is a gamma-ray photon, a packet of very high-energy electromagnetic radiation. The final stage in the process is

$$^3_2\text{He} + ^3_2\text{He} \rightarrow ^4_2\text{He} + 2^1_1\text{H}. \tag{3.2c}$$

Here, two helium-3 nuclei react to give a helium-4 nucleus, plus two protons.

The overall effect of two reactions of type (3.2(a)), followed by two reactions of type (3.2(b)), followed by a single (3.2(c)) reaction, is that four hydrogen nuclei (protons) are converted into a helium-4 nucleus. The mass of the helium nucleus is less than that of four protons and this extra mass is converted into energy according to Einstein's well-known equation $E = mc^2$, in which m is the mass converted into energy, c is the speed of light and E the amount of energy produced. The rate of loss of mass of the Sun due to this process is 4 million tonnes per second, which sounds horrendous but corresponds to a loss of a fraction 0.0003 of its mass over the less-than 5 billion years of its main-sequence existence to the present time.

3.3. Leaving the Main Sequence for Low- and Moderate-Mass Stars

The conversion of hydrogen to helium progresses most rapidly in the very centre of a star where the temperature is highest. Consequently, in the central region the concentration of hydrogen decreases with time. Paradoxically, this leads to an *increase* in the rate of energy production in the core. As the hydrogen is exhausted, the equilibrium of the star is maintained by a slow collapse of the central regions that increases both the temperature and density. Rates of nuclear reactions are highly sensitive to temperature and initially the increase in temperature overcomes the factor of the dilution of hydrogen to give an increased rate of energy production, and hence of hydrogen exhaustion. The increased pressure generated from the greater generation of heat at the centre leads to a slight expansion of the star and an increase in its surface temperature. This is illustrated in Fig. 3.3 for a solar-mass star as the passage from point A to point B on the H–R diagram. Hydrogen becomes exhausted in the core but the surrounding regions are still hydrogen-rich and so it continues to burn in a shell surrounding the core. As time progresses, this hydrogen-shell burning moves outwards as the hydrogen becomes exhausted at increasing distances from the

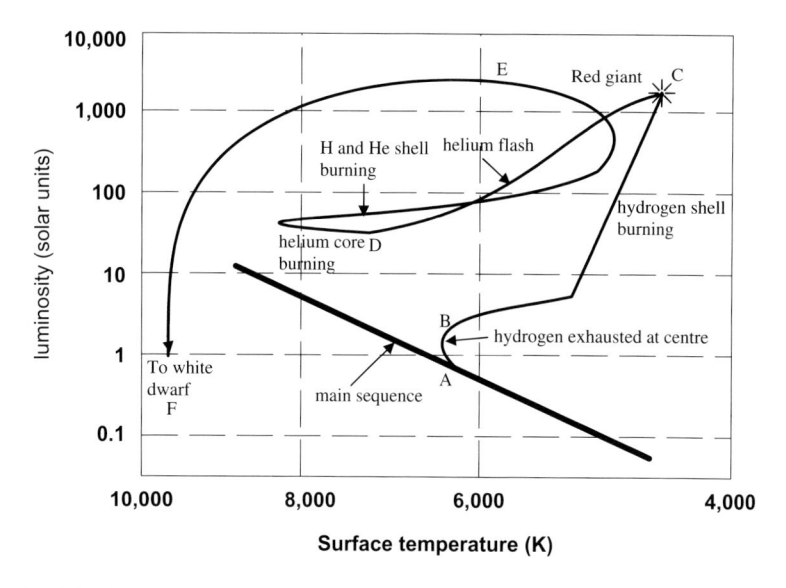

Fig. 3.3 The progress of a solar-mass star on leaving the main sequence.

centre. During this period there are no nuclear reactions going on at the centre of the star and the shell-burning applies pressure inwards, adding to the normal gravitational pressure and so compressing the core material to ever-higher density and temperature. At the same time the burning shell applies pressure outwards and so expands the star, which has the effect of reducing the surface temperature. This is represented in Fig. 3.3 as the path from B to C, and at point C the star has become a *red giant*, with radius of order 1 au and a surface temperature of around 4,500 K. When the Sun reaches this stage in five billion years from now, it will envelop the terrestrial planets out to Earth. During all this time the core of the star, now consisting mainly of helium, is increasing in density and temperature until, eventually, the conditions are reached for nuclear reactions involving helium to take place, which requires a temperature of order 10^8 K. The net outcome of these new reactions is that helium is converted into carbon, something that happens in two stages. In the first stage two helium nuclei react to give a nucleus of beryllium-8:

$$^4_2\text{He} + ^4_2\text{He} \rightarrow ^8_4\text{Be}. \tag{3.3a}$$

Beryllium-8 is an unstable nucleus that reconverts back into two helium nuclei with an extremely short half-life of 7×10^{-17} seconds, but in the very dense environment of the core at this stage there is a reasonable chance that some of the beryllium-8 will react further before it breaks down by the reaction

$$^{8}_{4}\text{Be} + ^{4}_{2}\text{He} \rightarrow ^{12}_{6}\text{C}. \tag{3.3b}$$

The net effect of this pair of reaction is that three helium nuclei have been converted into a carbon-12 nucleus. One type of particle emitted by radioactive materials is the α-particle, which is the same as a helium nucleus, so this pair of reactions is called the *triple-α process*.

At this stage the very dense helium core is in what physicists refer to as a *degenerate state*. Because the nuclei and the accompanying electrons have been squeezed so tightly together, the normal classical behaviour of matter has given way to behaviour governed by quantum mechanics. For a classical gas, as the temperature increases so does the pressure, but for degenerate material, the temperature can increase greatly with only a negligible increase in pressure. The triple-α process generates a great deal of energy but since the pressure does not increase much, there is no safety-valve effect in which pressure-induced expansion reduces both the pressure and the temperature. Thus there is a very large and very fast temperature increase and the path CD in Fig. 3.3 is referred to as the *helium flash*, which may last no longer than a couple of minutes. At the end of the helium flash the temperature in the core has increased enormously and the effect of this very high temperature is to remove the degeneracy. The star has now embarked on a state of helium-core burning that in some ways resembles the original main-sequence state when hydrogen was being consumed in the core. This leads to the evolution of the star towards a state resembling the original main sequence — although with a much higher core temperature, larger radius and higher surface temperature.

Eventually the helium in the core becomes exhausted with the core consisting mainly of carbon. Helium-shell burning then sets in, with hydrogen-shell burning occurring further out (Fig. 3.4). Once

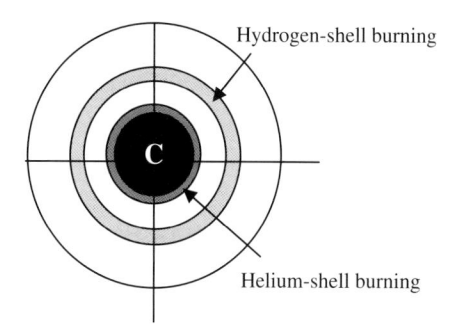

Fig. 3.4 Helium- and hydrogen-shell burning with a carbon core.

Fig. 3.5 A planetary nebula (WIYN/NOAO/NSF).

again the star evolves towards the red-giant part of the H–R diagram, region E, but this time, because energy generation in the star is so high, the star does not just expand but also expels material violently from its surface. This material is a complete shell but is seen through the telescope as a ring; it is called a *planetary nebula* (Fig. 3.5) although it has nothing to do with planets.

The inward pressure due to the helium-shell burning again leads to a degenerate core while more and more material is being lost from the surface. Eventually so much material is lost that all that is left is the degenerate core (position F) and the star has become a body known as a *white dwarf*. This consists mainly of carbon, has no nuclear reactions taking place, and the energy it radiates is that stored within it. A white dwarf can have the mass of the Sun in a body the size of the Earth, giving a density some two million times greater than that of water.

3.4. The Evolution of Higher-Mass Stars

Higher-mass stars have very large reservoirs of potential energy and so, when they collapse, the temperature in their cores becomes high enough for the triple-α reaction to occur before the density reaches the level at which the material becomes degenerate. For this reason, the normal pressure-based safety valve does operate and the helium flash does not occur. What does happen is that ever-heavier nuclei are produced by new reactions such as

$$^{4}_{2}\text{He} + ^{12}_{6}\text{C} \to ^{16}_{8}\text{O} \text{ (oxygen)},$$
$$^{12}_{6}\text{C} + ^{12}_{6}\text{C} \to ^{20}_{10}\text{Ne} + ^{4}_{2}\text{He} \text{ (neon + helium)},$$
$$^{12}_{6}\text{C} + ^{12}_{6}\text{C} \to ^{23}_{11}\text{Na} + ^{1}_{1}\text{H} \text{ (sodium + proton)},$$
$$^{12}_{6}\text{C} + ^{12}_{6}\text{C} \to ^{23}_{12}\text{Mg} + ^{1}_{0}\text{n} \text{ (magnesium + neutron)},$$
$$^{16}_{8}\text{O} + ^{16}_{8}\text{O} \to ^{28}_{14}\text{Si} + ^{4}_{2}\text{He} \text{ (silicon + helium)}.$$

At each stage, when a new energy-generating process is initiated in the core, the star reverts to a state closer to the main sequence and the path on the H–R diagram moves to and fro, first towards the red-giant region then back towards the main sequence, but with an ever decreasing size of swing. The reactions giving heavier elements can progress to the point at which iron is produced and then they can go no further. The reason for this is that reactions up to that stage are *exothermic*, that is, they generate energy, heat up the material and so enable new reactions to occur at an increasing rate. Reactions that produce nuclei heavier than iron are *endothermic*, that is, they require an input of energy for them to occur. A star with a mass in the range of 10–20 $M_\odot$ eventually reaches a stage where there is an iron core with several shells of burning taking place, with different major constituents in each shell, as shown in Fig. 3.6.

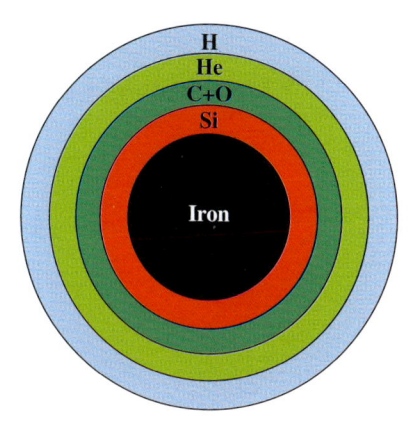

Fig. 3.6 Shells in a highly evolved massive star (not to scale).

Now something remarkable happens. The pressure on the iron core from the external material is so great that the material in the core is unable to support it. The protons and electrons in the core combine together to form neutrons so that the core consists of closely packed neutrons. The core shrinks in an instant to a density of 10^{18} kg m^{-3} and a radius of a few kilometres. The material outside the core rushes in to fill the space abandoned by the original iron core, strikes the neutron core, bounces outwards and collides with material still moving inwards. The result is an explosion of incredible energy — a *supernova*. Figure 3.7 shows the Crab Nebula, the debris from a supernova that was observed by Arab and Chinese astronomers in 1054.

The occurrence of supernovae is important in a chemical sense. The huge amount of energy generated enables nuclear reactions to take place, producing elements heavier than iron. All such elements that now exist in the Universe are the products of supernovae.

Another remnant of the Crab supernova is a *neutron star*. This is the neutron core of the exploding star — a sphere of closely packed neutrons, a few kilometres in diameter, wrapped in an iron crust. These neutron stars emit regular radio pulses, which were how they were first discovered, and this phenomenon gives them the name *pulsars*.

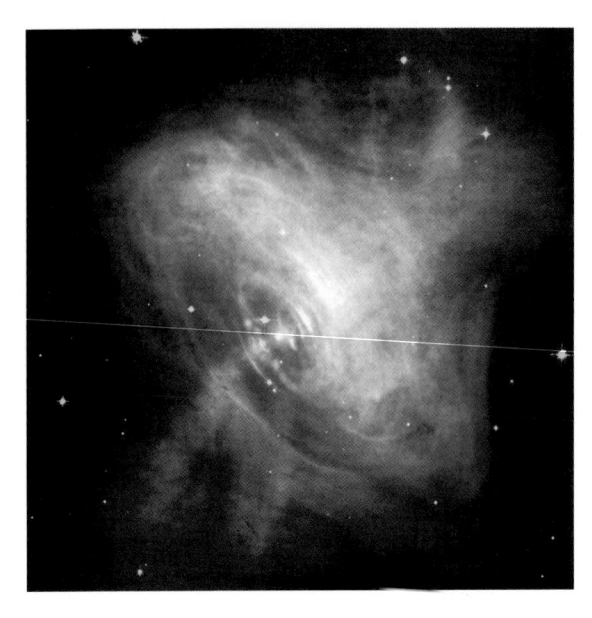

Fig. 3.7 The Crab Nebula — a supernova remnant (NASA/ESA).

The Indian (later American) astrophysicist Subrahmanyan Chandrasekhar (1910–1995) showed theoretically that there is a limiting mass to a white dwarf, $1.44 M_\odot$, and any remnant stellar core with a mass greater than that would have to be a neutron star. The same kind of theory also shows that there is a limited possible mass for a neutron star, somewhere in the region of 3–6 $M_\odot$, beyond which gravitational forces would crush even a neutron star. The final outcome in this case is a *black hole*, a mass existing at a point in space which is so massive that not even light can escape from it, and can only be detected by its gravitational influence.[2]

That completes my account of the life and death of stars. Now we must consider the process by which planets form, a story that is more concerned with stars at the point of their birth and in the heyday of their youth.

---

[2]The theory of the Chandrasekhar limit, white dwarfs, neutron stars and black holes has been given in G.H.A. Cole and M.M. Woolfson, 2002, *Planetary Science: The Science of Planets Around Stars* (Bristol: Institute of Physics), pp. 251–255.

3.5. Summary

The progress of a protostar to become a main-sequence star is illustrated in the Hayashi plot, Fig. 3.1. Initially the protostar slowly collapses under free-fall conditions. During this stage, when its radius decreases by a factor of 100, its temperature stays roughly constant because it is transparent, so that the energy of the collapse is radiated away. Eventually it becomes opaque, its temperature increases and pressure forces slow down the collapse. After a short bounce it takes up a state of quasi-equilibrium in which it slowly collapses under the control of the rate at which it radiates energy. During this Kelvin–Helmholtz stage of collapse, both its surface and internal temperatures increase. When the internal temperature reaches 15 million K, hydrogen-to-helium nuclear reactions occur and the star begins its main-sequence existence.

When hydrogen is exhausted in the core, hydrogen burning continues in a shell that gradually moves outwards. The pressure exerted by this shell burning compresses the core, and increases its temperature, and also expands the star to a red giant with a radius of order 1 au and a surface temperature of about 4,500 K. The very dense helium core eventually reaches a combination of density and temperature at which the triple-α process occurs, transforming helium into carbon. The core has become degenerate so it heats up quickly, accelerating the triple-α process, without a corresponding pressure increase to give a safety-valve effect. There is a runaway increase in temperature until the core material becomes non-degenerate; this helium flash stage lasts about two minutes. Helium-core burning then begins and continues until helium in the core is exhausted; during this time the star has moved towards the main-sequence region of the Hertzsprung–Russell diagram. When the helium in the core is exhausted, then helium-shell burning begins, with hydrogen-shell burning further out from the centre, and again the core is compressed and becomes degenerate.

For a star of low to medium mass, the large amounts of generated energy expel outer layers of the star, giving rise to a planetary nebula, until eventually only the degenerate carbon core remains — a white dwarf. For stars of higher mass, the core temperature increases to the point at which further nuclear reactions can take place, successively producing nuclei of increasing mass up to iron. Eventually the pressure on the iron core is so great that protons and electrons combine to give a core of tightly-packed neutrons. This happens almost instantly and material rushing into the vacated space bounces off the core, collides with oncoming material and generates energy to give a supernova event. The neutron core residue becomes a neutron star.

If the mass of the core exceeds a limit somewhere between three and six times the solar mass, then even the neutron core is crushed and the outcome is a black hole.

Chapter 4

The Evolution of a Galactic Cluster

The stars in galactic clusters are Population I stars, like the Sun, containing between 1–2% of their material in the form of elements heavier than hydrogen and helium. It is probable that all Population I field stars originated within a cluster in which other similar stars were forming; it is difficult to envisage a scenario in which an isolated Population I star could form separated by the ISM from all other stars. On the basis of that assumption, to understand the origin of the Sun and other stars, we must understand how galactic clusters form and evolve.

4.1. Embedded Clusters

I described the process by which a protostar would form in Chapter 2. The first step is the occurrence of a supernova that would both compress local ISM material and also inject coolant material into it. At the same time the local area would become enhanced in heavier elements, so increasing the previous metallicity of the local ISM material. The compressed material could be precipitated into collapse to form a DCC, and if the total mass of the collapsed material exceeded the Jeans critical mass, then it would continue to collapse under self-imposed gravitational forces.

A DCC should not necessarily be thought of as just the birthplace of a single galactic cluster. Within the DCC the generation of turbulence, fed by the collapse of the whole cloud, could produce sub-condensations, each of which could be the source of a different

Fig. 4.1 The Trapezium Cluster seen in visible light (left) and in the infrared (right) (HST, NASA).

galactic cluster. Such a hierarchical structure occurs in the Orion Nebula, a DCC within which there is the forming Trapezium Cluster. This cluster is shown in Fig. 4.1, taken both in visible light and in the infrared, the latter view showing much more since infrared radiation is better able to penetrate the dusty environment and also makes lower temperature objects seem brighter. The five brightest stars in the cluster have masses over 15 times the stellar mass.

While protostars are being formed here and there in a developing cluster, the whole cluster will be in a state of collapse; it is, after all, the collapse that feeds the turbulence which causes the protostars to form. In the early stages of collapse the cluster material retains its original temperature but eventually, as it becomes denser and more opaque, the temperature begins to rise, so slowing down and eventually halting, or even partially reversing, the inward motion. However, during the whole duration of the collapse the stars, YSOs and protostars within the developing cluster will also be falling inwards, with random motions of these bodies superimposed on the general inward motion. Consequently we have the situation where stars and their precursors exist in ever-increasing numbers in a smaller and smaller volume. It is clear that at some stage of the development of the stellar cluster there could be a very high number density of stars, objects due to become stars and higher density regions, produced by turbulent collisions, which will fail to become stars.

Clusters of stars forming within dense clouds of gas and dust are called *embedded clusters*, for the obvious reason that they are embedded in the dusty clouds. In the 1990s, infrared observations were showing that nearly all young stars were found in dense embedded clusters in which there was a high number density of stars, with stars much closer together than in a normal open cluster.[1] To illustrate this, in the vicinity of the Sun, a field star, there are about 0.07 stars per cubic parsec. Within a developed galactic cluster there can be between 0.3 to 30 stars per cubic parsec. Within the densest parts of an embedded cluster the number density can be from 100 to 10^5 per cubic parsec. It has been estimated that within the core of the Trapezium Cluster, the number density of stars is in the range of 10^4–10^5 per cubic parsec.

The gravitational forces of the stars and gas within a cluster region bind the cluster together and keep it in a state of equilibrium, or even of collapse. Indeed the mass of gas may be greater than the total mass of the stars — typically 50% greater. The earliest stars formed will tend to become more massive than later ones, a conclusion supported both by observation and theory[2] and they go through their evolutionary processes very rapidly (Table 3.1). After a few million years the most massive stars will have evolved to the supernova stage and the energy they release drives out the gas from the region. The gravitational cohesion of the cluster is lessened by the escape of the gas and the cluster begins to expand. It has been estimated that about 10% of the clusters have sufficient mass contained within their stars to continue to retain their cluster status after all the gas has been lost. The remainder will expand indefinitely with the stars — some single, some as binary pairs and a few as multiple systems — being released as field stars.

---

[1] Clark, C.J., Bonnell, I.A. and Hillebrand, L.A., 2000, *Protostars and Planets IV*. eds. V. Manning, A.P. Boss, and S.S. Russell (Tucson: Arizona Press), p. 151.
[2] Woolfson, M.M., 1979, *Phil. Trans. R. Soc.*, **A291**, 219.

However, even a cluster that initially retained its identity will not survive indefinitely. The individual stars, milling around within the cluster, sometimes have gravitational interactions with other stars that greatly affect their velocities. If a star near the boundary of the cluster happens to reach a sufficiently high speed, and if it is moving in an outward direction, then it will escape. In this way the cluster slowly evaporates; evaporation is an appropriate description because this is similar to the way that individual molecules leave the surface of a pool of liquid. Subrahmanyan Chandrasekhar calculated the time it would take for a galactic cluster to completely evaporate and this was in the range of 10^8 to 10^{10} years, but mostly at the lower end of that range. The same analysis applied to a globular cluster gives lifetimes of order 10^{12} years or more, many times the age of the Universe, so we can be sure that any globular clusters that formed in the early Universe still exist.

In the scenario presented here, there will be a stage in the lifetime of any Population I star when it will be in an environment densely populated with other objects. Some of these other objects, such as protostars or dense regions produced by colliding streams of material, will be of large extent so that significant interactions may occur. A description of two outcomes of such interactions follows.

4.2. The Formation of Massive Stars

Even if the original protostar were of more or less uniform density, it would not preserve this uniformity as it collapses towards the main sequence. The pressure forces are greatest at the centre and it is there that the collapse to high density proceeds most rapidly. Indeed it is possible that outer material could be stationary or even moving outwards while a dense core is forming in the collapsing protostar.

The collapse of the centre, which compresses the material there, causes it to increase in temperature, and radiation from this heated region streams out and traverses outer material that is still moving inwards to join the growing core. The force acting on incoming material, in general, has two main components — first, the gravitational force exerted by inner material pulling it inwards, and second, gas-pressure forces pushing it outwards. Radiation coming from the core is absorbed by the incoming material and exerts an additional outward pressure, combining with gas pressure to counteract the inward force of gravity. For small and medium mass protostars with correspondingly small core masses, the radiation pressure is trivial compared to the gas pressure and it does not substantially affect the motion of the inward moving material. However, the gravitational energy released by a collapsing core increases rapidly with the total mass of material involved and the temperature rise for a given proportionate collapse, say, by a factor of 10 in linear dimension, will be much larger for a massive core than for one of moderate mass. In Section 1.4 it was pointed out that the radiation emitted by a body at a temperature T is proportional to T^4 so that doubling the temperature gives 16 times the outwardly acting radiation pressure, dependent on the rate of energy output, whereas it only doubles the gas pressure. For a very massive protostar, once the mass of the central core reaches somewhere in the region of 8–10 solar masses, the radiation pressure would prevent any further material from falling in to join it. Since stars of considerably more than 10 solar masses actually exist, then there must be some mechanism for overcoming this effect.

The problem of how massive stars could form is one that has exercised astrophysicists for a considerable time. A suggested mechanism by which they can form is the coalescence of smaller mass stars or embryonic stars within a dense embedded cluster, a process first described by I.A. Bonnell, M.R. Bate and H. Zinnecker in 1998.[3] Even when an aggregation of 8 solar masses or more has formed, the radiation would not be able to prevent direct collisions with fairly dense and massive bodies, which could be one of collapsed protostars, YSOs or even lesser-mass main-sequence stars. For this mechanism to be effective it is clearly necessary that the environment is sufficiently crowded for direct collisions between bodies to occur fairly frequently.

---

[3] Bonnell, I.A., Bate, M.R. and Zinnecker, H., 2005, Massive Star Birth: A Crossroads of Astrophysics, *Proc. I.A.U. Symposium*, No. 227, eds. R. Cesaroni, M. Felli, E. Churchwell and C.M. Walmsley.

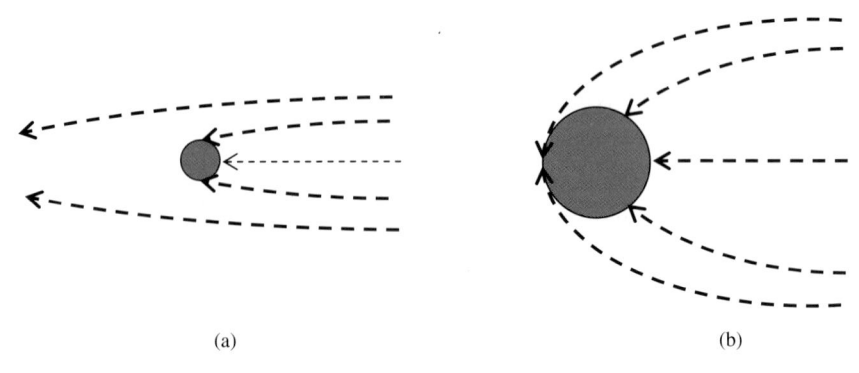

Fig. 4.2 Paths of bodies approaching a stellar core for (a) a light core (b) a massive core.

Once a massive body has formed, it will both be better able to attract other bodies towards itself and will also offer a larger target for collisions. The difference between a less and more massive stellar core in this respect is illustrated in Fig. 4.2 and described analytically in Appendix R. For the less massive core, shown in Fig. 4.2(a), the gravitational deflection of an oncoming body is small and the collision cross-section is little more than the geometrical cross-section of the core. For the massive core, shown in Fig. 4.2(b), there is considerable gravitational deflection and the collision cross-section is much greater than the geometrical cross-section of the core.

The more massive a core becomes, the more it is able to attract other bodies to join it, and the more bodies that join it, the more massive it becomes.

4.3. The Embedded Cluster Environment and Binary Star Frequencies

In Section 2.5 I described the fragmentation process by which a binary system would be produced. The final separation of the binary stars depends, amongst other things, on the angular momentum contained in the original protostar. If the stars of the binary system are far apart, then they are called *soft binaries*, in the terminology of astrophysicists, and are easily disrupted by the close passage of another star. Conversely, *hard binaries*, in which the stars orbit each other in close proximity, would require an extremely close, and hence highly improbable, passage of another star to effect a disruption.

The Czech astrophysicist, Pavel Kroupa, has made a special study of the frequency of binary systems in dense stellar environments. A study of the frequency of binary systems in two different star-forming regions — one called the Taurus–Auriga dark cloud and the other the Orion Nebula — has shown that the frequency of binary systems is significantly higher in the Taurus–Auriga region.[4] The stellar number densities in these two regions are very different, with a much higher one in the Orion Nebula. These observations suggest that there have been fewer disruptions of binary systems due to close stellar passages in the Taurus–Auriga region.

4.4. The Progress of Star Formation in a Galactic Cluster

Except for stars of very high mass, the approximate free-fall process that brings a protostar to the stage of becoming a YSO is very short compared with the duration of the Kelvin–Helmholtz collapse that takes it onto the main sequence. For a solar-mass star the life of the protostar may be 20,000 years or so compared with 50 million years for its YSO lifetime. For this reason it is possible to assign an age to a star as the time that elapses from when it first becomes a YSO; the error in ignoring the time it spends as a protostar is negligible. If we know that a star is approaching the main sequence, then its position on the H–R diagram tells us both its mass and its age. That this is so can be seen from Fig. 3.2. The theoretical curves for different star masses do not cross each other so every point to the right of the main-sequence line can be only on one curve corresponding to a particular mass. In addition the position on the curve indicates the period of time from when it started moving along the curve — in other words, the age of the star. However, there is a problem — from Fig. 3.3 we see that when a star leaves the main sequence it also moves off to the right so we need to determine whether a star that

---

[4]Kroupa, P., Petr, M.G. and McCaughrean, M.J., 1999, *New Astronomy*, **4(7)**, 495.

has been observed in such a position on the H–R diagram is in its birth stages or moving off to its death.

When we observe a stellar cluster we can be reasonably sure that the stars that formed within it have all done so within a fairly restricted time. Some stars that are produced will, by the process described in Section 4.2, have accumulated material to become very massive stars. If such a star is observed in its main-sequence state within a cluster, then we can be certain that it is a young cluster and hence that stars seen to the right of the main-sequence line are young, not old. The main-sequence lifetime of a 15 solar-mass star is 10 million years — a tiny fraction of the main-sequence lifetime of most stars of lesser mass (Table 3.1).

In 1966 I. Iben and R.J. Talbot examined some young stellar clusters, recognized as such by the presence of massive main-sequence stars, and found the masses and ages of the constituent YSOs. I.P. Williams and A.W. Cremin made a similar and rather more extensive examination of four young stellar clusters in 1969.[5] For one of these clusters, NGC2264, which was typical of the others, the general conclusions were that:

(i) Star formation began about 8×10^6 years ago.
(ii) The first stars produced had mass of about $1.35 M_\odot$.
(iii) There are two general streams of development of stars, in one of which stellar masses decrease with time while in the other stellar masses increase with time.
(iv) The overall rate of formation of stars increases with time.

I shall refer to this pattern of star development later as it has a bearing on planet formation.

4.5. Summary

The collapse of a region of a DCC within which stars are forming leads to a period when the evolving stellar cluster is in a state with high stellar number densities, held together by the embedding

---

[5]Williams, I.P. and Cremin, A.W., 1969, *Mon. Not. R. Astr. Soc.*, **144**, 359.

gas. During this stage in the development of a cluster, there are interactions involving protostars, YSOs and main-sequence stars, some of which can enable the formation of stars of greater than 8 solar masses and others that can lead to the disruption of soft binary systems.

A study of young stellar clusters shows that stars of somewhat more than a solar mass are first produced with separate streams of development, in one of which stellar masses decrease with time while in the other they increase with time.

Chapter 5
Exoplanets — Planets Around Other Stars

5.1. Planets Orbiting Neutron Stars

A neutron star is the final product of a supernova for a star that gives a core mass between the Chandrasekhar limit of 1.44 solar masses and the lower limit for the formation of a black hole, somewhere between three and six solar masses. Neutron stars are sources of strong magnetic fields and they emit electromagnetic radiation at radio wavelengths, although x-ray and γ-ray emissions have also been detected in some cases. This radiation is emitted in the form of fine beams along the directions of the magnetic axis of the neutron star that, in general, will be at some angle to the spin axis (Fig. 5.1). This means that as the neutron stars spin, the beams of radio waves trace out the surface of a cone; if one of the directions along the cone intersects the Earth, then pulses of radio signals will be detected at regular intervals corresponding to the spin period.

Neutron stars spin very rapidly with periods ranging from 1.4 milliseconds to a few seconds; because of their enormous density they are able to withstand the disruptive forces of such rapid rotation. Jocelyn Bell Burnell, at that time a research student of Anthony Hewish (b. 1924; Nobel Prize in Physics, 1974) at Cambridge, first detected these objects, which were given the name pulsars, in 1967. The regularity of the pulses suggested that they might be the product of some extraterrestrial intelligence and the first pulsar was labelled LGM-1, where the LGM stood for 'Little Green Men'. However, soon after the discovery of pulsars, Thomas Gold (1920–2004), an

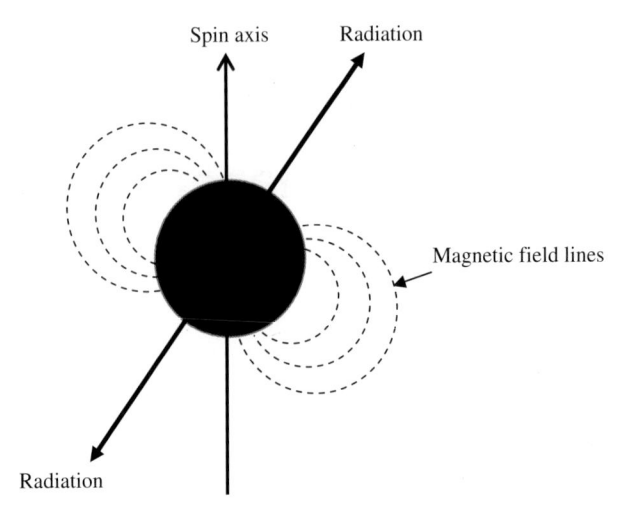

Fig. 5.1 The relationship of the radiation beams to the spin axis for a pulsar.

Austrian, later American, astrophysicist correctly identified them as neutron stars.

A star with a single planetary companion would be just like a binary-star system although with components of a very different mass ratio. A representation of the motions of the star and the planet around their centre of mass, not to scale, is given in Fig. 5.2.

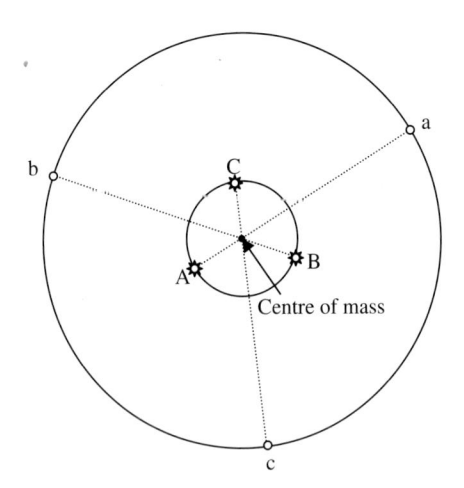

Fig. 5.2 The motion of a star and planet around the centre of mass.

Corresponding points on the orbits — shown as circles although they could be ellipses — are indicated by lower and upper case letters for the planet and star respectively. The scale of the motion of the star is much smaller than that of the planet; if the planet has one-thousandth of the mass of the star, then the motion of the star is one-thousandth that of the planet.

The first ever planets outside the Solar System were detected around neutron stars. The basis of the discovery is the precise regularity of the radio pulses. If the line of sight to the pulsar is in the plane of the motions of the planet and star, then, as the star moves round its orbit, it is sometimes moving away from, and sometimes towards, the observer. Whenever the star is moving away from the observer the interval between successive pulses is slightly longer than the average and, conversely, when the star moves towards the observer the interval between successive pulses is shortened. By analysing the pattern of the changes of intervals it is possible to estimate both the period of the orbit and also the speed of travel in the orbit. Some neutron stars occur as members of binary systems and estimates of their masses show a very small variation from the Chandrasekhar limit of 1.44 times the solar mass. This information is enough to estimate the mass of an orbiting planet, with the usual problem that also occurs with estimating the masses of stars in a binary system — the unknown inclination of the orbit to the line of sight (Appendix K).

In 1992 the Polish astronomer, Aleksander Wolszczan, reported the first detection of planets around a neutron star, PSRB1257+12. His first report, based on an analysis of the motion of the neutron star, was that there were two planets but later, more careful analysis showed that three planets were present with masses 3.9 $M_\oplus$,[1] 4.3 $M_\oplus$ and 0.25 $M_\oplus$, with orbital radii 0.46 au, 0.36 au and 0.19 au respectively. The difference in the intervals between pulses can be measured so accurately — to better than 1 microsecond — that even the gravitational effects of the planets on each other can be detected. There is a suspicion that there may be a fourth planet

---

[1]The symbol $\oplus$ represents the Earth so that $M_\oplus$ is the mass of the Earth.

around PSRB1257+12, of mass 0.0004 $M_\oplus$ and with an orbital radius of 2.6 au. This body, if it exists, has about one-thirtieth of the mass of the Moon.

Since a neutron star is a product of a supernova, the problem is to understand how the planets survived the event, assuming they were present around the original star. There seem to be two possibilities. The first is that the original star had no planets and the existing planets are accumulations of material from the debris that surrounded the neutron star after the supernova. The second is that they *were* present but with orbits so large that, although they lost their volatile outer layers, their cores survived the explosion. Subsequently their orbits decayed as they orbited in the medium left around the neutron star. There may be other possibilities but in the light of present knowledge any postulate is just that and cannot be tested.

5.2. The Characteristics of Orbits

I have been mentioning orbits in a rather general way but now it is time to be more specific about their characteristics. When there are two bodies orbiting around their mutual centre of mass, as for a binary system or a planet and star, then, in general, the orbit is an ellipse. An ellipse, shown in Fig. 5.3, is represented in a Cartesian co-ordinate system by the equation

$$\frac{x^2}{a^2} + \frac{y^2}{b^2} = 1 \tag{5.1}$$

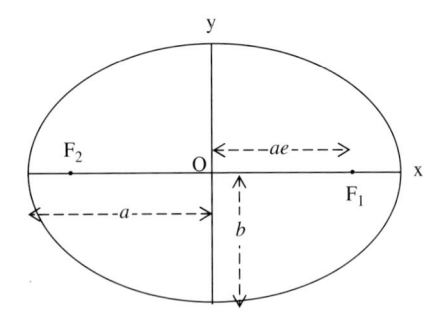

Fig. 5.3 The characteristics of an ellipse.

where a and b are the *semi-major* and *semi-minor* axes respectively. The foci, F_1 and F_2, are displaced from the geometrical centre of the ellipse by $\pm ae$ in the x direction, where e is the *eccentricity* of the ellipse. The semi-major axis is related to the semi-minor axis by

$$b = a(1 - e^2)^{1/2}. \qquad (5.2)$$

The eccentricities of most solar-system planetary orbits are quite small; the largest is that of Mercury, 0.2056. However, we shall find ourselves interested in orbiting bodies with much larger eccentricities than that of Mercury. For an ellipse the possible range of eccentricity is from 0, corresponding to a circular orbit, to 1.0, at which point the ellipse degenerates into a non-closed shape, a *parabola*.

Another characteristic of an orbit that may be of interest is its *inclination*. To define this we need some reference plane with which to compare the plane of an orbit. In the case of the Solar System this is taken as the *ecliptic*, the plane of the Earth's orbit. The most inclined planetary orbit is that of Mercury, at 7°.

The best way of defining the direction of an orbital motion or the spin of a body is by an arrow perpendicular to the plane of the orbit or along the spin axis. To give direction to the arrow, it points in such a direction that, with the arrow pointing towards you, the orbit or spin is seen to take place in a counter-clockwise direction. By this definition, with the ecliptic giving the reference direction, the tilt of the solar spin axis is 7°.

5.3. Planets Around Main-Sequence Stars; Doppler-Shift Detection

The detection of planets around main-sequence stars by the use of Doppler-shift measurements depends on the same kind of analysis by which the masses of binary stars are determined. However, there are important differences. First, something that makes the task more difficult is the large ratio of the stellar mass to that of the planet. The only body emitting detectable radiation is the star and the speed of the star is very small, a few metres per second, so the Doppler shifts in spectral lines are tiny and require the use of

Table 5.1 The contributions to the speed of the Sun due to three planets.

Planet	Eccentricity of orbit	Period (years)	Radius of solar orbit (km)	Speed (m s^{-1})
Jupiter	0.0483	11.86	7.4×10^5	12.4
Saturn	0.0560	29.46	4.1×10^5	2.7
Earth	0.0167	1.00	4.5×10^2	0.09

very precise optical spectrometers. The basis of this difficulty is illustrated in Table 5.1, showing the contribution to the speed of the Sun due to two major planets and the Earth. It is clear that the best conditions for having a large orbital speed of the star is if the planet is massive and in a close orbit. Early measurements were made with spectrometers accurate to about $10\,\mathrm{m\,s^{-1}}$ so the measurements were extremely noisy and unreliable; there have been improvements in instrumentation and now an accuracy approaching $1\,\mathrm{m\,s^{-1}}$ is attainable. A second difference, but one that simplifies the process, is that it is comparatively easy to determine the spectral class of the star and this gives a good estimate of its mass (Table 1.1).

We consider a situation where there is a single planet in a circular orbit around the star and the line of sight from the Earth is contained in the plane of its orbit (Fig. 5.4). At point S the star is moving

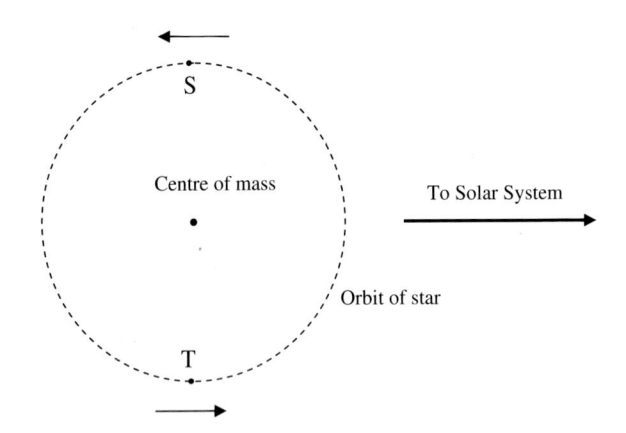

Fig. 5.4 The circular orbit of a star relative to the centre of mass.

away from the Sun and at point T towards the Sun, both motions being described relative to the centre of mass. The measurements are made from the Earth, a moving platform relative to the Sun, so it is necessary to correct for its motion if the velocity of the star relative to the Sun is to be found.

After correction for the motion of the Earth relative to the Sun, Doppler-shift measurements give the velocity profile shown in Fig. 5.5. The period of the orbit is the time for one complete cycle of the velocity profile and the difference between the maximum and minimum velocity is twice the orbital speed of the star. Knowledge of the mass of the star, its orbital speed and the period of its orbit is sufficient to give the mass of the planet and the radius of its orbit (Appendix S).

The above discussion, and also Appendix S, is predicated on the assumption that the line of sight is in the plane of the orbit, that the orbit is circular and that there is a single planet. If the line of sight makes an angle θ with the plane of the orbit, then the velocity measured is just the component of the velocity along the line of sight (Fig. 5.6). The actual velocity will be underestimated by a factor $\cos\theta$ and the mass of the planet will be underestimated by the same factor. For this reason, estimates of the masses of exoplanets are, in general, given as minimum masses.

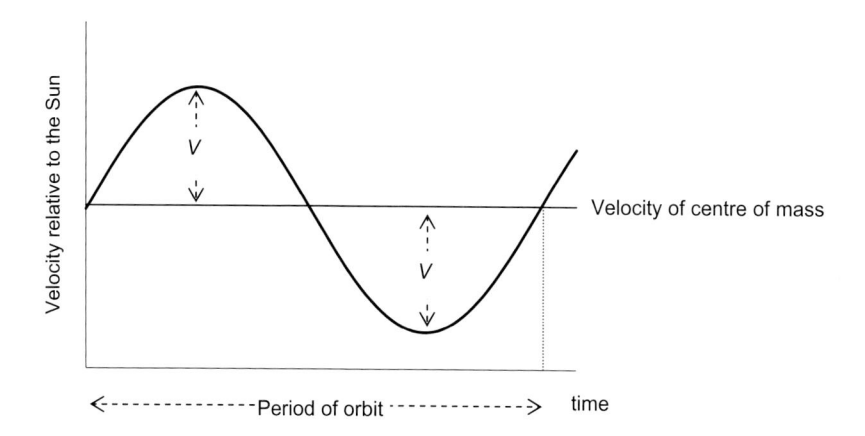

Fig. 5.5 An ideal star velocity profile with a single exoplanet.

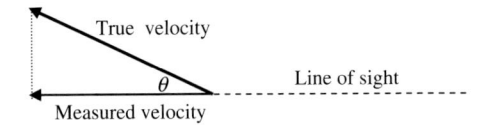

Fig. 5.6 The relationship between measured velocity and true velocity.

The profile shown in Fig. 5.5 is a calculated one but in practice the profile must be deduced from data, with experimental errors, gathered over a considerable period of time. The very first exoplanet detected was that around the star 51 Pegasus in 1995 by the Swiss astronomers Michel Mayor and Didier Queloz working at the Observatoire de Genève.[2] The star is very similar to the Sun, with a mass 1.06 $M_\odot$; the planet has a minimum mass 0.468 times that of Jupiter (remember the line-of-sight problem) and a circular orbit with radius 0.052 au. Since the planet had an orbital period of just 4.23 days, it was possible in a short time to observe several periods. By contrast, Fig. 5.7 shows the observations for the exoplanet detected around a main-sequence star, 47 Uma, by the American astronomers Paul Butler and Geoff Marcy. These were reported in 1996[3] although, perforce, the observations were started a long time before that. The figure shows how the velocity profile was fitted to the experimental measurements. The observations gave a period of 1,094 days for the planetary orbit and since the mass of the star is about 1.03 times that of the Sun this implies an orbital radius of 2.09 au. The minimum mass of the planet is 2.62 times that of Jupiter.

If the second assumption, that the orbit is circular, is untrue, then this manifests itself by the velocity profile being a distorted version of that shown in Fig. 5.5, as illustrated in Fig. 5.8. Given the variety of elliptical orbits and directions of the line of sight that are possible, the interpretation of the velocity profile is quite a complicated process. Figure 5.8 shows the profile for a Jupiter-mass planet in orbit around a solar-mass star with a semi-major axis of 0.33 au and eccentricity

---

[2] Mayor, M. and Queloz, D., 1995, *Nature*, **378**, 355.
[3] Butler, P. and Marcy, G., 1996, *ApJ. Letters*, **464**, L153.

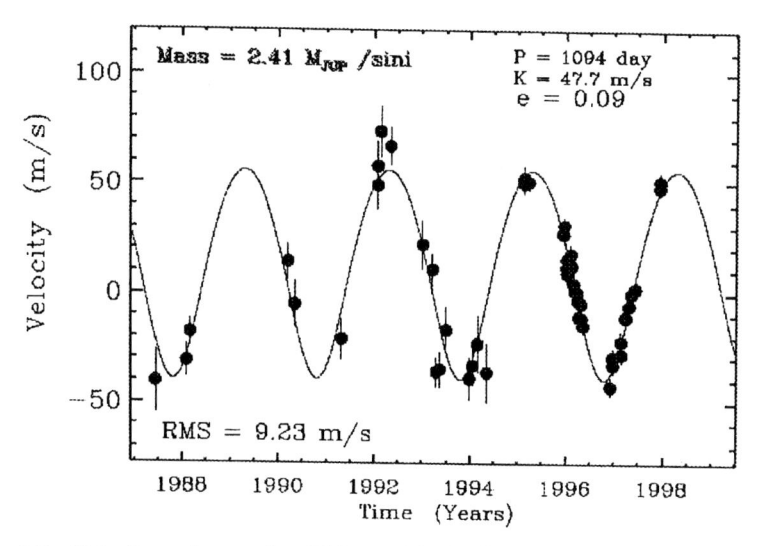

Fig. 5.7 Velocity estimates for 47 Uma with the fitted velocity profile (Butler & Marcy, 1996).

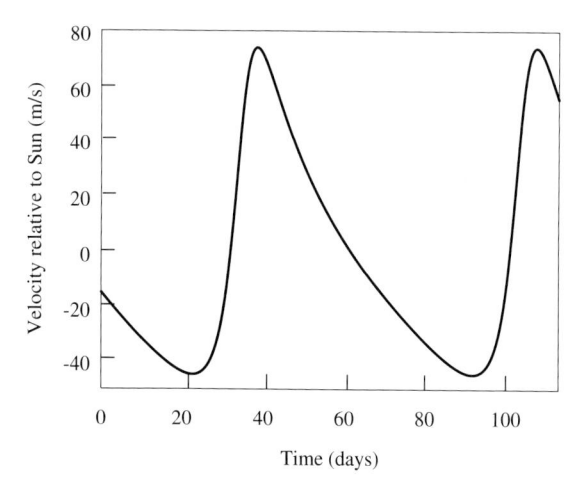

Fig. 5.8 A typical velocity profile for a star with a planet in an eccentric orbit.

0.328. The line of sight is in the plane of the orbit and makes an angle of 30° with the major axis. From measurements of the distortion the eccentricity and other characteristics of the orbit can be found.

The final complication in the detection and characterisation of exoplanets is when there is more than one of them orbiting a star. The overall motion of the star is then a superposition of the separate curves for the planets taken individually. Decomposing the observed velocity curve into separate periodic components allows the companion motions to be identified and the semi-major axes and eccentricities to be estimated. The mass estimates will still be of minimum masses, as explained previously.

The rate of discovery of exoplanets has greatly increased since the first one was detected and, at the time of writing, the characteristics of about 400 are known. Table 5.2 lists the major characteristics of some of them; they have been chosen to illustrate the range of masses, orbital semi-major axes and eccentricities. There are also planetary families that have been discovered where there are two or more planets associated with a star.

There are several interesting features of the exoplanets listed in Table 5.2. There is a family of five planets associated with the star 55 Cancri and there are three planets in the family of GJ581. The

Table 5.2 The characteristics of some Doppler-shift detected exoplanets.

Star	Minimum planet mass (M_J)[4]	Semi-major axis (au)	Eccentricity
HD41004B	18.4	0.0177	0.081
GJ436	0.0692	0.0278	0.159
55 Cancri	0.816	0.114	0.0159
	0.165	0.238	0.053
	3.84	5.84	0.063
	0.0235	0.0377	0.264
	0.141	0.775	0.002
HD72659	3.30	4.77	0.269
HD80606	4.31	0.468	0.9349
GJ581	0.0490	0.0406	0.000
	0.0159	0.0730	0.000
	0.0263	0.253	0.000

---

[4] M_J is the mass of Jupiter.

minimum masses of the planets in the table vary from 18.4 M_J, which is actually a brown-dwarf mass, to 0.0159 M_J — about five times the mass of the Earth. To appreciate the significance of the orbital semi-major axes we should compare them with 0.387 au, the semi-major axis of Mercury, the innermost solar-system planet, and 5.2 au, the semi-major axis of Jupiter, the most massive planet of the Solar System. The smallest semi-major axis of an exoplanet, 0.0177 au, is less than 5% of that of Mercury. The temperature on the exposed face of the orbiting body, which is really a brown dwarf, is extremely high and any much smaller body might have insufficient mass to retain its gaseous component in such an environment. Eccentricities also show a huge variation, from 0, corresponding to a circular orbit, to 0.9349 for the planet orbiting HD80606. At its closest this planet is distant — 0.03 au from the star, and at its furthest 0.9 au. It has been estimated that when it swings past the star at closest approach its surface temperature goes from 800 K to 1,500 K in the space of six hours.

So far in discussing the detection of exoplanets and their properties, we have assumed that we have no knowledge of the angle that the line of sight makes with the plane of the planetary orbit. However, there is a rare circumstance when it is definitely known that the line of sight is in the plane of the orbit, or very closely so. This is when the planet transits the star, that is to say, it passes over the face of the star as seen by the observer. Of course the observer will not directly see the planet's disk pass over the disk of the star, as for example when observing a transit of Venus over the Sun's disk, but what will be detected is a diminution of the intensity of the star as the planet blocks off some of the light coming from it. An early observation of such a transit, taken by a group of amateur astronomers in Finland, is shown in Fig. 5.9. The dip in intensity as the planet passes in front of the star can clearly be seen.

An extra bonus of observing the transit of an exoplanet is that, in addition to getting a precise estimate of mass of the planet, rather than its minimum mass, one can also estimate its radius. HD209458 is a main-sequence star and from its spectral class its radius can be deduced. The transit of the planet leads to a 1.5%

Exoplanet transit over HD209458
September 16, 2000 - Nyrölä Observatory, Finland

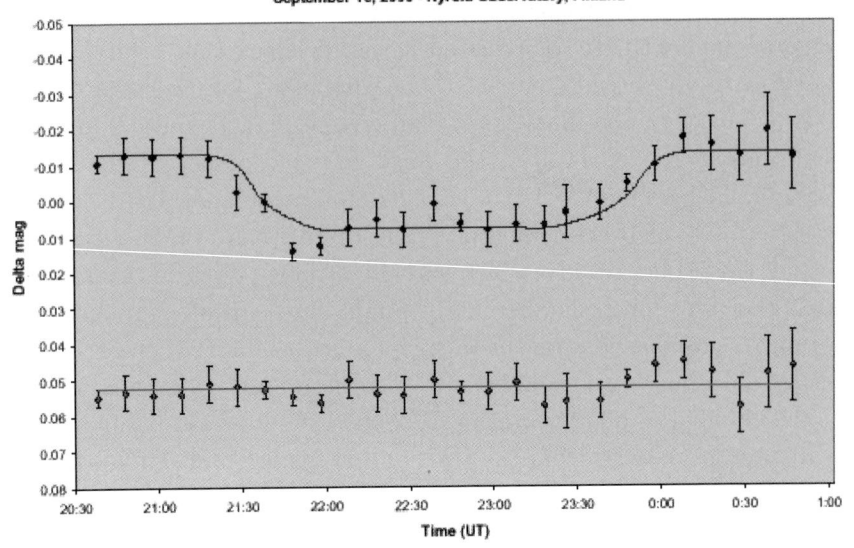

Fig. 5.9 The light curve from the transit of an exoplanet over the star HD 209458. The lower line is from a check star. This observation was made by a group of amateur astronomers in Finland at the Nyrölä Observatory.

drop in the intensity of the star from which it may be deduced that the radius of the planet is a fraction 0.122 of the radius of the star. The characteristics of this planet and its orbit have been very well determined. It has a mass 0.69 M_J and a radius some 35% greater than that of Jupiter. It is in a somewhat distended state because of its high temperature, about 1,250 K — related to its close, nearly circular, orbit of radius 0.047 au.

A characteristic of all the exoplanets detected by Doppler-shift and transit measurements is that they are fairly close to their parent stars. This does not mean that this is a normal characteristic of planetary systems; if it were so, then the Solar System would be an exception in having planets out to 30 au. For a given mass of planet, the closer it is to a star, the greater the speed of the star in its orbit. Another adverse factor for a large orbit is that the long orbital period would require a long period of observation to establish the velocity profile. For a given planetary orbit, the more massive the planet, the greater the speed of the star in its orbit. Clearly, the easiest planets to detect by Doppler-shift measurements are those that are most massive and in close orbits. However, there are now other ways of detecting planets where these constraints do not apply.

5.4. The Direct Imaging of Exoplanets

In the prologue to this book, 'An Historical Sketch', it was mentioned in the section relating to the Solar Nebula Theory that the theory was given a fillip by the discovery of dusty disks around very young stars. Actually, dusty disks around stars are very common — even the Sun, by no means a very young star, has a dusty disk. In parts of the world that are remote from urban environments and the light pollution they generate, it is possible to see the Sun's dusty disk with the unaided eye. It is best seen after sunset or just before dawn, when the sky is dark but the Sun is not too far below the horizon. It appears as a faint band of light in the sky called the *zodiacal light* (Fig. 5.10). The Solar System is heavily populated with dust, concentrated towards the ecliptic, the plane of Earth's orbit, and the zodiacal light is scattered sunlight, the directions of which define the extent of the dusty region. Most of the dust is tiny particles

Fig. 5.10 The zodiacal light (NASA/courtesy of nasaimages.org).

from micron-size up to the size of grains of sand. Under exceptional viewing conditions, by looking in the opposite direction from the Sun one may see another band of light in the sky corresponding to back scatter from dust that is further out from the Sun than the Earth. This is normally called the *gegenschein*, but is sometimes translated into English as the *counterglow*.

The presence of all this dust tells us something very important about what must be happening in the Solar System. There is a phenomenon known as the *Poynting–Robertson effect* in which solar radiation acting on small particles draws them inwards towards the Sun (Appendix T). A grain of sand orbiting the Sun at 1 au would be absorbed by the Sun in a million years or so. What this indicates is that there must be one or more sources of constant replenishment of the dust, otherwise during the lifetime of the Solar System, 4.6×10^9 years, it would all have disappeared. In the case of the Solar System we know what these sources are. The well-known appearance of a comet is as a bright ball with a tail, and this tail contains a great deal of dust that is being continuously shed as the comet moves through the Solar System. It is when Earth moves through the region of the orbital paths of some of these shed grains that we observe meteor showers as the dust enters the atmosphere and becomes incandescent. Another source of dust is the occasional collisions of asteroids, bodies orbiting the Sun made of stone and/or iron, which can vary in size from metres to hundreds of kilometres. Larger fragments from these collisions produce the meteorites that land on Earth but there is also a great deal of fine debris produced.

Seen from outside the Solar System, this dusty layer could, in principle, be detected by means of the infrared radiation it emits. However, in practice this would be difficult because of the small amount of dust in the Sun's disk. If the disk contained much more material, say, equivalent to the mass of the Earth, then it could be very easily detected. Although the dust is cold, and hence radiates very little energy per unit area, the total area of radiating dust particles is so large that the total radiation emitted would be easily detected. That, after all, is the explanation of the detection of bumps in the radiation curves of young stars, which reveal the presence of accompanying disks. The same material concentrated into a body the size of the Earth, even at a higher temperature, would be undetectable since the emitting area would be smaller by a factor of order 10^{10}.

There are many stars — not young stars but considerably younger than the Sun, which possess easily detectable dusty disks, 1,000 or more times denser than that of the Sun. Since the Poynting–Robertson effect has universal application, it follows that these stars must also possess some accompanying bodies that are rich sources of dust. From the last few years of the 20th century, there have been extensive infrared studies of a number of stars with dusty disks carried out by a Scottish group headed by Jane Greaves of the University of St Andrews.

This group examined the star Fomalhaut, with a mass of 2.1 $M_\odot$ and situated some 25 light years from the Sun. This star has a dusty disk that is cleared out in its middle region, suggesting the presence of planets that are sweeping up the dust. There is also a bright ring at about 40 au from the star and it was suggested that this could be from a region rich in comet-like bodies.

The star Vega, also with a mass of 2.1 $M_\odot$ and 25 light years from the Sun, again has missing emission close to the star. But an even more interesting observation is a bright concentrated source at about 80 au, a possible interpretation of which is that it is emission from a dust cloud surrounding a major planet. If such a planet exists, then it would not be detectable by Doppler-shift methods because of the long orbital period, about 500 years.

The star β-Pictoris, with a mass of 1.75 $M_\odot$ and 63 light years from the Sun, has the distinction of being the first star to have its dust disk imaged. A great deal of structure has been found in the disk with bright rings at 6, 16, 28, 52 and 82 au from the star. It has been suggested that the spaces between these rings have been swept clean by planets and orbital radii of 10, 25 and 44 au have been suggested.

The final star we mention in this context is ε-Eridani, with a mass of $M_\odot$ at a distance of 10.5 light years. There have been claims of the detection of a planet around this star by Doppler-shift measurements but these measurements are not good enough to reliably support the claim. There is a belt-like structure in the dust disk that has been interpreted in terms of perturbations by planets.

Another kind of exoplanet observation was reported by Jane Greaves at the National Astronomy Meeting of the Royal Astronomical Society in Belfast in April 2008.[5] Radio imaging of the star HL Tau, which is surrounded by a dusty gaseous disk, revealed the presence of a 14-Jupiter-mass companion (Fig. 5.11), a mass just above the formal limit for a planet. The observations suggested that a nearby star, XZ Tau, may have passed by HL Tau about 1,600 years ago and it was suggested that this passage had 'triggered' the formation of the companion. XZ Tau is a binary, or even possibly

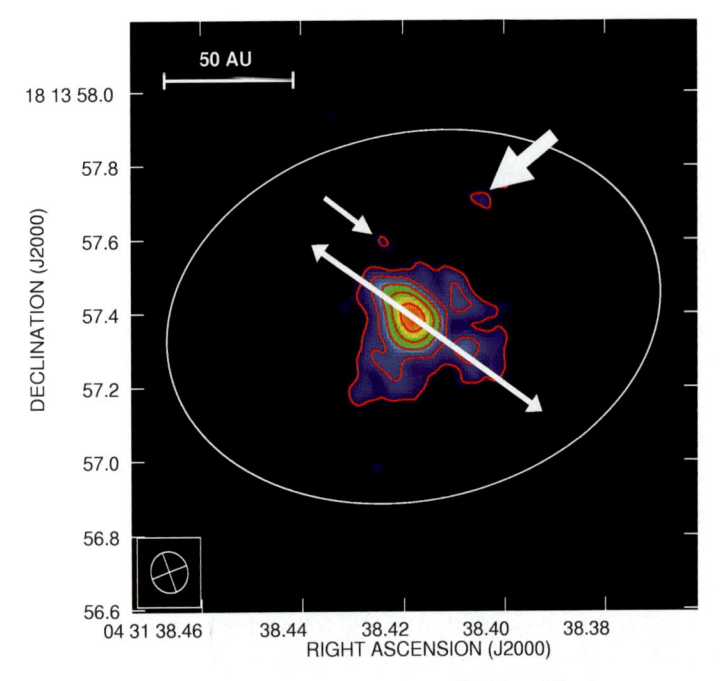

Fig. 5.11 A 1.3 cm false-colour map centred on HL Tau. The large arrow marks the reported body. The smaller arrow may indicate the presence of a smaller body. The oval represents the limit of the gaseous disk surrounding the star.

---

[5]Greaves, J.S., Richards, A.M.S., Rice, W.K.M., and Muxlow, T.W.B., 2008, *MNRAS Letters*, **391**, L74–78.

Exoplanets — Planets Around Other Stars

a three-star system, one member of which is probably at the pre-YSO protostellar stage. The image shows a smaller, fairly bright condensation, indicated by a small arrow in the figure, although this was not identified as a possible planet.

A new era of exoplanet investigations began in 2008 when the American astronomer, Paul Kalas, showed the first optical image of a planet orbiting Fomalhaut. In 2005 it was inferred from a sharp inner boundary of the dust disk that a planet was present. Photographs of the dust disk taken in 2004 and 2006 were then carefully examined and showed the planet, which had moved in the interval (Fig. 5.12). Available information does not give the characteristics of the planet or its orbit with great precision. Its mass could be anywhere between one-twentieth and three Jupiter masses. The best estimates of its orbital characteristics are that its semi-major axis is about 115 au, corresponding to a period of 872 years and an eccentricity of approximately 0.11 — but these values are highly uncertain.

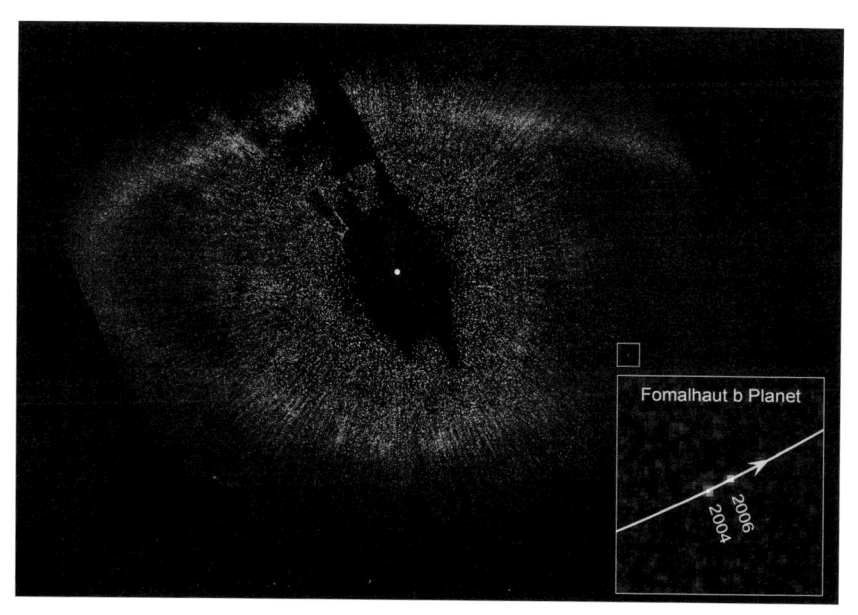

Fig. 5.12 The dust disk around Fomalhaut. The enlarged image of the box shows the planet in 2004 and 2006 (NASA/ESA).

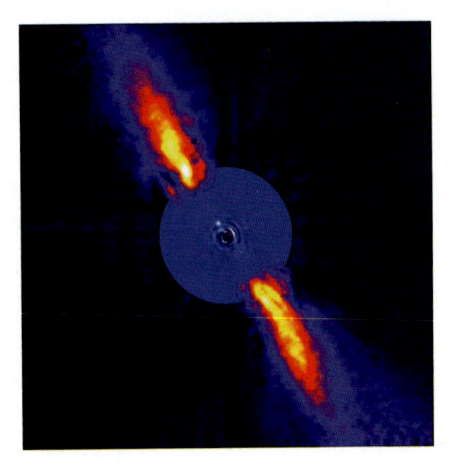

Fig. 5.13 The composite image of β-Pictoris. The planet is the bright spot near the centre (ESO).

At the same time as the Fomalhaut announcement, an image was released by the Gemini Observatory of three planets around HR8799, a star of mass 1.5 $M_\odot$ in the constellation Pegasus. These planets, with estimated masses of $10M_J$, $10M_J$ and $8M_J$ are at distances 25, 40 and 70 au from the star respectively.

Because of distortions in the disk of β-Pictoris, it was believed that there ought to be a major planet orbiting the star at a distance of around 10 au. At the end of 2008 a team of French astronomers produced a picture that seems to show a planet at a distance of about 8 au (Fig. 5.13). It was not a straightforward image as for Fomalhaut and HR8799. The outer part of the image shows the reflected light from the dust disk. The inner circular region was imaged in the near infrared, at a wavelength of 3.6 microns — about five times the wavelength of the red end of the optical range. In the circular region the contribution of the glare of the star was artificially subtracted and what was left was the bright spot indicating the presence of the planet.

5.5. Exoplanets and the Solar System

At the time when theoreticians were first considering how the Solar System began, it was not known whether the Solar System was unique, a rare example of a planetary system or one of many. It is now clear that it is one of many. Had the Solar System been unique or rare, then a theory that proposed an unlikely and exotic theory for its production could not have been excluded — indeed, it might almost have been preferred. However, we now know that there are many planetary systems and so any mechanism proposed for their formation must involve processes that are both commonplace and robust in their outcomes. We also know that the Solar System is not unusual in its characteristics. The latest infrared and visible-light observations, combined with those from Doppler-shift measurements, have revealed exoplanets in orbits with semi-major axes much less than that of Mercury and much greater than that of Neptune. Orbital eccentricities of exoplanets have a much greater range of values than those of the Solar System. Only in the range of planetary masses is the Solar System outside the observed exoplanet range, but this is probably an observational constraint that the skill and ingenuity of astronomers will overcome in due course.

5.6. Summary

The first planets to be detected were those around neutron stars. Because the radio pulses from these stars are so regular, it was possible to determine the minimum mass of the planets and their orbits with some precision. The presence of these planets suggested that they might have been present when the star was in its main-sequence stage.

Doppler-shift measurements have confirmed that there are planets around many main-sequence stars. There is a wide range of estimated minimum masses and orbital parameters, including eccentricities up to a value of 0.935. Several planetary systems have also been detected with several planets orbiting a star. Because of measurement difficulties this kind of detection preferentially reveals planets of higher mass in closer orbits. Consequently, Doppler-shift observations are unable to detect planets resembling either the terrestrial planets of the Solar System or major planets in orbits with semi-major axes much larger than that of Jupiter. A rare but useful observation is when the planet transits the star, which allows both a true estimated mass of the planet, rather than the minimum mass, and also an estimated radius to be determined.

Infrared observations of dusty disks around some young stars gave indications of the presence of planets, some at distances from their stars much greater than that of Neptune from the Sun. This conclusion has been confirmed by the direct optical imaging of planets around some stars — observations that complement those of Doppler-shift measurements in that they preferentially detect planets that are distant from their stars.

The weight of evidence is that the Solar System is a typical planetary system, so any theory explaining the presence of exoplanets will also provide an explanation for the formation of the Solar System.

Chapter 6
The Formation of Planets: The Capture Theory

It is clear that any theory of planetary formation should involve processes that commonly occur in all parts of the galaxy and, in particular, in all star-forming regions. It seems unlikely that a star could form, could subsequently become a field star without planetary companions and then somehow acquire planets. I suppose it is possible to envisage some rather unlikely process for this to happen, say, by the capture of odd planets that happened to be floating around the galaxy, but this would hardly satisfy the condition of requiring a common and robust mechanism for forming planetary systems.

The theory that I am going to describe here is one that takes place in an environment we have already met — a dense embedded cluster — that appears to be a part of the process of forming Population I stars. The loss of the embedding gas then leads either to a complete dispersion of the stars, the most likely outcome, or, for about 10% of cases, to the formation of a galactic cluster. We have seen that interactions that might take place in a dense embedded cluster have been invoked to explain the variation in the frequency of binary systems, related to different stellar-density environments, and also to explain the formation of stars with masses greater than $8M_\odot$. Now we shall consider other kinds of interactions that explain the formation of planetary systems.

6.1. The Interaction of a Star with a Protostar

The study of young stellar clusters, described in Section 4.4, showed that some stars on the main sequence, and many others at the YSO stage, will be present while lesser-mass protostars are still being formed. We now consider an interaction in which a protostar in an extended state — which will be for most of its free-fall time — closely approaches a compact star, which can be either a star on the main sequence or a well developed YSO. In this context, 'closely' means close enough for the protostar to be affected by tidal forces due to the compact star and this would involve a closest approach within several hundred au. The method used for investigating this interaction was smoothed-particle hydrodynamics (SPH), described in Appendix L. In this particular instance the usual form of SPH used by most researchers was deemed to be inadequate. In many astronomical applications of SPH, material undergoes expansion or contraction and cools or is heated as a result of these changes of state. However, in this case we have a very cool protostar passing close to a very luminous star and being heated in the process. The behaviour of the protostar is going to be different from what it would be if the star were a cold non-radiating object of the same mass. For this reason Steven Oxley and I introduced a new feature into SPH that closely simulated the transfer of energy by radiation.[1] This required the consideration of other properties of the materials in the model, for example, *opacity*, which describes the extent to which it absorbs and transmits radiation. It was a very expensive addition to SPH in terms of computing time — increasing it by a factor of about 10 — but one could have far more confidence in the validity of the computational results.

The first simulation[2] I showed was done with the following parameters:

Characteristics of the star
Mass of the star, $M_* = 2 \times 10^{30}$ kg $\approx M_\odot$
Luminosity of the star, $L_* = 4 \times 26$ W $\approx L_\odot$

---

[1] Oxley, S. and Woolfson, M.M., 2003, *Mon. Not. R. Astr. Soc.*, **343**, 900.
[2] Oxley, S. and Woolfson, M.M., 2004, *Mon. Not. R. Astr. Soc.*, **348**, 1135.

Characteristics of the protostar
Mass of the protostar, $M_P = 7 \times 10^{29}$ kg $\approx 0.35 M_\odot$
Initial radius of the protostar, $R_P = 800$ au
Initial temperature of the protostar, $T_P = 20$ K
Mean molecular mass of protostar material, $\mu = 4 \times 10^{-27}$ kg

Characteristics of the protostar orbit
Initial distance of centre of protostar from star, $D = 1{,}600$ au
Closest approach of protostar orbit to star, $q = 600$ au
Eccentricity of the orbit, $e = 0.95$

SPH simulation of the protostar
The protostar had a uniform density and was represented by 5,946 SPH particles

Figure 6.1 shows the SPH particles projected onto the plane of the orbit at four different times. The protostar started as a sphere but after 6,000 years it has closely approached the star and is highly distorted. It will be noticed from the parameters that the closest approach of the orbit to the star is less than the radius of the protostar so that an actual collision of the two bodies might be expected. In practice this does not happen since the protostar is collapsing and is also being stretched by tidal forces. After 12,000 years the protostar has been drawn out into a long filament and, from the theoretical work by Jeans on the gravitational instability of filaments (Appendix C), it should be expected to break up into a string of condensations. The simulation results at 18,000 years show that such instability does actually occur; these condensations are potential *protoplanets*, entities that may evolve into planets.

Five of the protoplanet condensations are retained in orbit around the star. These protoplanets, with their masses (in Jupiter units) and the semi-major axes (a) and eccentricities (e) of their orbits, are given by:

Mass $= 4.7 M_J$, $a = 1{,}247$ au, $e = 0.835$
Mass $= 7.0 M_J$, $a = 1{,}885$ au, $e = 0.772$

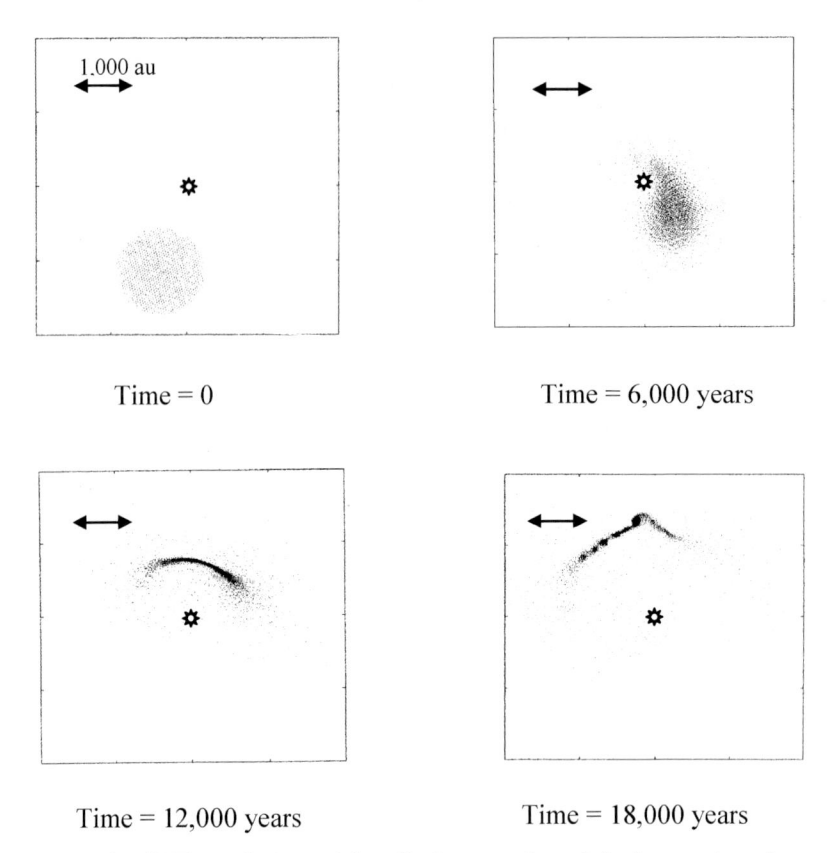

Fig. 6.1 An SPH simulation, with radiation transfer, of the interaction of a star and a protostar.

Mass = $4.8 M_J$, $a = 1509$ au, $e = 0.765$
Mass = $6.6 M_J$, $a = 1{,}325$ au, $e = 0.726$
Mass = $20.5 M_J$, $a = 2{,}686$ au, $e = 0.902$

The mass of the final condensation is outside the range of planetary masses — it is in the range for being a brown dwarf but the remainder are within the range of observed planetary masses, although at the upper end of the range.

Some of the condensations are not retained by the star but are released into the general star cluster. They will probably end up just like field stars except they will be of planetary or brown dwarf mass. In 2000 two British astronomers, P. Lucas and P. Roche, were using infrared observations to detect brown dwarfs within the Orion Nebula. They did find many brown dwarfs but what they also found were some objects of planetary mass, which they called *free-floating planets*.[3] When these were first reported they were noted as an unexpected outcome of the 'standard', *i.e.* Solar Nebula, model but they are a natural outcome of the mechanism that has just been described.

The general appearance of the condensations in Fig. 6.1 indicates that they will eventually condense into giant planets. To verify this, Fig. 6.2 shows the collapse of a protoplanet — one taken from another simulation.

Figure 6.2 shows the motion of the protoplanet and the projected SPH particles at intervals of 100 years. About one half of the mass goes into the collapsing core while the remainder forms an extensive disk. This means that the masses given for the condensations appearing in Fig. 6.1 will overestimate by a factor of two the masses of the resultant planets. Later we shall discuss what happens to the disk.

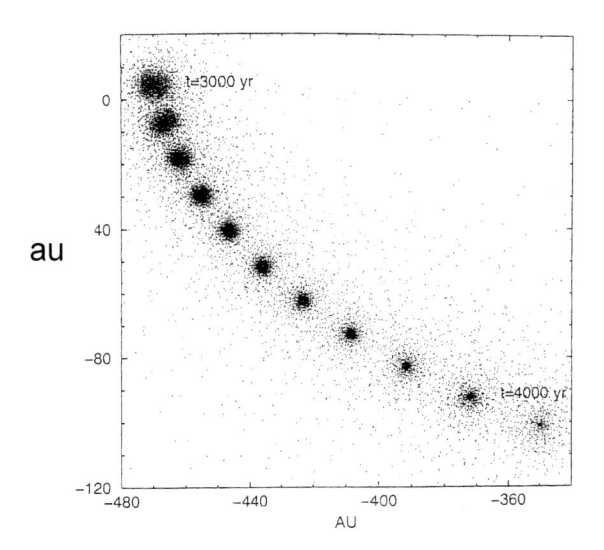

Fig. 6.2 The collapse of a protoplanet at 100-year intervals.

---

[3]P. Lucas and P. Roche, 2000, *Mon. Not. R. Astron. Soc.*, **314**, 858.

I have stressed the requirement that any plausible mechanism for producing planets should be robust, which means that it will give the required outcome over a wide range of parameters. If planets could only be produced with parameters that had to be fine-tuned, the mechanism would clearly be unacceptable. Many simulations, all with star characteristics like those of the Sun, have been made with this model of planet formation. Protostar masses have varied between 0.35–$0.65\,M_\odot$, their radii between 600–1,300 au, nearest approach distance from 600–800 au and orbital eccentricities from 0.9 to 1.1. In all cases filaments were produced within which were condensations with masses in the range of 5 to 23 Jupiter masses. In another calculation on a completely different scale the protostar had a mass of 0.05 $M_\odot$, radius 100 au and had a nearest approach distance to the star of 60 au. This gave rise to two planets, each with a mass of about 5 M_J. It is clear that the robustness criterion for this process is well-satisfied.

The basic feature of this mechanism is that the material to form planets is captured by a star from a passing protostar. This aspect of the model has given rise to the name Capture Theory (CT) for this process of planet formation.

6.2. The Interaction of a Star with a High-Density Region

In Chapter 2, I described how protostars originate as high-density regions produced by the collision of streams of gas in a turbulent star-forming cloud. Not all such regions will go on to produce a protostar; for this to happen the region must not be heavily disturbed by further collisions during the early part of its free-fall collapse. High-density regions are being formed and disrupted continuously during the star-forming period and at any time their number would probably considerably exceed the number of collapsing protostars. For this reason it is interesting to see what would be the outcome of an interaction between a star and a forming high-density region. Now, in place of protostar parameters to define the conditions of the interaction, we must define the characteristics of the two streams of

The Formation of Planets: The Capture Theory

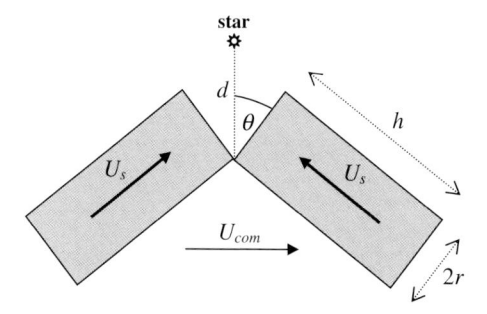

Fig. 6.3 The collision of two cylindrical streams of gas in the vicinity of a star.

material and how they are moving. The geometry that has been used is a symmetrical one and is illustrated in Fig. 6.3.

The geometry that was adopted placed the star in the plane defined by the axes of the two cylinders; two gas streams and a star allow far greater variation in a geometrical sense than the scenario involving a star and a protostar. The parameters that were taken to define the simulation are:

The mass of each stream, M_s
The density of the gas, ρ
The mean molecular mass of protostar material, μ
The ratio h/r for the cylindrical streams
The distance d as shown in Fig. 6.3
The angle θ as shown in Fig. 6.3
The stream speed U_s relative to the centre of mass
U_{com} is calculated such that in combination with U_s the centre of mass of the two streams will move in an orbit with closest approach, q, and eccentricity, e
The number of SPH particles used in the simulation, n

Figure 6.4 shows a simulation with:

$M_s = 10^{30}$ kg $= 0.5 M_\odot$: $\rho = 4 \times 10^{-15}$ kg m^{-3}; $\mu = 4 \times 10^{-27}$ kg; $h/r = 2$; $d = 2{,}000$ au; $\theta = 36.1°$; $U_s = 1$ km s^{-1}; $q = 800$ au; $e = 0.8$; $n = 7{,}990$

92   On the Origin of Planets

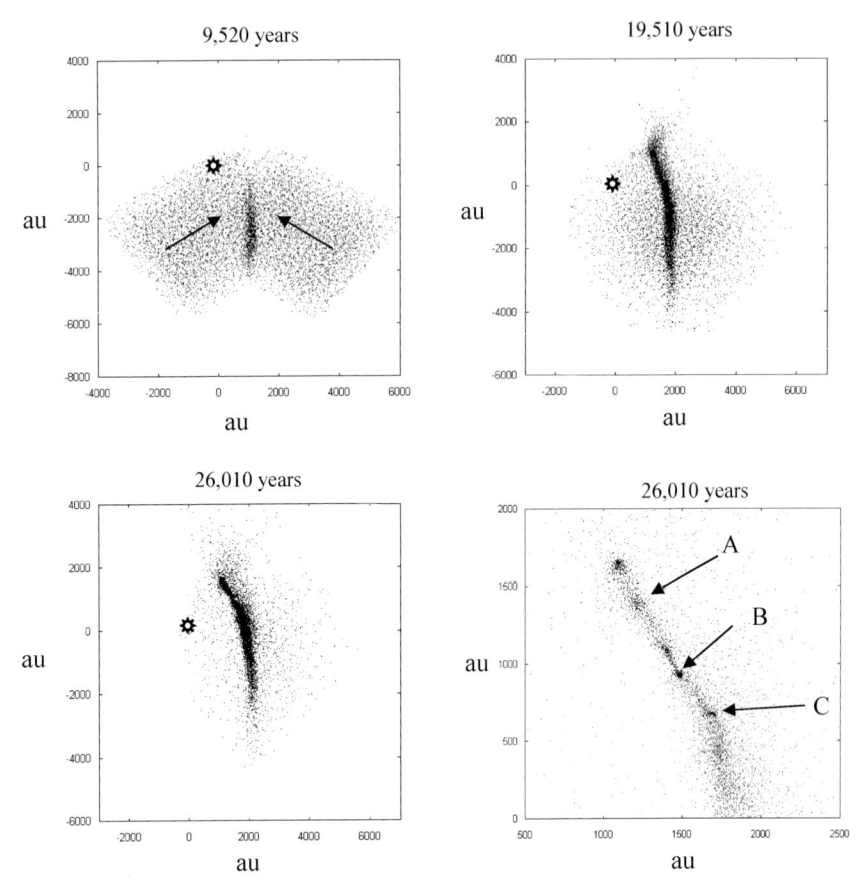

Fig. 6.4 Simulation of colliding gas streams at times 9,520 years, 19,510 years and 26,010 years. The final image shows a higher resolution view at 26,010 years.

The three condensations indicated by arrows are captured with characteristics:

Condensation A; mass 1.00 M_J, semi-major axis 4,867 au, eccentricity 0.768
Condensation B; mass 1.6 M_J, semi-major axis 1,703 au, eccentricity 0.381
Condensation C; mass 0.75 M_J, semi-major axis 1,736 au, eccentricity 0.818

Other simulations of colliding streams, with a wide range of parameters, give outcomes of a similar kind with some condensations captured and others released to give free-floating planets.

6.3. Summary

The embedded state of a developing stellar cluster, with a high number density of stars, is an environment that most, possibly all, Population I stars occupy at some stage during their formation process. Within such an environment, a compact star, either on or approaching the main sequence, might interact with either a diffuse protostar or a newly formed high-density region produced by colliding streams of gas. Under a wide range of conditions the tidal forces due to the star lead to the formation of an unstable filament of material, condensations within which become potential protoplanets. The star captures some condensations while others escape from the gravitational influence of the star to become free-floating planets. The collapse of a protoplanet produces a dense core surrounded by a substantial dusty disk having about the same mass as the core.

The initial orbits of the captured condensations have semi-major axes of one to several thousand au and usually very high eccentricities. If the final orbits are to correspond to those observed for solar-system planets and exoplanets, then these orbits must be substantially modified.

Chapter 7
Orbital Evolution

In Chapter 6 I described how planets could be produced and I also pointed out that the capture-theory process gives planets with very extended and eccentric orbits. The protostars taken for the calculations had masses in the range of 0.35–0.65 $M_\odot$, which is equivalent to 350–650 M_J. Only a small part of the protostar mass goes into planet formation, certainly less than 100 M_J in all the simulations that have been carried out and usually much less. The question then arises of what happens to the remainder of the protostar mass. The calculations show that, normally, most of it is lost into space but some is retained and forms a disk around the star. Sometimes a small part directly impinges on the star and is absorbed by it. The disk material constitutes a medium within which the newly formed planets move, either for all of their orbits or just the inner part of their orbits if the disk has a limited extent.

In the Historical Sketch I explained the need of the SNT for a migration mechanism to explain the outermost planets of the Solar System. There are two types of migration process. In type II migration, the planet is massive enough to open up an annular gap in the medium in which it moves. The gravitational force due to the planet adds angular momentum to the more slowly moving exterior medium material and so propels it further outwards. The reaction to this effect is to cause the planet to move inwards. Conversely, the planet's gravitational force removes angular momentum from inner faster-moving material and so pushes it inwards while the reaction process moves the planet outwards. For type I migration, the planet has too little mass to open a gap and hence the planet moves in contact with the medium. The action of type I migration is always to move the planet inwards; in principle, type II migration could move a planet outwards but it is comparatively slow and inefficient.[1]

The type I and II migration processes are both predicated on the circular orbits of the planets, which is what is expected from the SNT. However, the CT produces planets in highly eccentric orbits and we must now consider how the medium will affect their motions.

7.1. The Nature of the Disk

When the term 'disk' is used to describe the arrangement of material surrounding a young star, the image that comes to mind is something like a coin, which has a uniform thickness, or even an athlete's discus that is thicker at the centre than at the edge. A stellar disk has a more complicated shape. The areal density of the disk, *i.e.* the mass per unit area if mass were projected on to the mean plane, will depend on the way that the disk material is deposited in the neighbourhood of the star and also on the temperature of different parts of the disk due to heating by the star. The amount of material close-in will be reduced by outward forces due to radiation pressure (Appendix T) and the effect of a stellar wind — the emission of very fast particles, mainly protons and electrons, by the star (Appendix U). At some distance from the star the disk will necessarily have to terminate so that the areal density profile will first rise, moving out from the centre, reach a peak and then fall off.

A requirement for the physical stability of the disk is that the maximum density at any point in the projection of the disk will be in the mean plane. If the density in the mean plane of the disk is ρ_0, then at a distance z from that point, in a direction perpendicular to the mean plane, the density ρ_z falls off in the way shown in Fig. 7.1. This shape is a Gaussian curve, the spread of which is described by its *standard deviation*, σ, shown in the figure. In this case, theory gives the standard deviation as

$$\sigma = \frac{\pi c}{2\sqrt{2}v_o} \qquad (7.1)$$

---

[1] A fuller description and analytical treatment of type I and type II migration is given in Cole, G.H.A. and Woolfson, M.M., 2002, *Planetary Science: The Science of Planets Around Stars* (Bristol: Institute of Physics).

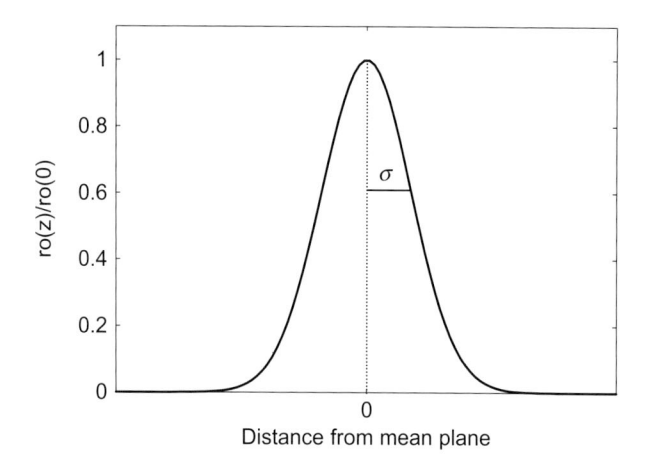

Fig. 7.1 The form of fall off of density with distance from the mean plane.

where c is the speed of sound in the medium, depending on the nature of the material and its temperature, and v_o is the Keplerian speed (orbital speed of a body in a circular orbit) at the distance of the point to the star. The orbital speed factor will tend to dominate in Eq. (7.1) and, since orbital speed falls with distance from the star, the value of σ, and hence the spread away from the mean plane, is larger at greater distances. To summarize — the material is more diffuse further out from the star and less concentrated close to the mean plane. A more detailed description of the structure of a disk is given in Appendix V.

The capture-theory process leads to the formation of a disk around the star but that is not to say that new stars would not have disks if they did not participate in a capture event. It is unlikely that a forming star would incorporate within itself all the material in the condensation that produced it. Indeed, for protostars formed from high-density regions produced by the collision of streams of gas, some peripheral material would probably be retained in orbit. Figure 7.2 shows in schematic form an offset head-on collision of two streams of gas. In the region where the gas streams meet, the gas is compressed. Outside that region the gas is uncompressed and some of it could be retained in orbit around the central compressed region. The actual physical arrangement would not be as clear-cut as is portrayed in

Fig. 7.2 (a) An offset head-on collision of gas streams. (b) The formation of a denser compressed region with two uncompressed streams of gas.

the figure but the conclusion that potential disk material is available is valid.

Whether or not a star had a pre-existing disk at the time of a capture-theory event would depend on how long after the formation of the star the event took place, since we know from observations that disks have a finite duration. The estimated masses of disks around new stars are in the range of 0.01–0.1 $M_\odot$ or 10–100 M_J. From capture theory simulations it is estimated that the mass of material captured by the star to give a disk has a similar range but normally towards the upper end. If there were a pre-existing disk at the time of a capture-theory event, then the final disk surrounding the star would be the sum of what was there originally and what the event added.

It is incumbent on scientists always to veer towards conservatism in their assumptions about the conditions for testing a theory. What I mean is that one should not choose parameters at the extreme limit of possible values that would tend to give the sought-for outcome — rather, one should choose in the other direction for then any positive result would have greater credibility. It is also important not to exclude factors that may act against obtaining the desired result. Thus in testing the CT, as described in Chapter 6, Oxley and I spent a great deal of time and effort in designing an algorithm to incorporate radiation transfer into SPH because it acted *against* the capture process; the capture of collapsing protoplanets happens much more readily with cold material. So, in the same spirit, in testing for the action of a resisting medium on the orbits of planets, I did not use a mass of medium at the upper end of what is feasible so as to amplify the effect being sought. Most calculations have been carried out with resisting mediums of modest mass, usually 25–50 M_J.

To define the medium within which the planet moves, it is necessary to define the way that the areal density varies with distance from the star and also the value of σ, which dictates the fall-off away from the mean plane. However, another factor should be included. Disks are not permanent features and they decay with time; they are under bombardment from stellar radiation and the stellar wind and gradually evaporate into space. Infrared observations suggest that disks become undetectable after periods varying from 10^6 to 10^7 years but mostly at the lower end of that range; about one-third of young stars lose their disks in under 3×10^6 years.[2] In the calculations that are about to be described, disk decay is simulated so that the density at every point falls off exponentially with time (Appendix V). This means that if the density at any point falls to one-half of its original value in time t, then it will have one-quarter of its original value in time $2t$ and one-eighth of its original value after time $3t$, and so on. Thus the effect of the medium on a planetary orbit declines with time and eventually its effect becomes negligible so that the orbit stabilizes.

7.2. The Force on a Planet Due to the Medium

We all have an instinctive acceptance, based on our experience, of the fact that a body moving within some substance — a gas or a liquid — will experience a retarding force and we know from news reports that from time to time Earth satellites crash to earth because of atmospheric resistance. However, the way that a resisting medium influences a body moving within it is greatly affected by the size and mass of the body and the nature and extent of the medium. I shall now describe the different types of resistance that can occur.

7.2.1. *Viscosity-based resistance*

We are all acquainted with the idea of resistance due to motion relative either to the air or to a liquid. Riding a bicycle, or even walking, in the face of a strong wind requires much more effort

---

[2]Haisch, K.E., Lada, E.A. and Lada, C.J., 2001, *Astron. J.*, **121**, 2065.

than carrying out the same motions in calm conditions. However, resistance is experienced even if the motion is in still air; it is the relative speed of the air and the mover that is the critical factor. Car manufacturers put a great deal of effort into designing vehicles that offer as little air resistance as possible so as to improve fuel economy. What applies to motion through the air applies even more to motion through a liquid such as water. This resistance is readily felt by moving a hand through water like a paddle; the faster the hand is moved, the greater the resistance felt. This kind of resistance is due to *viscosity*, the property of a fluid — a gas or liquid — that measures its resistance to flow. Air has a very small viscosity, water a small viscosity, engine oil a higher viscosity, treacle an even higher viscosity and pitch, seemingly almost solid unless it is hot, an extremely high viscosity. The fluid in contact with a body moving through it moves with the body, so setting up variations of speed within itself. It is the friction between layers of fluid moving at different speeds that is responsible for the resistance to the motion of the body.

Viscosity-based resistance is very dependent on the size of the moving body. If the dimensions of the body were increased by a factor of two — for example, a sphere of radius 1 unit was changed into a sphere of radius 2 units — then the viscous resistance for a given speed, which is proportional to the cross section in the direction of motion, increases by a factor of four. However, the mass goes up by a factor of 8 so that the resisting force per unit mass is halved. The larger the body, the less important is viscosity-based resistance; for planetary-sized objects it is insignificant. We must look elsewhere for an explanation of how a planet experiences resistance as it moves through a medium.

7.2.2. Mass-based resistance

The type I and II migration mechanisms depend on the gravitational forces between the moving body and the medium. Figure 7.3 gives a simple illustration of the way that gravitational forces can affect a body moving through a medium. In Fig. 7.3(a) we see a body approaching a region occupied by a system of solid particles

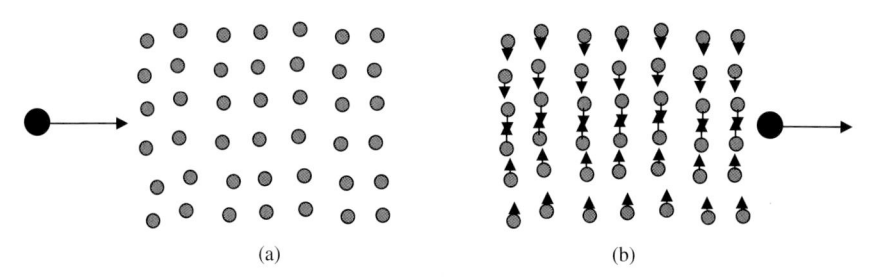

Fig. 7.3 (a) A body approaching a region with a stationary system of particles. (b) The particles are moving after the body moves through the region.

separated from each other but initially stationary. In Fig. 7.3(b) the body has passed through the region without touching any of the particles but, as a result of its gravitational field, during its passage the particles are all set in motion as shown by the arrows. The particles now have kinetic energy (energy of motion) that previously they did not have. Since energy must be conserved, this energy has to come from the moving body that now moves more slowly.

There is no property of viscosity involved in this process of removing energy from the body; the 'medium' is not even a fluid. It is purely a gravitational effect and although it is most easily visualized as the way the body affects the particles, it is clear from the net effect of the passage that each individual particle must also have influenced the body. This indicates that, in order to simulate the passage of a planet through a disk, it is necessary to represent the distribution of mass within the disk, for this is the property that is going to affect the motion of the planet.

7.3. Modelling the Medium and Details of the Calculation Method

The method chosen to represent the medium for the purpose of the calculations was somewhat like that used in Fig. 7.3. A system of particles was set up with variable masses so as to reproduce the desired variation of density within the disk. These particles were first placed in a series of rings in the mean plane of the disk and given velocities that put them into circular orbits around the star.

So that the mass distribution should remain constant if undisturbed, the particles were non-interacting, that is to say, they did not interact gravitationally with each other.

When the motion of the planet is elliptical, as it certainly is at the beginning of the simulation, then its distance from the star changes throughout the orbit and it moves faster the closer it is to the star. For this reason it is desirable to have a higher density of disk particles close-in so that the variation of density over small distances can be better represented. This is achieved by making the first ring of points closer in than the planet will ever reach and then relating the radius of the $(n + 1)$th ring to that of the nth ring by

$$r_{n+1} = \phi r_n \quad (\phi > 1). \tag{7.2}$$

With this arrangement the gaps between the rings gradually become wider as the distance from the star increases. The particles were then uniformly arranged around each ring at distances as close as possible to the average distance to the rings on either side. This gave a distribution of particles as seen in Fig. 7.4(a).

In order to simulate the distribution of particles away from the plane, the particles were randomly displaced perpendicular to the plane between $+\sigma$ and $-\sigma$, where σ is given by Eq. (7.1). To determine c, which then defines σ at any distance from the star, requires assignment of the mean molecular mass of the medium and also its temperature. Although this distribution, displayed in Fig. 7.4(b), is not the one shown in Fig. 7.1, it is a sufficiently good representation for the purpose of the simulation.

The next quantities to be fixed are the masses associated with the individual particles, the variation of which could give any desired radial distribution. The capture-theory computational results give the captured medium with a variety of configurations. Often it has a maximum density somewhere in the range of 100–300 au from the star, with density tailing off on either side. Where this type of distribution is taken, the form adopted for the variation of the areal density of the medium with distance from the star was a Gaussian distribution (Fig. 7.1). The medium is then fully defined by the distance of the peak from the star, the standard deviation,

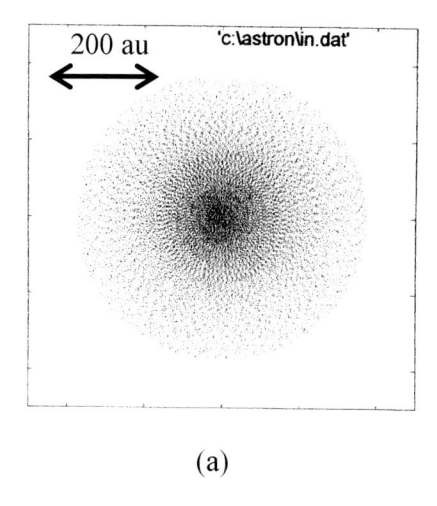

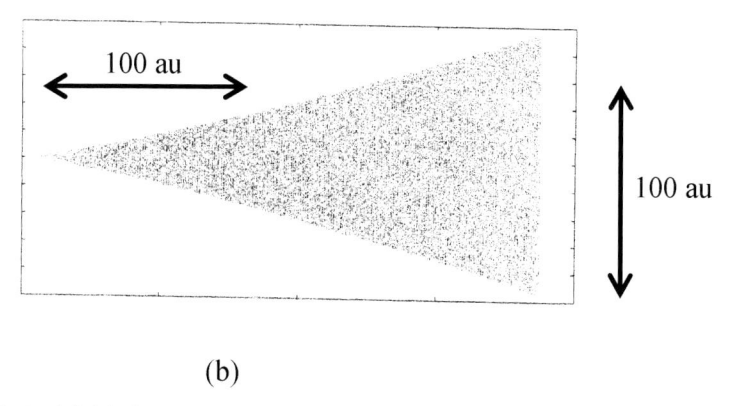

Fig. 7.4 (a) Medium particles seen in plan view. (b) The distribution perpendicular to the plane.

σ_r, and the total mass of the medium. The other type of distribution that occurs is where the density is highest very close to the star and falls off with increasing distance. This distribution is modelled by a declining exponential function (Appendix V), the form of which is shown in Fig. 7.5.

The final characteristic of the medium that must be defined is the rate at which it declines with time, bearing in mind that disks become unobservable on a timescale of a few, up to 10, million years at

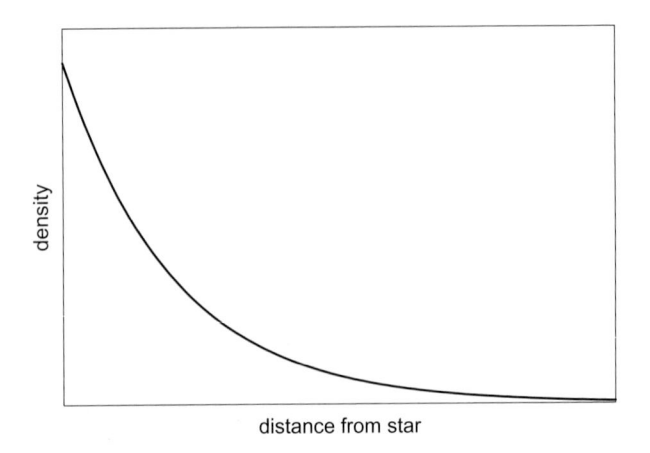

Fig. 7.5 A declining exponential distribution.

most. Introducing an exponential decline with time of all the particle masses does this.

With the medium thus established and with its particles set moving in circular orbits, the planet is introduced, moving in the proposed initial orbit derived from a capture-theory origin. In the calculation the planet and medium particles act gravitationally on each other but, as previously stated, gravitational interactions between the medium particles are excluded. This gives a stable medium in the absence of the planet, and the extent to which the planet disturbs the medium also gives the reaction that influences the planet's motion. However, in a real medium the disturbance caused by the passage of the planet would gradually disappear, just as the wake of a boat disappears at some distance behind it. The gap in the gaseous medium produced by the passage of the planet would create a region of lower density and pressure, and neighbouring gas would flow in to restore equilibrium. To take account of this, the medium particles were reformed into their original configuration from time to time. For a protoplanet with an orbital period of more than 30,000 years, the medium was reformed after each orbit. When the orbital period was between 15,000 and 30,000 years, it was reformed after two orbits, and for less than 15,000 years, it was reformed after every three orbits. Variations of this formula for restoring the medium

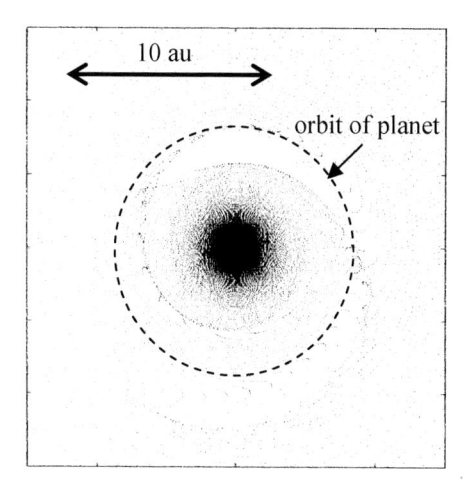

Fig. 7.6 Spiral waves generated by a Jupiter-mass planet in a gaseous medium.

gave minor changes in the computational results, but these changes were more of detail than substance.

Before giving the results of running the model with extensive and highly eccentric planetary orbits, I show in Fig. 7.6 the appearance of a medium, of mass 25 M_J, with a Jupiter-mass planet in a circular orbit of radius 6.3 au. The interest here is that it shows the spiral waves travelling inwards and outwards from the planet that have been suggested by SNT theorists as a way of giving migration for outer planets, as explained in the Historical Sketch. It is clear that the scale of the spiral waves is such that the quantity of energy imparted to a small target such as a planet could be only an infinitesimal fraction of the energy in the waves — energy that derives from the decay of the planet's orbit.

7.4. Calculations of Orbital Decay and Round-off

Simulations of orbital decay and round-off have been carried out over a wide range of parameters.[3] The following example of a simulation is illustrated in Fig. 7.7.

---

[3] Woolfson, M.M., 2003, *Mon. Not. R. Astr. Soc.*, **340**, 43.

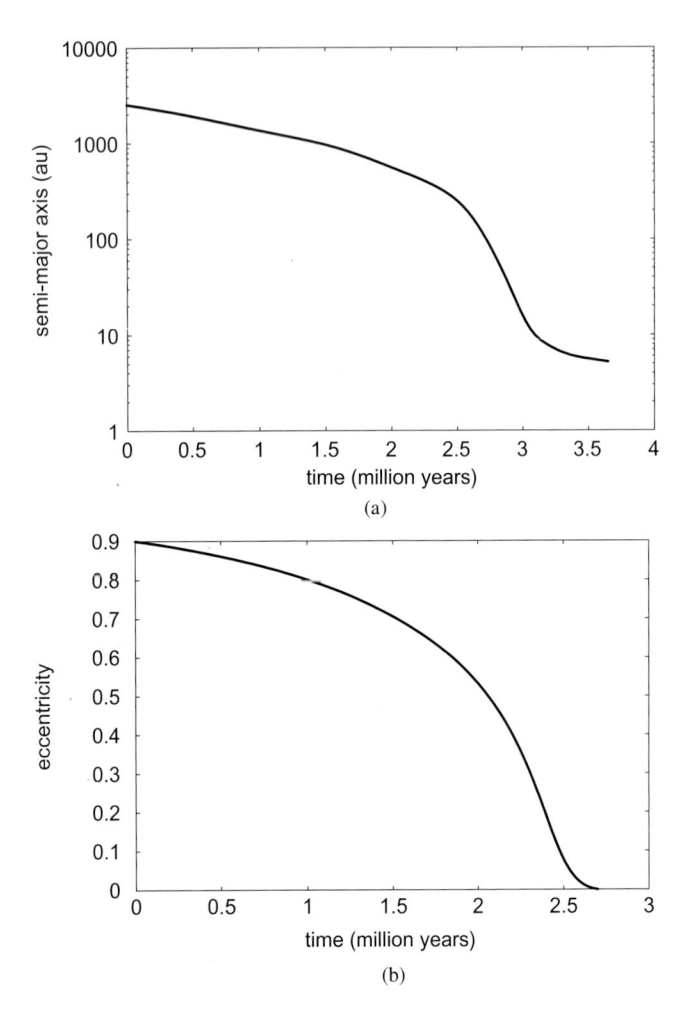

Fig. 7.7 The variation with time of (a) semi-major axis and (b) eccentricity.

Mass of planet = M_J; Initial semi-major axis = 2,500 au; Initial eccentricity = 0.9 Mean molecular mass for medium = 2×10^{-27} kg; Temperature of medium = 20 K; Total medium mass = 50 M_J; Exponentially declining density, halving in 139 au; Medium halves every 693,000 years. Medium represented by 77,408 particles.

In about 2.7 million years the orbit rounded off and after another million years the orbit decayed to a semi-major axis of

Table 7.1 The variation of the final a and e with protoplanet mass for a given medium.

Mass of planet	Semi-major axis (au)	Eccentricity
M_J	2.980	0.0067
$2M_J$	1.549	0.0065
$3M_J$	1.056	0.0064
$4M_J$	0.787	0.0064
$5M_J$	0.633	0.0062

about 5.2 au, which happens to be the orbital radius for Jupiter. Similar simulations with different parameters show some obvious characteristics. For example, increasing the overall mass of the medium, or decreasing its rate of decay with time, reduces the times for round-off and decay and also increases the total decay of the orbit. A less expected result is that the effect of a medium of given characteristics *increases* with the mass of the planet. In Table 7.1 the final semi-major axes, a, and eccentricities, e, are given for protoplanets of different masses moving in a medium of mass $50\,M_J$ with an exponentially declining distribution of the type shown in Fig. 7.5.

In all the cases so far considered, the final orbits have been close to circular. However, as seen in Table 5.2, there are a few observed exoplanets with very high eccentricities, so we must now consider how these could come about.

7.5. Orbits of High Eccentricity

In the calculations described in the previous section, it was assumed that each element of the medium orbited the star with Keplerian velocity, the same assumption of the SNT theorists who have worked on planet migration. Their migration mechanisms assume that the planet is in a circular orbit with medium material outside it moving a little slower and that inside it moving a little faster. For a planet in an *elliptical* orbit, this relationship of planet motion to medium motion will no longer apply. At the extremes of its orbit — when it is closest to the star, at *periastron*, and furthest from the star, at *apastron* — it is moving parallel to the local medium but not at the same speed. The medium is impinging on the planet, either slowing it down or speeding it up, depending on the direction of impingement relative to the planet's motion. Whatever the mechanism by which the medium applies force on the planet, the proposition is always true that the force is in a direction that will oppose the motion of the planet relative to the medium and the force is greater for a greater relative speed.

We now consider the effect when the medium at any point is moving with Keplerian velocity. Although the medium exerts a force on the planet at all points on its orbit, we can understand the overall effect on the planet by just considering what happens at periastron and apastron. In Fig. 7.8(a) the full line shows an elliptical orbit and the solid arrow the speed of the planet at periastron. The medium is moving more slowly, with a speed shown by the dashed arrow, so it is going to resist the motion of the planet at that point and slow it down. The slowing down will produce a new orbit like that shown dashed, which has the same periastron position but a smaller semi-major axis and a smaller eccentricity than the original one. Now we consider what happens at apastron, shown in Fig. 7.8(b). Since the medium is now the faster moving, it speeds up the planet, so producing a new orbit, shown dashed, which has the same apastron as the original orbit but a larger semi-major axis and a smaller eccentricity. At both extreme positions the effect is to round-off

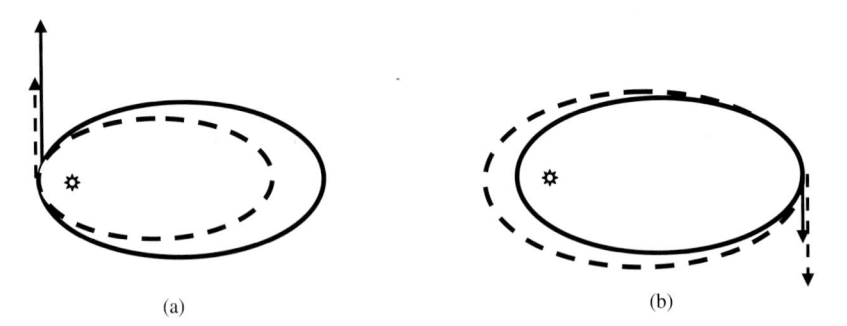

Fig. 7.8 The change in orbit due to medium forces at (a) periastron and (b) apastron. Shown are original orbits (full line) and modified orbits (dashed line). Protoplanet speeds are shown as full arrows and medium speeds as dashed arrows.

the orbit but the semi-major axis is decreased at periastron and increased at apastron. Normally the medium is denser and therefore more effective closer in, so that the semi-major axis decreases and the orbit decays. However, whatever the distribution of the medium happens to be, if at all points it is freely orbiting as a planet would do, then the eccentricity will decrease. Round-off in this case is inevitable. Another factor is required to explain the highly elliptical orbits observed for some exoplanets.

It is known that young stars go through an active stage in which they have much stronger stellar winds and are much more luminous than when they reach the main sequence. For example, in 1998 two Italian astrophysicists, D'Antona and Mazzitelli, estimated that the early Sun could have been 60 times as luminous and could have had solar winds between 10,000 and 100,000 times as strong as at present.[4] Luminosity-based radiation pressure and bombardment by stellar wind particles apply an outward force on the medium that counteracts the gravitational attraction of the star (Appendix U). We are going to consider the effect of having the gravitational field of the star neutralized to some extent by one or both of strong radiation and a powerful stellar wind. The effect on the medium is as though the star has a smaller mass but the massive planet would be unaffected and so experience the full gravitational force of the star. In this case all parts of the medium are still in circular orbits but they are moving more slowly — perhaps much more slowly — than the Keplerian speed. We now consider the possible effects at periastron and apastron in the case where the medium is heavily slowed down, as illustrated in Fig. 7.9.

Figure 7.9(a) shows that, just as in the previous case, the planet at periastron moves faster than the medium so it is slowed down, changing the orbit to one of smaller semi-major axis and lower eccentricity. However, at apastron (Fig. 7.9(b)) the medium has been slowed down so much that it is moving *more slowly* than the planet. Now the planet is slowed down, decreasing the semi-major

---

[4]Dantona, F. and Mazzitelli, I., 1998, http://www.mporzio.astro.it/~dantona/prems.html.

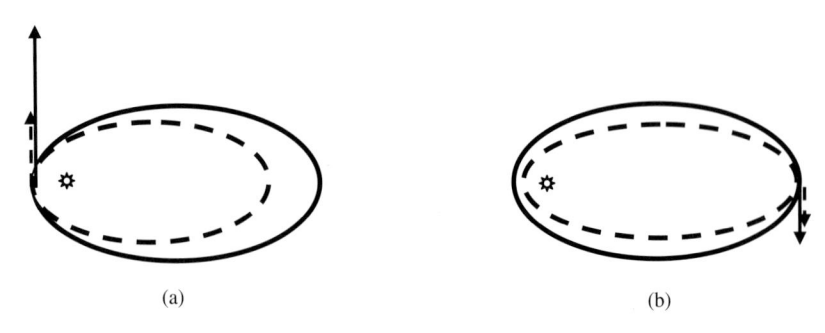

Fig. 7.9 The effect of a slowed-down medium on a protoplanet orbit at (a) periastron and (b) apastron.

axis but *increasing* the eccentricity. At both extremes the semi-major axes are reduced, giving decay of the orbit, but the effect on the eccentricities is different. In most cases the medium will be much denser at periastron and the effect there will be the stronger so the eccentricity will be reduced. However, in many capture-theory simulations the captured medium takes on a strong doughnut-like form (Fig. 7.10) and this can lead to dominance of the effect at apastron and hence to an increase in eccentricity.

In considering the influence of an active star in affecting the medium, it is necessary to introduce a new factor — the relationship

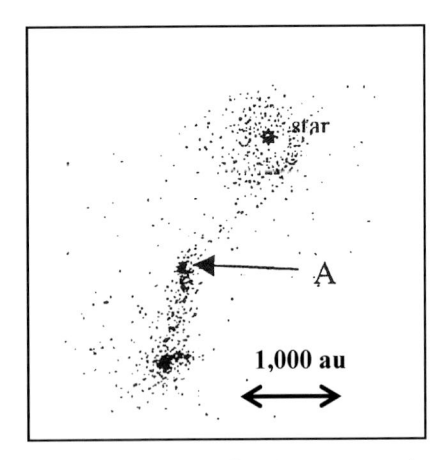

Fig. 7.10 A capture-theory simulation showing a strong doughnut-like captured medium.

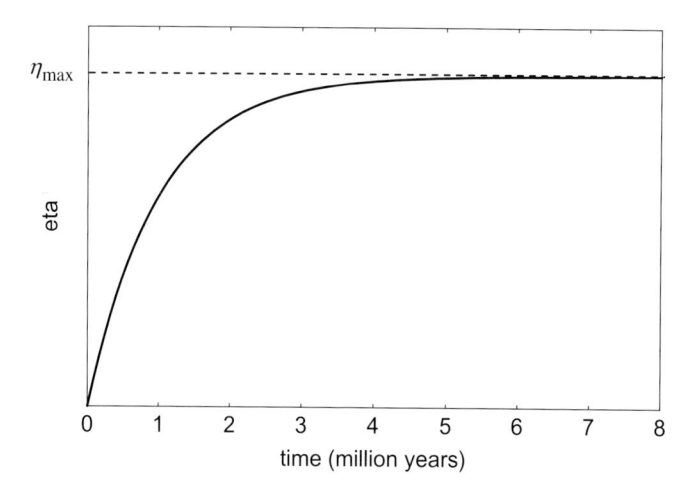

Fig. 7.11 The proportion of the stellar gravitational field neutralized by stellar activity.

of the period of the star's enhanced activity to the period during which the medium is acting. In the computational results that I report here, the proportion, η, of the gravitational field of the star that is removed by the stellar activity is modelled as a function of time as in Fig. 7.11. As the stellar activity increases, then so does η until it reaches a value η_{max}.

In the following simulations a planet of mass $5\,M_J$ starts in an orbit with semi-major axis 1,500 au and eccentricity 0.9. The medium, of total mass $50\,M_J$, with mean molecular mass 2×10^{-27} kg and temperature 20 K, has the form shown in Fig. 7.1 with a peak at 200 au from the star and with $\sigma = 100$ au. The simulations are run with different values of η_{max} and give the results in Table 7.2. With $\eta_{max} = 0.5$ or less, the orbits are circular, but for higher values of η_{max}, the final eccentricity steadily increases with η_{max}. The variation of semi-major axis and eccentricity with time for the runs A, F and H are shown in Fig. 7.12.

There is a wide range of parameters that are possible for simulations of this kind. By taking stellar activity that almost eliminates the gravitational attraction of the star on the medium, even higher eccentricities can be obtained.

Table 7.2 Variation of final orbital parameters with η_{max}.

Run	η_{max}	a_{final}	e_{final}
A	0.0	0.870	0.000
B	0.2	0.652	0.000
C	0.4	0.438	0.000
D	0.5	0.613	0.000
E	0.6	1.877	0.268
F	0.7	3.690	0.302
G	0.8	5.782	0.442
H	0.9	7.571	0.566

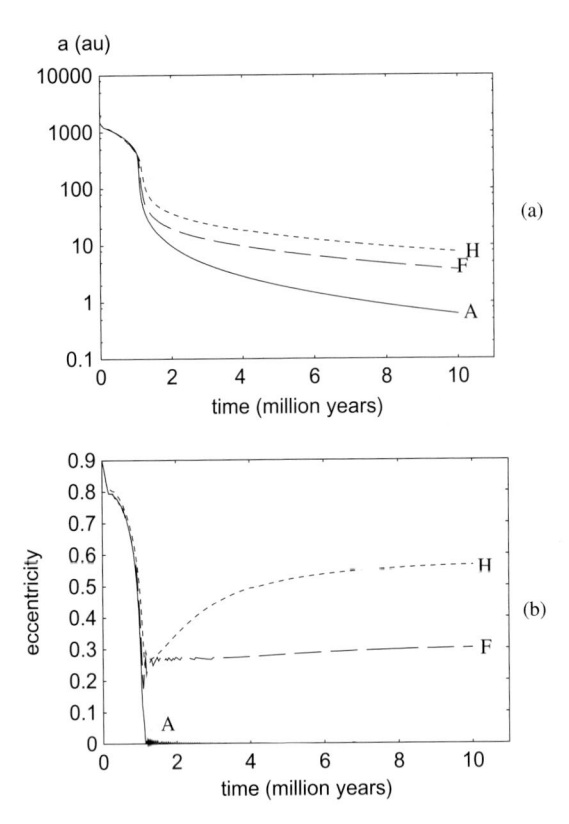

Fig. 7.12 Three simulations of orbital evolution, two of which give eccentric orbits.

7.6. The Range of Semi-Major Axes

Infrared and direct optical imaging of exoplanets has greatly extended the range of observed distances of planets from stars, from a few percent of that of Mercury to several times that of Neptune. Even semi-major axes of several hundred au, were they to be observed, would be allowed by the CT model. Planets are produced with much greater semi-major axes and the extent to which they decay and round off depends on the density and distribution of the resisting medium and its duration. A low-density medium of short duration could leave a planet at a considerable distance from its star. It should also be noticed from Figs. 7.7 and 7.12 that round-off is a faster process than decay so a planet could be left in an orbit of large semi-major axis with a small eccentricity.

A rather more challenging observation is of exoplanets with very small semi-major axes, which raises the question of how they can get so close to the star but avoid being drawn into it completely. Rather curiously, this question is related to the behaviour of the Moon in relation to Earth. If a planet is orbiting a star closely enough for it to appreciably distort the star by tidal forces, then, if the spin period of the star is less than the orbital period of the planet, energy is transferred from the star to the planet. When the rate of energy transfer equals the rate at which energy is lost by motion in the resisting medium, the orbit is then stabilized. The position reached in this way is quite stable. If the planet moves a little closer to the star, the rate of energy transfer from the star increases, giving a net increase of energy with time, so it moves out again. Conversely, if the planet moves a little further out, the rate of transfer of energy from the star is reduced, giving a net loss of energy per unit time, and so the planet moves inwards. The mechanism by which the energy transfer takes place, in the context of the Moon and Earth, is described more fully in Section 12.1.

7.7. Simple Ratios of Orbital Periods

When orbits round-off and decay, protoplanets are influenced not only by the star and the medium but also by each other. The ratio of the orbital periods of some pairs of major planets of the Solar System is very close to the ratio of small integers. For example,

$$\frac{\text{Orbital period of Saturn}}{\text{Orbital period of Jupiter}} = \frac{29.46\,\text{years}}{11.86\,\text{years}} = 2.48 \approx \frac{5}{2}$$

and

$$\frac{\text{Orbital period of Neptune}}{\text{Orbital period of Uranus}} = \frac{164.8\,\text{years}}{84.02\,\text{years}} = 1.96 \approx \frac{2}{1}.$$

Mario Melita, an Argentinian astrophysicist, and I investigated this topic of *commensurate orbital periods* in 1996.[5] We found a mechanism that comes into play when the orbits of pairs of planets have become circular and are decaying at different rates. When the orbits become *commensurate*, that is, the ratio of their periods equals the ratio of two small integers, an energy exchange takes place between them. The effect of this is that the orbits continue to decay but they do so linked together in such a way that the ratio of the orbital periods remains constant. This was confirmed by computational means; the effect is a fairly subtle one that is possible, but difficult, to explain. It might be wondered why the neighbouring planets Saturn and Uranus do not have periods connected in this way; the ratio of their orbital periods is 2.85. An explanation could be that, just by chance, the pattern of orbital decays of the pairs Jupiter:Saturn and Uranus:Neptune gave simple ratios while the medium was still very effective, but that a simple ratio between Saturn and Uranus did not become established before the medium decayed. Alternatively, it could be that the factors locking the two pairs of planets in commensurate orbits inhibited the establishment of commensurability between one of each of the established pairs.

7.8. Stellar Spin Axes

Although it is not directly related to the topic of this chapter, the evolution of planetary orbits, the direction of stellar spin axes do have an indirect relationship and so is introduced here. An argument

---

[5]Melita, M.D. and Woolfson, M.M., 1996, *Mon. Not. R. Astr. Soc.*, **280**, 854.

that has been raised against the SNT is that the spin axis of the Sun is not perpendicular to the mean orbital plane of the planets but is inclined at 7° to that direction. This angle does not seem to be particularly large, and hence one might tend to disregard it, but in fact it is very significant. It would be difficult to explain how, in the solar-nebula scenario, the spin axes of the central blob that formed the Sun and that of the outer material that formed the disk could differ by that amount. But then the CT also has some explaining to do. There can be no systematic relationship between the orbital plane of the protostar (or dense region) and the spin axis of the star. The probability that the angle between two random directions is 7° or less is about 0.004, so it would be stretching credulity to say that it is just by chance that the solar spin axis is so close to being perpendicular to the mean plane of the planets.

The Sun spins very slowly and its angular momentum is equivalent to that of one-fifth of the mass of Jupiter in free orbit around its equator. The estimated mass of the disks around young stars, which we have identified as the resisting medium, is between 10 and 100 times the mass of Jupiter. It is possible that when the protostar, or dense region, is disrupted, some part of it would be directly absorbed by the Sun and, indeed, the capture-theory simulations show that this is sometimes so.

Any material joining the Sun in this way would pull the spin axis closer to the perpendicular to the plane of the disk, which we identify as the mean plane of the planetary orbits. Another factor to consider is that the total mass of solids in the disk — between 1 and 2% of the whole — is between 0.1 and 2 times the mass of Jupiter. The Poynting–Robertson effect (Appendix T) will draw into the Sun any dust particles larger than about a micron in diameter. Information about the distribution of the sizes of dust particles in the interstellar medium is mostly derived from measurements of the way light is scattered and absorbed, and is only reliable up to sizes of about 1 micron. There are particles larger than 1 micron and, although they contribute very little to the absorption or scattering of light by dust, they may account for a substantial proportion of the mass of the dust (Appendix F). Any dust that is drawn inwards by the

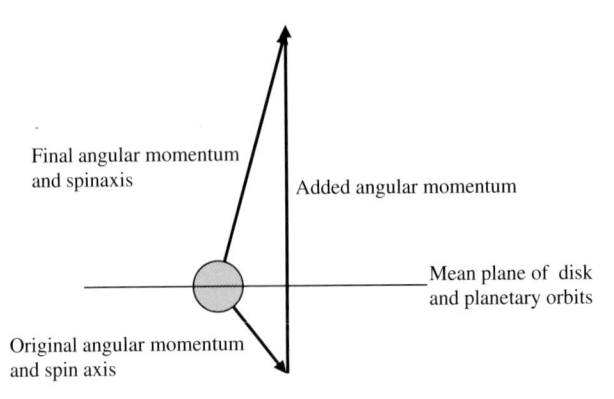

Fig. 7.13 Adding disk material to the Sun pulls the spin axis towards the normal to the disk.

Poynting–Robertson effect gradually spirals inwards and, just before it is absorbed by the Sun, it is in free orbit just above the surface. This addition of material would also have gradually pulled the solar spin axis towards the perpendicular to the mean plane of the planetary orbits. Figure 7.13 shows how the angular momentum added to the Sun could have pulled the spin axis to its present position.

In considering the spin of stars in general, and of the Sun in particular, we should bear in mind that if the star goes through an active period after its final spin axis has been established, its spin may be slowed by the mechanism described in Appendix O. However, this will reduce the rate of spin but not the direction of the spin axis.

In August 2009 the British astronomer Andrew Collier Cameron observed an eclipsing exoplanet orbiting a star, WASP-17, about 1,000 light years away from the Sun, which had a retrograde orbit with respect to the spin of the star. The inclination of the orbit was 150°, taking the plane perpendicular to the star's spin axis as the reference plane. While the orbits of some other exoplanets had quite large inclinations, none had previously been found greater than 90°, beyond which angle the orbit could be described as retrograde. As usual, the first reaction to this observation was to envisage ways in which it could be accommodated to the SNT, for which it represented a considerable difficulty. It was postulated that its path had been reversed by an interaction with an undetected companion planet or by some passing object. This explanation cannot be ruled out.

The interpretation of this observation in terms of the CT is quite straightforward. Although the addition of material to a star will nearly always pull its spin axis towards parallelism with the vector representing the planetary orbit, if the initial spin and orbital vectors are close to anti-parallel, if the stellar spin speed is high and if the amount of absorbed material is low then a retrograde planetary orbit will be the result. Since there is no systematic relationship between the orbital plane of the protostar (or dense region) and the stellar spin, then, through the combination of circumstances as described above, a retrograde orbit, although uncommon, should occasionally occur.

7.9. Summary

The capture-theory mechanism not only gives protoplanet formation but also leaves a medium in the form of a disk around the star. This would supplement any disk that existed before the capture took place and disks of mass $50\,M_\mathrm{J}$ would be quite typical. These disks would have a flared structure in cross-section, being more spread out at greater distances from the star.

The influence of the medium on a planet is due to the gravitational forces between the planet and the distribution of mass in the medium. For purposes of numerical simulations, the medium can be represented by a distribution of particles with masses and positions that reproduce the required density distribution of the medium in three dimensions. The most common areal densities, expressed as a density variation with distance from the star, resemble either a humped Gaussian distribution or an exponentially declining density with the highest density close to the star. To simulate the gradual evaporation of the disk, an exponential decline with time of the particle masses is introduced. Simulations that start with semi-major axes in the range of 1,500–2,500 au and eccentricities of 0.9 or thereabouts typically give decay and round-off on timescales of a few million years. The rate and extent of orbital change depend on the mass and distribution of the medium and its rate of decay. Counter intuitively, the extent of the orbit modification *increases* with increasing planet mass.

If the orbital modification takes place in a period of great activity of the star, with high luminosity and powerful stellar winds, then the gravitational force of the star on the medium can be partially counterbalanced, with the effect of slowing down each part of the medium in its circular orbit. If the slowdown is sufficiently large, and if the medium has a doughnut-type distribution, then this can lead to final orbits of high eccentricity — as is observed for some exoplanets.

A pair of protoplanets, with orbits decaying at different relative rates, can reach a stage when the orbits are commensurate, that is to say, the ratio of the orbital periods is close to the ratio of small integers. The decays of the orbits are thereafter coupled together so that while both decay, they do so in such a way that the ratio of periods is maintained. This explains the close commensurability of the orbital periods of the pairs of solar-system planets Jupiter:Saturn and Uranus:Neptune.

The original spin axis of the Sun was at some unknown angle to the plane of the disk. Absorption of protostar, or dense region, material by the Sun and of dust, through the operation of the Poynting–Robertson effect, pulled the solar spin axis to within 7° of the normal to the disk. In some capture-theory interactions, the vectors representing the stellar spin and the protostar orbital plane will correspond to retrograde motion of planets relative to the stellar spin. If little material is added to the star, this can lead to an outcome similar to that of the planet orbiting the star WASP-17.

Chapter 8
The Frequency of Planetary Systems

I have described how, in the environment of a dense imbedded cluster, the tidal interaction of a compact star with either a protostar or a dense region of the cloud can lead to the formation of planets, some captured by the star but others escaping to become free-floating planets. The capture of other material by the star in the form of a disk gives a medium within which the orbits of the newly formed protoplanets evolve. Depending on the total mass, distribution and rate of decay of the medium, the final orbit of a planet can have a wide range of semi-major axes and eccentricities, such as are found for exoplanets.

A plausible theory of planetary formation must involve a mechanism that is both common and robust. There can be no doubt about the robustness of the capture-theory mechanism; it operates over a very wide range of parameters. Within a dense embedded cluster it also seems likely that capture-theory events would be common — but 'seems likely' does not constitute proof in a scientific context. To reinforce the plausibility of the CT theory it is necessary to show that the proportion of Sun-like stars that would acquire a family of one or more planets is consistent with observations. So that is the question that I must now address.

8.1. Observations and Observational Constraints

The exoplanets that are most easily detected by Doppler-shift methods, as described in Chapter 5, are those that are the most massive and closest to the star. These give a combination of the greatest stellar speeds — the quantity that is measured — and the shortest orbital periods, meaning that measurements can be made over many orbital periods, thus giving more precise results within a reasonable time frame. Since the first detection of exoplanets, the precision of the Doppler spectroscopy method has greatly increased, thus enabling exoplanets of ever-smaller mass to be detected. In April 2009, by the use of an instrument called HARPS (High Accuracy Radial velocity Planet Searcher) mounted on European Space Agency telescopes in Chile, two quite small exoplanets were detected around Gliese 581, a close star at a distance of 20.3 light years that has a mass just one-third that of the Sun and is designated as a *red dwarf*. There are at least four planets orbiting this star. One of them, designated as Gliese 581d, with an estimated mass of 7.7 $M_\oplus$,[1] orbits at a distance that could allow liquid water on its surface, giving rise to speculation that it might support some form of life. Another planet, Gliese 581e, with an estimated mass of just 1.9 $M_\oplus$, is much closer to the star so that, although it may be more Earth-like in mass, it is unlikely to possess a biosphere.

There is a claim for the detection of a planet with even less mass than Gliese 581e, although the detection was made in a very different way. One conclusion from Einstein's general theory of relativity is that light passing close to a heavy object is deflected. The British scientists Arthur Eddington and Frank Dyson confirmed this in 1919 when they observed the apparent displacement of stars when the light from them passed close to the Sun during a solar eclipse. This bending of light can give rise to what is called *gravitational microlensing*, as illustrated in Fig. 8.1. Normally a detector — for example, the eye — observes a star by the light that enters it having travelled in a straight line from all parts of the star's surface. However, if there is a massive object between the star and the detector, then light that has been deflected by the object will enter the detector and the star will appear to be brighter (Fig. 8.1(a)). The microlensing star does not actually act like a true lens; as will

---

[1] The symbol $\oplus$ indicates the Earth, so $M_\oplus$ is the mass of the Earth.

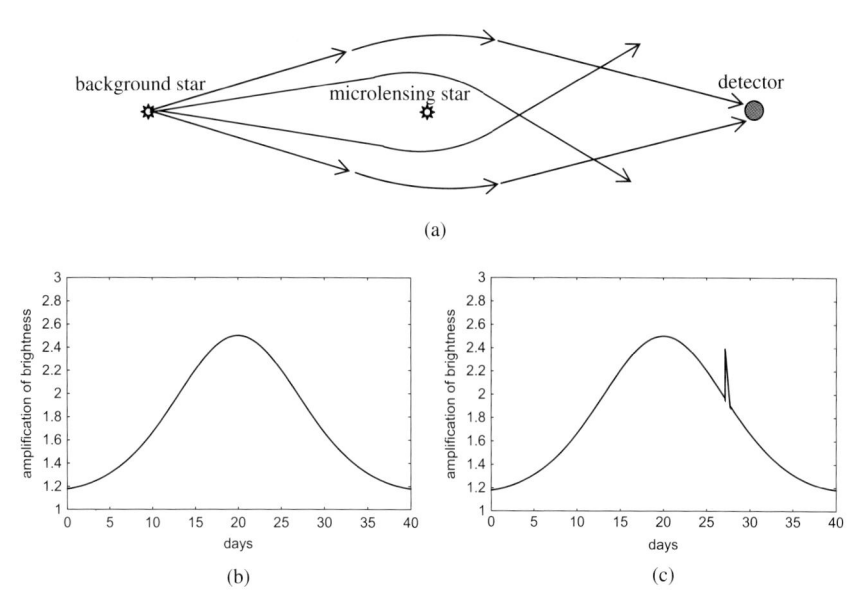

Fig. 8.1 An illustration of microlensing. (a) The paths from the background star to detector. (b) The variation of brightness as the microlensing star moves across the field of view. (c) The spike shows the effect of the planet.

be seen in Fig. 8.1(a), only the light leaving the background star in certain directions will reach the detector. However, there are many more paths for the light to reach the detector than if the microlensing star were absent, so the background star appears brighter. As the microlensing star moves across the field of view, so the background star will first increase in brightness, reach a peak and then return to its normal appearance (Fig. 8.1(b)). If the microlensing object is a star with a planet, then, if the planet happens to pass close to the line between the background star and the detector, there will be a sharp rise in brightness (Fig. 8.1(c)), normally lasting a few hours. From the height of the sharp rise it is possible to estimate the mass of the planet. It was claimed early in 2009 that a planet, detected in this way and orbiting a red dwarf, has a mass of 1.4 $M_\oplus$.

The first estimate of the proportion of Population I stars with planetary companions was about 3%, an estimate that has steadily risen as techniques for detecting exoplanets have improved. More accurate Doppler-shift spectroscopy and microlensing have drastically decreased the minimum mass of planets that can be detected. We have also seen that it is possible to directly image planets at some distance from their stars — planets that would be impossible to detect by present Doppler-shift methods. With about 400 exoplanets detected by late 2009, it has become clear that early estimates of the proportion of Sun-like stars with planets were far too low. Some astronomers have suggested that the proportion may be as high as 30%, although the observational evidence for this is not clear. What can be said with some confidence is that any plausible theory should predict the proportion of Sun-like stars with planetary companions as 10% or more. We shall now check to see if the capture-theory mechanism satisfies this criterion.

8.2. Initial Formation Statistics

The number of variable parameters associated with a capture-theory event is quite large. The only important parameter concerning the compact star is its mass — it just acts as a source of gravitational field. Its luminosity is also a relevant factor but a less important one when it comes to the initial formation of a tidal filament. Although the luminosity of the star was incorporated into the capture-theory calculations described in Chapter 6, because it was a potentially adverse factor, it turned out to have only a marginal effect on the overall outcome of the simulations. Of far greater importance is the nature of the body that gives rise to the tidal filament. It can be either a protostar or a dense region formed by the collision of streams of material. If it is a protostar, then its mass and radius are important, as is the density distribution within it. The state of the protostar most likely to give planets is when it is in the early stages of collapse and hence of largest radius. It will then approximate most closely to having a uniform density. Alternatively, if the filament-providing body is a high-density region, then its mass and shape will be relevant. However, whatever the filament-providing body happens to be, its motion relative to the star will be of critical importance in deciding whether or not protoplanets are formed. If the closest approach distance is too large, no filament will form; if it is too small, there will be a collision and no orderly formation of a filament. A final, and very important, factor in determining the proportion of stars with planets is the number density of the stars in the embedded cluster. This rises to a peak as the cluster collapses and then reduces again when supernovae drives out the binding gaseous component of the cluster, thus causing it to expand. A model of the way that the stellar number density changes with time, which I have used in various computations, is shown in Fig. 8.2.

One of the difficulties in estimating the expected number of planetary systems from the CT is that, although we can estimate the number of protostars being formed — given that they are not disrupted by a capture interaction they turn into stars and we know the number of various kinds of stars that exist — we have no idea how many high-density regions form but do not become protostars. In 2004, at the end of the paper describing the CT, Stephen Oxley and I calculated the proportion of stars expected to have planets by a purely analytical process.[2] We only considered protostars in our analysis and, even without high-density regions,

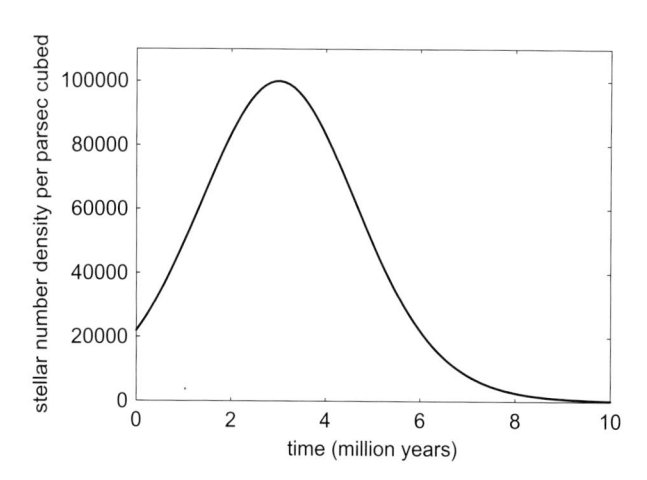

Fig. 8.2 A model of the rise and fall of stellar number density in an embedded cluster.

---

[2]Oxley, S. and Woolfson, M.M., 2004, *Mon. Not. R. Astr. Soc.*, **348**, 1135.

our conclusion was that the predictions of the frequency of planetary systems from the CT were consistent with observations. At that time the general view was that about 7% of Sun-like stars had planets and we found that protostars alone could give that proportion. The addition of high-density regions as potential producers of filaments meant that a higher observed number of exoplanet systems could also be accommodated.

Here I shall take a different non-analytical approach to estimate the proportion of stars with planets, using a numerical model instead. The steps in setting up this model are given here in a rather general way and explained numerically in Appendix W.

(i) A star number density in the embedded cluster, n, is randomly chosen in the range of 10^4–10^5 pc^{-3}.

(ii) The radius, R, of a spherical region of the cluster is chosen that will accommodate 1,000 stars.

(iii) The 1,000 stars are randomly placed within the sphere.

(iv) The mean speed of stars in an embedded cluster is estimated to be in the range 500–2,000 metres per second.[3] For each star a random speed is chosen within that range and the star is given that speed in a random direction.

(v) Each star is assigned a mass, randomly chosen within the range of 0.8–2.0 $M_\odot$.

(vi) The protostar is placed at the centre of the sphere with a velocity selected as in (iv).

(vii) The mass of the protostar is selected randomly in the range of 0.25–0.75 $M_\odot$.

(viii) Given a temperature of 20 K and a mean molecular mass of 4×10^{-27} kg, the density and radius of the protostar is found based on the assumption that when formed it is of a Jeans mass (Fig. IX).

(ix) From the density of the protostar the free-fall time is found (Eq. M7 in Appendix M). The simplifying and conservative assumption is taken that the protostar is only capable of giving

---

[3] Gaidos, E.J., 1995, *Icarus*, **114**, 258.

a filament for a time, t_i, that is half the free-fall time, during which its radius is constant at 0.9 of its initial radius.

(x) By numerical analysis the motions of the stars and the protostar are followed for a time up to t_i. If during this time the centre of the protostar passes a star at a distance closer than 1.5, but further than 0.5, of its radius then it is deemed to have produced a filament and planets.

(xi) Steps (i) to (x) are repeated 1,000 times and the proportion of protostars giving planets is found.

The result of running this program is shown in Fig. 8.3 in the form of a histogram — in which each block gives the number of closest passages in ranges of one-half of the radius of the protostar. There are 92 passages within the prescribed distance so that 9.2% of the protostars end up as the producers of planets, the majority of the remainder collapsing to produce stars.

What has been found in the above computational analysis is the proportion of protostars giving planets. However, for the known distribution of stellar masses, there are five times as many protostars with masses in the range of 0.25–0.75 $M_\odot$ as there are stars in the mass range of 0.8–2.0 $M_\odot$. Assuming that no star experiences more than one capture-theory event, this means that the proportion of

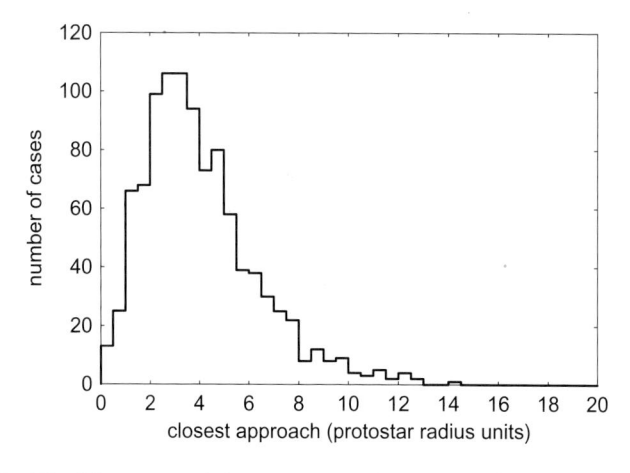

Fig. 8.3 The histogram of closest approach distances for 1,000 protostars.

stars expected to have planets is $0.092 \times 5 = 0.46$, *i.e.* just under one-half.

The model that has been used here is by no means a perfect one. The arbitrary distinction between stars and protostars on the basis of mass is clearly flawed since some exoplanets orbit stars with masses well within the range we have taken for protostars. However, the approximations in the model have not been of the kind that would unduly exaggerate the estimated proportion of planetary systems and there is plenty of slack between the results found here and what observations so far indicate. For example, if we took the necessary approach distance between protostar and star to be the radius of the protostar, rather than 1.5 times the radius, then this would still indicate that 3.8% of protostars produce planets and hence that 19% of stars acquire a planetary system. In addition, we have not taken into account interactions between stars and dense regions that will contribute an unknown, and probably large, number of additional planetary systems.

While it is impossible to put a hard figure on the proportion of stars that should have planets based on this model, what is clear is that the CT should lead to a large proportion of stars with planets — as observations now suggest.

8.3. The Disruption of Planetary Systems

So far, in considering the CT as a source of planetary systems, we have enjoyed the benefit of a high number density of stars within an embedded cluster that has given a large number of planet-forming interactions. However, once a planetary system has been formed, with planets on extended orbits carrying them up to a few thousand au away from the stars, then the dense environment becomes an adverse factor. The nature of extended orbits of high eccentricity is that the planets spend a very high proportion of each orbital period moving slowly near apastron, the furthest distance from the star. While there, a planet is in danger of being torn away from its parent star by the near passage of other stars in the cluster and becoming a free-floating planet. What was a good environment for producing planets in orbit around stars has become a bad one for retaining them.

After the planet has been produced, two things are changing, both of which greatly influence its eventual destiny — the planet's orbit and the density of stars in the embedded cluster. The original orbit of a planet is very extended. If it remained unchanged and if the stellar density in the embedded cluster also did not change, then it is certain that, sooner or later, the planet would be removed from the star. At the other extreme, if the planet achieved its final orbit and the stellar density in the embedded cluster had fallen to a small fraction, say 0.01% of its peak value, then the star would retain the planet almost indefinitely. So what is important is the proportion of planets that survive in orbit during the period from the initial state, when they are extremely vulnerable to being removed, to the final state, when they are tightly and permanently bound to the parent star.

To estimate the probability that a planet will be removed from its parent star, the method adopted was to carry out a large number of individual calculations. In each calculation a planet in an orbit corresponding to a particular stage of the decay and rounding-off process, with semi-major axis a and eccentricity e, is subjected to the perturbation of a star passing by on an orbit, taking it to a closest distance D from the parent star. The situation with the two stars at their closest approach is illustrated in Fig. 8.4.

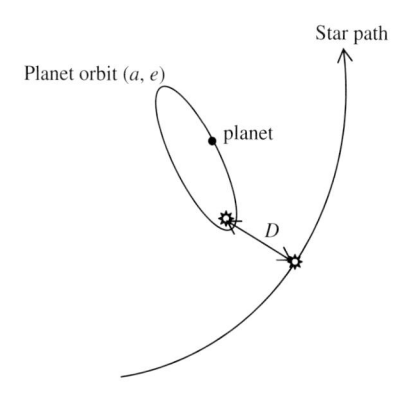

Fig. 8.4 The passing star at closest position to the parent star.

Defining the conditions for the calculation requires other quantities to be defined. The figure is two-dimensional and we may think of it as defining the plane of the stellar orbit. However, the planetary orbit will not, in general, be in this plane and its orientation in space needs to be defined. Taking the line joining the two stars at closest approach as an x-axis and the x-y plane as the plane of the diagram, the orientation of the planetary orbit can be defined mathematically in terms of three angles. Other quantities that need to be defined are the masses of the two stars and the position of the planet in its orbit at closest approach of the stars. For a given stellar passage, planets at some positions in the orbit are lost while at other points they are retained. The time during which the passing star exerts significant perturbation on the star is short enough to ignore any decay and round-off of the planetary orbit during the interaction.

Once the parameters have been defined, then an individual calculation is easily and quickly carried out. One of three outcomes can be recorded. The first is that the planet is removed from its orbit to become a free-floating planet. The second is that the planet is retained in orbit but the semi-major axis is increased so that it is less firmly bound to the parent star. In the language of planetary scientists, the orbit has *softened*. The third outcome is that the semi-major axis is reduced and the planet is more firmly bound — *i.e.* the orbit has *hardened*.

Turning the results of single calculations into an overall estimate of the proportion of planets removed for the whole period when planetary orbits are evolving is a very complex process and cannot be done with any precision. Conditions in the embedded cluster are constantly changing with respect to stellar density and the stellar density affects the probability of getting close stellar passages during the period a planetary orbit evolves. During the time a planetary orbit evolves — a million years or so — the stellar density in the cluster can greatly change, either increasing or decreasing (Fig. 8.2). To turn this complicated problem into a form that can be tackled requires some simplifications and the acceptance that the resultant estimate of the proportion of planets retained can only be a rough estimate. The way that the problem was tackled is described in

Appendix X and in somewhat more detail in a paper I wrote in 2004.[4]

The result of the calculation indicated that over the complete range of conditions, taking into account their relative likelihood, the proportion of planets being retained is somewhere in the range one-third to two-thirds. However, the proportion of retained *planetary systems* is higher. For example, if there were four initial planets and the probability of retaining each of them was 1/3, then, taking the probabilities for retaining the different planets as independent of each other, the probability that all four planets were lost would be $(2/3)^4 = 0.1975$, *i.e.* there would be a four in five chance that at least one planet would be retained.

In the previous section it was estimated that nearly one-half of Sun-like stars should have undergone a capture-theory event and so acquired one or more planets. The subsequent stellar interactions will reduce the number of retained planets, perhaps by a factor of three, but will not so greatly reduce the proportion of stars with at least one planetary companion. The combination of formation and disruption probabilities indicates that up to 40% of solar-type stars may have planetary companions. Another outcome of the calculation is that in about one-third of the stellar interactions in which the planets were retained, the planetary orbits were hardened, and in the remaining two-thirds they were softened. This had no substantial effect on the proportion of stars with planets; the main effect was to modify the time it took for the planetary orbits to evolve.

8.4. Summary

The Doppler-shift method of detecting exoplanets has steadily improved since the first exoplanet was found in 1995, and the pace of exoplanet detection is increasing. Exoplanet masses of less than 2 $M_\oplus$ have been found in association with low-mass red dwarf stars. The range of methods available to detect exoplanets has also expanded. In addition to the direct optical imaging, described in Chapter 5, the

---
[4]Woolfson, M.M., 2004. *Mon. Not. R. Astr. Soc.*, **348**, 1150.

technique of microlensing can reveal planets with masses similar to that of the Earth. The estimated proportion of stars with planets has risen from an initial estimate of 3% and is almost certainly higher than 10%; some estimates put the proportion as high as 30%.

A model of protoplanet formation, which involves computationally following the path of a protostar within an embedded cluster, indicates the possibility that nearly one-half of stars could initially acquire planetary systems. Although the actual proportion depends on the details of the model, which like all models involves approximations, the conclusion that many planetary systems should form is strongly indicated.

The dense embedded environment that favours planet formation by the capture-theory mechanism is also favourable to the disruption of planetary systems once they have formed. When a new protoplanet is created, its initial orbit may extend out to a few thousand au from its star. It is then vulnerable to being removed by the gravitational pull of a passing star. When its orbit has evolved to its final state and the stellar number density in the cluster has greatly diminished, then the planet will remain attached to its star virtually indefinitely. Computational analysis shows that between one-third and two-thirds of planets may be lost by stellar perturbation during the period when the orbits are evolving. However, the proportion of complete planetary systems lost will be much smaller. It is concluded that the combined effect of forming and disrupting planetary systems may leave up to 40% of stars with planetary companions.

Chapter 9
Satellite Formation

Current observations of exoplanets have revealed stars with families of several planets, some of which orbit their stars at both smaller and larger distances than do the planets of the Solar System. Exoplanets have been found within 0.02 au of stars whereas the closest solar-system planet to the Sun is Mercury with a mean distance of 0.387 au. Direct imaging of exoplanets has revealed distances of more than 100 au from stars compared with the 30 au orbital radius of the furthest major planet from the Sun — Neptune. Again, the masses of many exoplanets are considerably greater than that of Jupiter, the most massive solar-system planet, and some exoplanets have masses greater than, but close to, that of the Earth. The message contained in this information is that the Solar System is in no way unique or unusual except, perhaps, in being able to support life on one of its planets; even that may not be unusual — we have only explored a tiny part of our own galaxy in the hunt for exoplanets.

A characteristic of *all* the major planets of the Solar System is that they possess extensive families of satellites. This suggests that whatever the mechanism is for forming planets it will give the conditions for producing accompanying satellite families. If we reverse the argument we have just employed to say that, not only is the Solar System like other planetary systems but also that other planetary systems are like the Solar System, then we could argue that the process of producing exoplanets is similar to that of producing solar-system planets and hence satellite systems should be expected for major exoplanets. Here I shall describe a likely mechanism for satellite formation, based on indications from the capture-theory model described in Chapter 6 and using theory that has been developed by solar-nebula theorists to explain planet formation.

9.1. Angular Momentum Considerations

Following the invention of the telescope in 1608 by Hans Lippershey, a German spectacle-maker working in Holland, Galileo Galilei made a similar instrument shortly afterwards and began to use it for astronomical observations. An early observation was of the planet Jupiter around which he discovered four large satellites — Io, Europa, Ganymede and Callisto — now collectively called the *Galilean Satellites*. Galileo was a supporter of the Copernican theory, which held that the Earth and all other planets orbited the Sun, and he saw in the Jupiter satellite system a Solar System in miniature. The idea that satellites are related to planets in the same way as planets are related to the Sun is one that has persisted until quite recently, and perhaps persists even today. In 1919 James Jeans expressed the view that any theory that proposed a mechanism for producing satellites that differed from that for producing planets would be 'condemned by its own artificiality'. More recently, in 1978, the Swedish astrophysicist, Hannes Alfvén, who was awarded the Nobel Prize in Physics in 1970, stated that,

> 'We should not try to make a theory of the origin of planets around the Sun but a *general theory of the formation of secondary bodies around a central body*. This theory should be applicable both to the formation of satellites and the formation of planets.'[1]

Three very famous and able scientists have favoured the idea that the similarity in the visual relationship of satellites to planets to that between planets and the Sun signifies that one system is just a scaled-down version of the other and that the formation processes of planets and satellites should be the same. Indeed, that is also a

---

[1] Alfvén, H., 1978, *The Origin of the Solar System*, ed. S.F. Dermott, (Chichester: Wiley), p. 27.

basic tenet of the SNT where, just as planets are believed to form in a disk around a star, so satellites form in a disk around planets. We now critically examine this belief.

In the Historical Review at the start of this book, it was noted that Laplace's nebular theory foundered on the problem of the angular momentum distribution in the Solar System. The evolution of an isolated collapsing nebula could not possibly give an outcome where 0.14% of the mass, residing in the planets, should end up with 99.5% of the angular momentum of the system. One way of looking at this division of angular momentum, which does not directly involve the masses of the planets and the Sun, is to compare two types of *intrinsic* angular momentum — that is, angular momentum per unit mass. That of the Sun's spin can be characterized by the value for the material at its equator, while that of a planet's orbit by the value for the whole body. For a satellite system the corresponding quantities are the intrinsic angular momentum of material at the equator of the spinning planet and that of the whole satellite in its orbit. Table 9.1 shows the ratio of the intrinsic angular momentum for the orbiting secondary body to that of equatorial material of the central body for various planet–Sun and satellite–planet combinations. From this table it is clear that, whatever the superficial resemblance between the two types of system, they do differ in at least one very important respect.

The smaller values of S for planet–satellite systems make it much more plausible that they could both have been derived from

Table 9.1 The ratio, S, of the intrinsic angular momentum of the secondary orbit to that of the spin of the central body at its equator.

Central body	Secondary body	Ratio S
Sun	Jupiter	7,800
Sun	Neptune	18,700
Jupiter	Io	8
Jupiter	Callisto	17
Saturn	Titan	11
Uranus	Oberon	21

the same body of material. For the model of the Sun produced as a condensation at the centre of an unconstrained, freely rotating nebula, there is the problem that, when initially formed — if it could form at all — it would be at the point of rotational disruption. However, the planet and its surrounding disk, as shown in Fig. 6.2, originate in a different environment, as a condensation locked within a filament

We can explore this situation numerically. From Figs. 6.1 and 6.4 it can be deduced that the filament within which the condensations form have a rotational speed of order $\omega_0 = 5 \times 10^{-12}$ radians s^{-1} and this is taken as the rotational speed of the protoplanetary blob when it breaks free of the filament. The same figures show that the initial radius of the blob is of order 30 au. From the distribution of SPH particles in the simulation it is found that about one-half of the mass of the blob, which collapses to form the planet, is contained within a spherical volume of radius of about 5 au (7.5×10^{11} m). If it collapses to a radius of 10^8 m — somewhat larger than that of Jupiter but then this part of the blob has mass 2.5 M_J — the final angular speed of the planet's spin is

$$\omega_f = 5 \times 10^{-12} \left(\frac{7.5 \times 10^{11}}{10^8}\right)^2 \text{ radians s}^{-1}$$
$$= 2.81 \times 10^{-4} \text{ radians s}^{-1} \qquad (9.1)$$

This corresponds to a final spin period of 6.2 hours, of the same order as the spin periods of the major planets of the Solar System.

The material further out will fall inwards until its angular speed maintains it in Keplerian orbit around the central mass, which we take as $M_P = 2.5$ M_J.[2] If its original distance from the centre of the blob was r_0 and it falls to r that, because of the nature of the disk, it would do by spiralling inwards, then its angular speed becomes

$$\omega_r = \omega_0 \left(\frac{r_0}{r}\right)^2. \qquad (9.2)$$

---

[2]The disk will also gravitationally influence disk material but we do not take this into account.

If this is the Keplerian angular speed at distance r then

$$\omega_r^2 = \omega_0^2 \left(\frac{r_0}{r}\right)^4 = \frac{GM_p}{r^3}$$

or

$$r = \frac{\omega_0^2 r_0^4}{GM_{\mathrm{P}}}. \tag{9.3}$$

If the disk had an initial radius of 30 au (4.5×10^{12} m), then, from Eq. (9.3) its final radius once it had settled down would be 3.07×10^{10} m (~ 0.2 au), a distance that is considerably greater than the orbital radii of all satellites that one would expect to be formed from a disk.

The situation we have here for our forming planet is just that envisaged in the Solar Nebula Theory, a collapsing central body and a surrounding disk. A body of theory has been developed to produce planets from disk material, giving a mechanism that is theoretically quite sound but requires more time to operate than observations of disk lifetimes will allow. Now we are going to see whether, within the smaller scale of a protoplanetary disk, satellites could form in a reasonable time.

9.2. The Form of the Disk

In Chapter 7, which discussed the evolution of planetary orbits, the nature of the medium surrounding a star was described. For satellite formation the form of the disk surrounding a planet is of great importance and will be similar to that previously described in Chapter 7. The first stage of satellite formation requires dust to form a carpet in the mean plane and the time this process will take depends on how diffuse the disk is in any region.

A disk is predominantly a gaseous structure and does not have a sharp boundary like a solid object. Figure 7.1 shows how the density of the gas varies with distance from the mean plane, with the width of the curve designated by the quantity σ, defined in Eq. (7.1). Actually, Eq. (7.1) is only valid if the mass of the disk is negligibly small compared to the mass of the central body, which

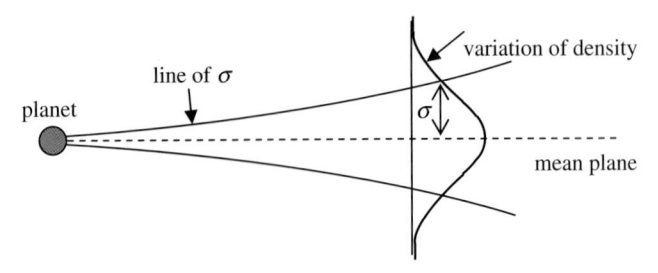

Fig. 9.1 A representation of the cross-section of a protoplanetary disk showing the mean plane, line indicating the variation of σ and the variation of density with distance from the mean plane.

is approximately true for a disk around a star, where the ratio of masses is between 0.01 and 0.1, but is not true for a disk around a planet where the two masses are similar. A high-mass disk pulls inwards on itself by gravitational forces and hence will be less diffuse than would be indicated by the values of σ given by Eq. (7.1). In Fig. 9.1 an impression is given of the shape of the cross-section of the disk, similar to that shown in Fig. 7.4(b), and the way that the density varies through its width.

A feature of the disk not illustrated in Fig. 9.1 is the variation in the density of the material with increasing distance from the planet. This is best expressed in terms of *areal density*, which is the mass of material, both gas and dust, per unit area if it were all projected onto the mean plane. A common form of distribution, which can be deduced from Fig. 6.2, is that the areal density is greatest closest to the star and steadily falls with distance. In the analysis of satellite formation, that will now be described, the areal density distribution was modelled in the form of a declining exponential, as shown in Fig. 7.5. To summarize, the picture we have is of a protoplanetary disk in which the areal density steadily falls with distance from the planet, with the material at any distance stretched out from the mean plane as illustrated in Fig. 9.1, with the extent of the stretch increasing with distance. We must now consider how the solid component of the disk gradually drifts down towards the mean plane and how long that process would take.

9.3. The Settling of Dust

Appendix F described the distribution of dust particle radii, which vary from small fractions of a micron up to about 5 μm. The very fine dust is the most easily detected by the way it scatters light to cause reddening of objects at great distances. However, it is the larger particles, of radii 2 μm and above, that constitute the bulk of the mass (Fig. F.3). The larger particles in the protoplanetary disk will fall faster towards the mean plane and it is those that will determine the time of the formation of a dust carpet. When the formation of a dust carpet was being considered in relation to the SNT, it was assumed that dust particle size was limited to 1 μm so that forming a dust carpet in a reasonable time depended on the growth of initially sub-micron particles to a larger size by accretion.[3] This growth is mainly due to their thermal motion, related to the temperature of the gaseous medium in which they are suspended. Very tiny solid particles suspended in a dense gaseous medium will be seen to execute a jittery motion. This is due to a phenomenon called Brownian motion, discovered by the Scottish botanist Robert Brown in 1827 when he observed it occurring for pollen particles floating on water. Brownian motion is due to the asymmetric bombardment of the particle by gas molecules. As the dust particles move they collide with other particles and stick together, so producing a particle of greater mass. The theory of this process is described in Appendix Y.

If the gas is very diffuse, such that the distance travelled by gas molecules between collisions is much greater than the radii of the dust particles, then the dust particles will be affected by the gas in a fairly predictable way and they will move through the gas at a fairly steady speed under the influence of any external forces, *e.g.* gravity. Figure 9.2 illustrates the force acting on a dust grain in a disk due to the gravitational force of the planet. The force has a component towards the mean plane that causes the dust particle to move in that direction. The much larger component parallel to the

---

[3]Weidenschilling, S.J., Donn, B. and Meakin, P., 1989, *The Formation and evolution of Planetary Systems*, eds. H.A. Weaver and I. Danley, (Cambridge: CUP), pp. 131–150.

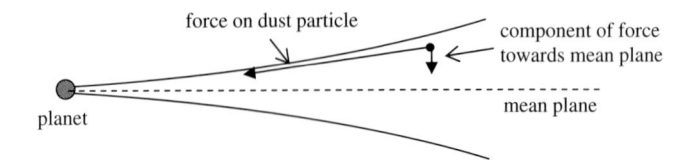

Fig. 9.2   The force on a dust particle towards the mean plane.

mean plane is the central force that enables the dust particle, with the rest of the disk, to orbit around the planet. The force on the dust particle impelling it towards the mean plane will, in reality, be larger than that due to the planet alone. The mass of the disk itself will contribute to that force and just taking that from the planet alone will give a considerable underestimate; ignoring the disk contribution opposes the desired outcome of a fast dust-settling time and so, once again, is an acceptable approximation.

There is an opposing force due to its motion through the diffuse gas that balances the gravitational force on a dust particle, so causing it to fall at a steady speed. When it moves through the gas the *average* relative speed of the gas molecules striking it in a direction opposing its motion is greater than that striking it in its direction of motion (Fig. 9.3). This gives a net force opposing the motion. The theory based on this idea, given in Appendix Y, indicates that the speed of a dust particle in a particular gravitational field within a very diffuse gas is proportional to its radius.

This speed dependence on radius gives a second mechanism for the growth of dust particles. Larger dust particles move towards the mean plane faster than smaller ones and when a larger particle collides with a smaller one moving more slowly, then the two will tend to stick together forming a larger particle that will fall faster still. The trend will be for small particles to be lost by combining

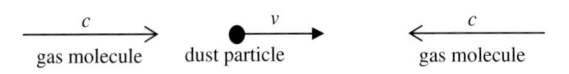

Fig. 9.3   A gas molecule moving at the average speed c in the same direction of motion as the dust particle strikes it with relative speed $c - v$. For a gas molecule moving in the opposite direction the relative speed is $c + v$.

with larger ones and the dust particles that finally reach the mean plane will be very much larger than those initially present. However, in relation to the growth of dust particle sizes by aggregation, I must remind the reader of the CODAG experiment, mentioned in the Historical Sketch, which suggested that particles would form wispy strings rather than compact spheres and so might not settle much more quickly than the original small particles. For that reason I shall just deal with the settling of larger particles of radius $2\,\mu$m or more. If there is effective particle growth that gives faster settling, despite the CODAG result, then this will make the formation of a carpet even more effective.

I have used the theory developed the American planetary scientist Stuart Weidenschilling to calculate the settling times for dust at various distances from a planet. The example I take here has the following parameters:

Mass of planet $= 2.0 \times 10^{27}$ kg (approximately the mass of Jupiter)
Mass of disk $= 2.0 \times 10^{27}$ kg
Mean molecular mass of gaseous material $= 4 \times 10^{-27}$ kg
Temperature of disk material $= 20$ K
Density of dust grains $= 3 \times 10^3$ kg m^{-3}
Ratio of principal specific heats of gas $= 5/3$

The declining exponential fall-off of the areal density of the disk with distance from the planet was taken so that the density reduced by a factor of e for every 2×10^9 m.[4] The settling time was found for particles of radius 2, 3, 4 and $5\,\mu$m from distances of 3σ from the mean plane at distances of between 3×10^9 m and 6×10^9 m from the planet. The results, shown in Fig. 9.4, indicate that although particles further out have a greater distance to travel before they reach the mean plane, this is more than compensated for by the lower density so that the time of settlement is less. Although we have not included the case where dust particles grow, because of the doubt

---

[4]The exponential e (=2.7182818) occurs in a natural way in mathematics and is used to describe exponential decline. The function e^{-x} reduces in value by a factor $1/e$ for each increase of x by 1.

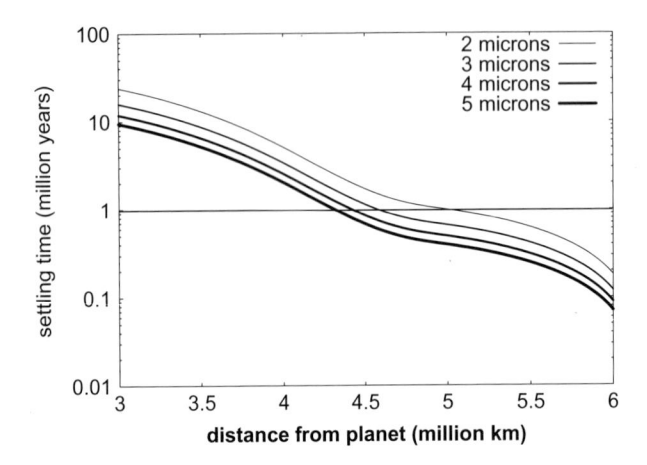

Fig. 9.4 The settling times for dust particles at different distances from the planet.

due to the CODAG results, it is interesting to note that analysis for this situation shows that *closer-in*, the settlement time is less. This is because in the denser parts of the disk the dust particles grow rapidly and by the time they reach the mean plane they have grown to a radius of more than 50 μm.

For the dust particles of fixed size there is no carpet formation at distances from the planet of less than 4×10^9 m for any reasonable disk lifetime. At distances beyond 5×10^9 m, for a reasonable disk lifetime in which most of the disk would endure for one million years, settlement into a carpet would be complete. However, there are other criteria to be satisfied. The total mass of the solid material in the carpet must be sufficient to provide the material for satellites. If we take the case of Jupiter, the total mass of its four large satellites — the *Galilean satellites* — is about 4×10^{23} kg. For the disk we have specified, the total mass beyond 4×10^9 m is over 8×10^{26} kg and if just 0.5% of that is deposited dust, the total mass of the carpet would be 4×10^{24} kg — an ample source of solid material to form the satellites.

Another criterion that the disk must satisfy is that it must survive the disruptive tidal effects of the star during periastron passages. The planet will be in an elliptical orbit which will pass through its periastron position several times during a period of a million years. From the orbital decay curves shown in Fig. 7.7, it seems that the closest approach of the planet to the star in this period will be of the order of 100 au. Theoretically it can be shown that the planet could not retain material at a distance greater than

$$r_p = \frac{r_{PS}}{2.5} \left(\frac{M_P}{M_S}\right)^{1/3} \tag{9.4}$$

where r_{PS} is the planet–star distance and M_P and M_S are the masses of the planet and star respectively. For a ratio of masses of 0.001 — that of Jupiter and the Sun — and with a planet–star distance of 100 au, the value of r_P is about 4 au, well outside the distance of significant density of the dust carpet.

The conclusion from this analysis, described in more detail in Appendix Y, is that there seems to be no theoretical reason why a dust carpet of sufficient total mass should not form around the protoplanet — the first step to forming satellites.

9.4. The Formation of Satellitesimals

Having formed a dust carpet, the next stage in the progress towards forming satellites is for the carpet to break up and condense into a number of roughly spherical solid bodies. This is an essential part of the SNT for which the bodies are called planetesimals. I have followed that terminology and called the solid bodies *satellitesimals*.

Gravitational instability is an important mechanism in astrophysical processes over a wide range of scales. It may have been the way in which the original material formed by the Big Bang broke up into isolated regions, eventually leading to the formation of galaxies. It was invoked by early workers in the SNT to produce planets by the spontaneous break-up of the circumstellar disk into planetary-mass blobs — an approach subsequently abandoned. In the Jeans tidal theory it was the process by which the filament drawn out of the Sun broke up into a series of planetary condensations — a theoretical conclusion supported by the computational results in Figs. 6.1 and 6.4 illustrating the capture-theory process. Just as for these situations, in which either a filament or a large volume of gas has spontaneously broken up, so gravitational instability can happen to an approximately two-dimensional structure like a dust carpet. The Russian planetary scientist Victor Safronov first gave a theory for this kind of instability in 1972; the Americans Peter Goldreich and William Ward subsequently modified it in 1973.[5]

A first step in looking at the formation of blobs by instability in the carpet is to impose the condition that they must be able to withstand disruption by the tidal effects due to the planet. The forces at play on a spherical blob of material due to self-gravitational forces and gravitational forces due to a planet, P, are illustrated in Fig. 9.5.

The self-gravitational forces act to pull points A and B towards each other while the unequal forces due to the planet, constituting a tidal effect with F_{PA} greater than F_{PB}, act to pull them apart. For the stability of the blob, so it is not pulled apart by tidal forces, the condition to be satisfied is

$$F_{gA} + F_{gB} \geq F_{PA} - F_{PB}. \tag{9.5}$$

It is shown in Appendix Y that this leads to

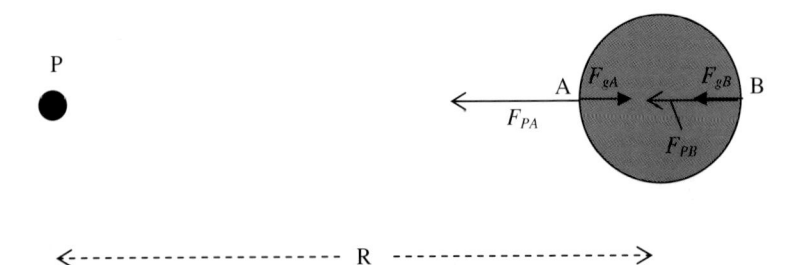

$$\rho_B \geq \frac{3M_P}{2\pi R^3} \tag{9.6}$$

Fig. 9.5 The forces per unit mass at points A and B of a spherical blob of material due to self-gravity F_{gA} and F_{gB} acting towards the centre of the sphere and those due to the planet F_{PA} and F_{PB} acting towards the planet.

---

[5]Goldreich, P. and Ward, W.R., 1973, *Astrophys. J.*, **183**, 1051.

in which ρ_B is the density of the spherical blob, M_P the mass of the planet and R the distance to the centre of the sphere. Once the dust carpet reaches a thickness, h, such that its density reaches the critical level indicated by Eq. (9.6), then gravitational instability will set in and the carpet will begin to break up. If the areal density (mass per unit area) of the dust component of the disk is ρ_{ad}, then this thickness will be given by

$$h = \frac{\rho_{ad}}{\rho_B}. \tag{9.7}$$

According to Safronov's theory, the area of the disk in each of the condensations will be about $60\,h^2$ so the total volume of a condensation, which will form a satellitesimal, is about $60\,h^3$ with a mass $m_B = 60\,h^3\rho_B$.

If we use the planet and the surrounding disk that gave the results illustrated in Fig. 9.4, then the masses of satellitesimals as a function of distance from the planet are shown in Fig. 9.6.

It is interesting to compare the masses of the satellitesimals indicated by the figure with those of the satellites of Jupiter, since the mass of the planet we have taken is similar to that of Jupiter. The satellite family of Jupiter has 63 members, most discovered by spacecraft observations, and is dominated by the Galilean satellites,

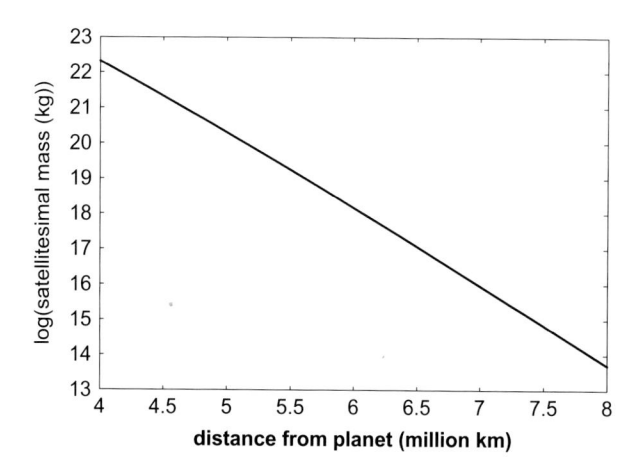

Fig. 9.6 Satellitesimal masses at various distances from the planet.

Io, Europa, Ganymede and Callisto, with masses 8.93×10^{22}, 4.88×10^{22}, 1.497×10^{23} and 1.068×10^{23} kg respectively. At a distance of 4.5×10^9 m, where settling times are well within the expected disk lifetime, a satellitesimal mass is about 2×10^{21} kg, between about one-twentieth to one-seventieth the mass of the Galilean satellites. The next most massive satellite of Jupiter is Amalthea, closer-in than the Galileans and with a mass of 2×10^{18} kg, within the range of masses of satellitesimals in the figure.

9.5. Satellite Formation

In his 1972 work Safronov described a theory for the formation of planets from planetesimals. This process was described in the Historical Sketch and is briefly repeated here. The planetesimals were produced in nearly circular orbits but gravitational interactions between them increased random motions in the system. As the random motions built up and planetesimal orbits became ever less circular, so the probability of collisions increased; the general tendency of collisions is to damp down the randomness of motion. A balance between the effects that increased and increased randomness gradually became established, at which stage the mean speed of the planetesimals in any region would be less than, but of the same order as, the escape speed from the largest planetesimal.

If two bodies, attracted to each other by gravitational forces, collide, and if all the energy of motion is retained, then the bodies cannot remain together but must fly apart again, perhaps in a fragmented form. However, in general, energy of motion will be lost, either in producing heat or in breaking up one or both of the bodies. In this case the energy of motion is less after the collision than before and it is possible for the bodies to coalesce. This is particularly likely to happen if one of the bodies is much more massive than the other. In the Safronov model one planetesimal in each region of the disk becomes dominant — of larger mass than all others — and it begins to accumulate other planetesimals through collisions (Appendix Z). The more bodies it accretes, the larger it becomes and the larger it becomes, the more it attracts other bodies towards itself to be accreted. This is the proposed mechanism for the formation of terrestrial planets and the solid cores of major planets according to the SNT. The expression found by Safronov for the time taken to produce a planet or planetary core is

$$\tau_P = \frac{4 r_L \rho_m P}{3 \pi \rho_{ad}(1+\beta)} \quad \text{(Appendix Z)} \tag{9.8}$$

where r_L is the radius of the planet or planetary core, P is the period of a circular planetary orbit in the region of formation of the planet, β is a constant somewhere in the range of 4 to 10, ρ_m is the density of the body being formed and ρ_{ad} is as previously defined. For a reasonable disk surrounding the early Sun, with a mass of 0.1 solar mass and with a surface density varying as $1/R$ (except very close to the Sun), the areal density of solids at 1 au would have been about 470 kg m^{-2}. For $\rho_m = 6 \times 10^3$ kg m^{-3} (mean density of the Earth), $r_L = 6.4 \times 10^6$ m (radius of Earth), $P = 3.156 \times 10^7$ seconds (1 year) and $(1+\beta) = 8$, this gives the time for the formation of the Earth as 4.4 million years. This formation time poses no theoretical problems. However, at the distance of Neptune (30 au) with $\rho_{ad} = 16$ kg m^{-2}, $P = 5.200 \times 10^9$ seconds, $\rho_m = 3 \times 10^3$ kg m^{-3} and r_L the same as that for the Earth — although it should certainly be larger — we find the time of formation as 10,500 million years — more than twice the age of the Solar System. Given that they must eventually accumulate gaseous envelopes, and hence that their cores must be produced within the few-million-year observed lifetimes of disks, the other major planets also present severe timescale problems.

Safronov's mechanism is clearly unsatisfactory when applied to solar-system planets so we must now see what it gives when applied to satellite formation. In Section 9.3 we found that, for the model disk being considered, it was necessary to go out to a distance of at least 4×10^9 m to have a dust carpet formed in a sufficiently short time. However, the orbital radius of Callisto, the outermost Galilean satellite, is 1.88×10^9 m so we must ask how it is that satellitesimals are available close enough to Jupiter to produce the Galilean satellites. The answer is that satellitesimals are moving in a resisting medium and their orbits are constantly decaying. Thus

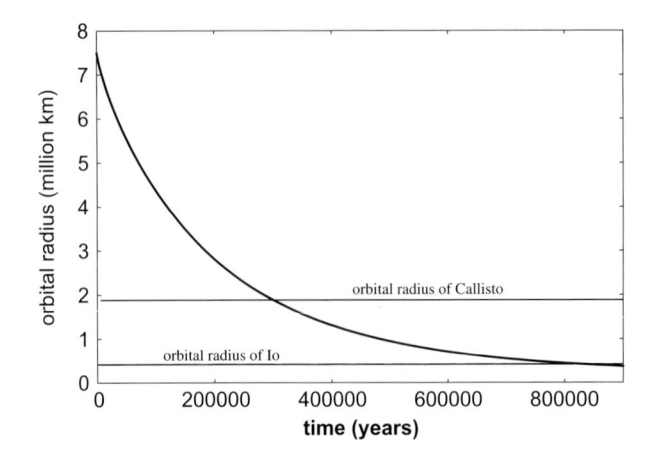

Fig. 9.7 The decay of the orbit of a satellitesimal of mass 10^{22} kg.

the orbit of the embryonic satellite is steadily decaying as it grows and, indeed, when the satellite is finally formed, the orbit will still continue to decay for the duration of the planetary disk. This point is illustrated in Fig. 9.7 that shows the decay of a partially formed satellite, with a constant mass of 10^{22} kilograms, moving in a resisting medium based on the disk that gave Fig. 9.4 but decaying so that the density falls everywhere by a factor of e every million years. The method of calculation used was as for the decay of planetary orbits shown in Fig. 7.7. Starting with a radius of 7.5×10^9 m, the orbit decays to the orbit of Callisto, the outermost Galilean satellite, in about 3×10^5 years and to the orbit of Io, the innermost one, in about 8×10^5 years. The orbit is close to circular for the whole of the decay period.

Taking averages for a satellite like Callisto, forming as its orbit decays inwards towards Jupiter, we take the characteristics of the disk at 5×10^9 m, giving a dust carpet with areal density $\rho_{ad} = 32650 \, \text{kg m}^{-2}$ and $P = 6.08 \times 10^6$ seconds that, together with $r_L = 4.8 \times 10^6$ m and $\rho_m = 2 \times 10^3 \, \text{kg m}^{-3}$, give a formation time of about 3,000 years. This is extremely short, but that is what is indicated by the Safronov equation (9.8). Given that it is a rather optimistic figure and may be several times larger, it still shows that at the smaller scales involved in satellite formation there are no timescale problems.

In the case of the Galilean satellites, subsequent to their formation and while the resisting medium was still present, their orbits evolved by the process described in Section 7.6 to give strong commensurabilities linking the periods of Io, Europa and Ganymede, the three innermost members of the quartet. Their periods are very closely in the ratio 1:2:4 but there is a precise relationship linking their *mean angular motions*, indicated by the symbol n, which is the technical term for their mean angular speeds. This relationship is

$$n_{\text{Io}} - 3n_{\text{Europa}} + 2n_{\text{Ganymede}} = 0. \tag{9.9}$$

Commensurabilities also link many other satellites of solar-system planets.

9.6. Comments

As with many modelling exercises, simplifications have had to be made in order to be able to deal with the process of satellite formation. It has also been necessary to make estimates of various quantities that are basic to the whole progress of the calculation. For example, the SPH simulations shown in Figs. 6.1 and 6.4 indicate an initial slow rotation of the protoplanetary blob of about 5×10^{-12} radians s^{-1}; if it were faster than that, the whole modelling would follow a different path with a more extensive disk and longer timescales. However, unless the rotation of the protoplanetary blob was considerable faster than I have assumed, then the process of satellite formation would still occur within an acceptable time, *i.e.* within the expected lifetime of the disk.

Another significant simplification in the model is to assume that the initial blob breaks up into a separated planetary condensation in which matter, supported by internal pressure gradients, rotates at less than Keplerian speed and a flared disk with all parts in Keplerian orbit. The reality would be far more as schematically illustrated in Fig. 9.8. In the neck between the central bulge, which will form the planet, and the main disk, within which a dust carpet forms,

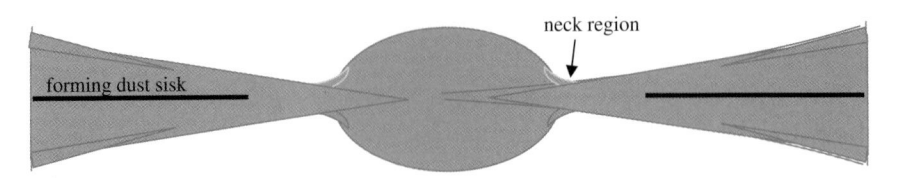

Fig. 9.8 A schematic illustration of a forming protoplanet with its surrounding disk.

there will be an intermediate region of transition between below-Kepler angular speed and Kepler angular speed. As the planetary condensation collapses, so it will become hot; in 1982, Nick Schofield and I estimated that for some considerable period the luminosity of a developing proto-Jupiter would be about 10^{22} W,[6] only 2.5×10^{-5} that of the Sun but still enough to promote evaporation of disk material. As time progressed, so the neck region would diminish and finally disappear, as would the disk, leaving satellites in the orbits to which they had decayed.

9.7. Summary

All the major planets of the Solar System possess substantial satellite families, which suggests that satellite formation is a natural, concomitant process to that of planet formation and that exoplanets probably also have accompanying satellites. The relationship of satellites to planets in terms of the angular momentum distribution in the two types of body is completely different to that between stars and planets. Whereas the Sun contains only 0.5% of the angular momentum of the Solar System, the planet Jupiter contains 99% of the angular momentum of its smaller-scale primary-secondary system. Despite suggestions by many eminent planetary scientists to the contrary, this suggests that the mode of formation of satellites in relation to planets may differ from that by which planets are produced in relation to stars.

The simulations of planet formation by the capture-theory mechanism indicate that planets form with surrounding extensive

---

[6]Schofield, N. and Woolfson, M.M., 1982, *Mon. Not. R. Astr. Soc.*, **198**, 947.

disks with masses similar to those of the planets themselves. The process of planet formation, according to the SNT, give times of formation greatly exceeding the observed lifetimes of circumstellar disks, but, nevertheless, the process has a sound theoretical basis. The application of the same theory to satellite formation gives times that are reasonable compared with the expected lifetime of disks around new planets — of the order of one million years. For a model resembling Jupiter, the settling of dust into a carpet takes from a few hundred to one million years, decreasing with distance from the planet. The fragmentation of the carpet through gravitational instability eventually to give satellitesimals is a rapid process; typical masses of the satellitesimals are in the range of 10^{14}–10^{22} kilograms, corresponding to the masses of some of the smaller satellites of the Solar System, but considerably less than the masses of the Galilean satellites. The amalgamation of satellitesimals by the process described by Safronov, which is the source of the timescale problem for planet formation for the solar nebula theory, is comparatively fast for satellite formation — just a few thousand years for the model under consideration.

The evolution of satellite orbits in the resisting medium provided by the disk led to commensurabilities between the periods of satellite orbits by the process that Melita and I described in 1996 (Section 7.6).

Chapter 10
Features of the Solar System

Although there have been tremendous advances in exoplanet detection, and large numbers have been found and continue to be found, our knowledge of them is limited. Orbits can be determined reasonably well but, with few exceptions, we can only estimate a minimum mass. In the rare cases where they transit the parent star, an estimate can be made of the actual mass and also the radius of the planet. By contrast, when we come to the Solar System, we have an abundance of detailed knowledge about its various members, gained from measurements and observations from Earth, equipment on surface-landers and instruments in spacecraft.

Given that there is a plausible theory for the origin of the Solar System just as a system of planets orbiting the Sun — the same theory as that for the exoplanet systems — the next interesting challenge is to attempt to interpret what we know about the detailed structure of the Solar System in terms of some theory of its evolutionary history. This is the topic for the latter part of this book. However, before we consider in subsequent chapters the processes giving rise to the detailed structure and properties of the various members of the Solar System, I shall first give a broad-brush description of the system and what it contains.

10.1. The Planets

The planets of the Solar System fall naturally into two groups distinguished both in their physical characteristics and their orbits.

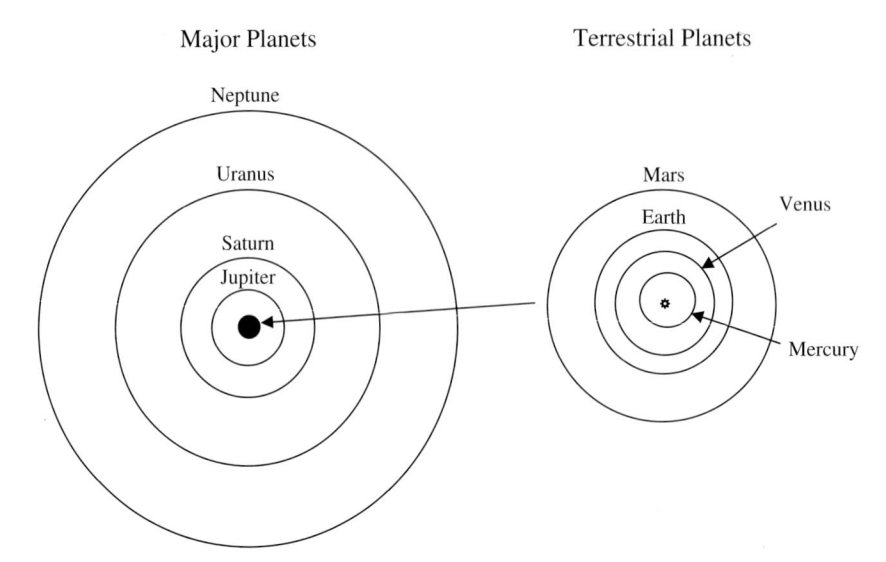

Fig. 10.1 Planetary orbits. The orbits of the terrestrial planets are contained within the black circle.

The orbits of these two groups, each containing four members, are shown in Fig. 10.1.

10.1.1. *The terrestrial planets*

As seen in Fig. 10.1, the *terrestrial planets* — Mercury, Venus, Earth and Mars — are closely packed together in the inner part of the Solar System. Their positions, relative sizes and structures are illustrated in Fig. 10.2 with their masses, in Earth units (5.974×10^{24} kg), given in parentheses following the planet name. They fall into two pairs — Earth and Venus, which are similar in size, and the smaller pair, Mars and Mercury.

All the terrestrial planets and the Moon, also represented in Fig. 10.2, have a core which is mostly iron with about 10% of nickel. Surrounding the core is a silicate mantle with a thin surface crust of less-dense silicates. The overall density of the planet depends mainly on the relative proportion of iron and silicate but also on the mass of the body since self-gravitational forces compress larger bodies. The densities of the four planets, from Mercury outwards,

Features of the Solar System 153

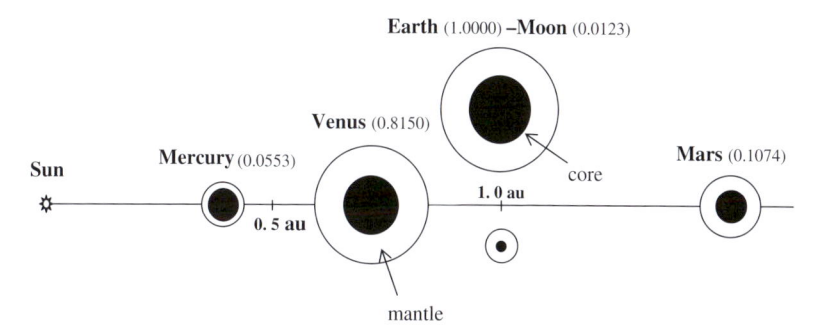

Fig. 10.2 Orbital radii, relative sizes, masses and structures of the terrestrial planets. The linear scale for the planets (Earth radius 6,378 km) is 3,000 times that for the semi-major axes.

in units of 10^3 kg m^{-3}, are 5.43, 5.24, 5.52 and 3.94, respectively; if uncompressed densities were taken, then Mercury would be denser than Earth. Mercury breaks the rule that density increases with mass — a matter we shall consider in due course.

The general appearance of the four terrestrial planets is shown in Fig. 10.3. The image of Venus is derived from a radar survey of its surface. Venus has a thick atmosphere, with a ground pressure 100 times that of Earth, and normal images with visible light only show unbroken cloud cover.

Fig. 10.3 The four terrestrial planets (to scale). From left to right: Mercury, Venus, Earth and Mars (NASA).

The smaller members of this quartet have the highest eccentricities for planetary orbits: 0.093 for Mars and 0.206 for Mercury. Mercury also has the highest orbital inclination. The plane of Earth's orbit is called *the ecliptic* and a planet orbiting in the ecliptic has zero inclination. The inclination of Mercury's orbit, 7°, is twice that of Venus, the planet with the next highest inclination.

10.1.2. *The major planets*

The *major planets* are widely spaced in the outer region of the Solar System. These planets all contain iron-silicate cores but with volatile material comprising the majority of their volume and mass. Their positions, relative sizes and structures are illustrated in Fig. 10.4, again with their masses in Earth units in parentheses following the planet name.

These planets fall into two pairs, the members of each pair being similar but the pairs being dissimilar. Jupiter and Saturn are *gas giants*. Apart from their iron-silicate cores, probably of between 5 and 10 Earth masses, they are mostly hydrogen and helium. The pressures in their deep interiors are so great that the hydrogen takes on the form of *metallic hydrogen* in which the hydrogen atoms form a regular lattice just like the atoms in a crystalline solid. Above that region, there is liquid hydrogen that gradually merges with the gaseous outer atmosphere. The proportion of helium, as detected in the near surface regions, is quite different for the two planets — about

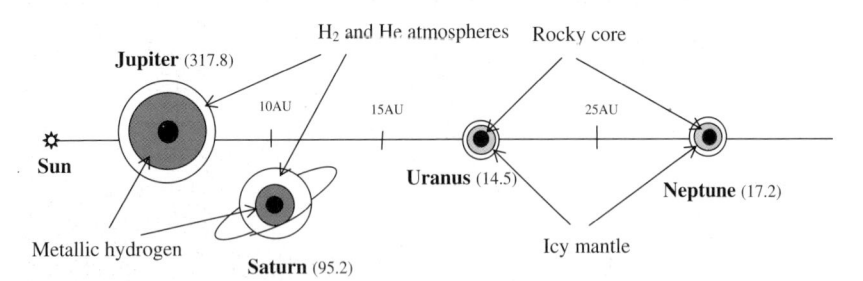

Fig. 10.4 The orbital radii, relative sizes, masses and structures of the major planets. Planets are shown at about 5,000 times their linear dimensions relative to their depicted orbital radii. The mean radius of Jupiter is 71,492 km.

14% for Jupiter but only 3% for Saturn. Both planets contain traces of other gases such as methane, ammonia and water vapour. The density of Jupiter, $1{,}326\,\text{kg}\,\text{m}^{-3}$, is much lower than that of the terrestrial planets, as would be expected from its composition, but is considerably higher than Saturn's density, $687\,\text{kg}\,\text{m}^{-3}$, since Jupiter is much more compressed by self-gravitational forces. Both the gas giants show marked flattening along the pole-to-pole diameter due to the effects of their spin. Jupiter spins slightly faster, with a spin period of 9 h 55 m compared with Saturn's 10 h 32 m, but Saturn's polar flattening is much greater because of its low density. The polar diameter of Saturn is 10% less than the equatorial diameter and is easily seen in images of the planet.

The other two major planets, Uranus and Neptune, are sometimes referred to as the *ice giants* because they contain considerable quantities of methane, ammonia and water that form icy mantles around the silicate-iron cores (Fig. 10.4). They differ from the gas giants in being much less massive, considerably smaller and having longer spin periods — 17 h 14 min for Uranus and 16 h 7 min for Neptune.

Images of the major planets are shown in Fig. 10.5. A prominent feature is the extensive ring system that encircles Saturn. Actually, the other three major planets also have ring systems, but so tenuous that they are not visible from ground-based telescopes. Visible on Jupiter is the feature known as the *Great Red Spot* (lower left) that is a storm that has been raging on Jupiter for at least the last 350 years, from when it was first observed. A feature of similar origin, the *Great Dark Spot*, can be seen on Neptune.

10.1.3. *Tilts of planetary spin axes*

A property of the planets that has something to say about their evolution, if not about their origin, is the tilt of their spin axes. These are shown in Fig. 10.6. A planet for which the spin axis is perpendicular to the orbital plane and the spin is *prograde, i.e.* in the same sense as the planetary orbits (anticlockwise, looking from the north) has zero tilt and this is shown in the figure by

Fig. 10.5 The major planets (not to scale). Top: Jupiter (NASA, Voyager image) and Saturn (NASA/JPL/Space Science Institute). Bottom: Uranus (NASA), Neptune (NASA).

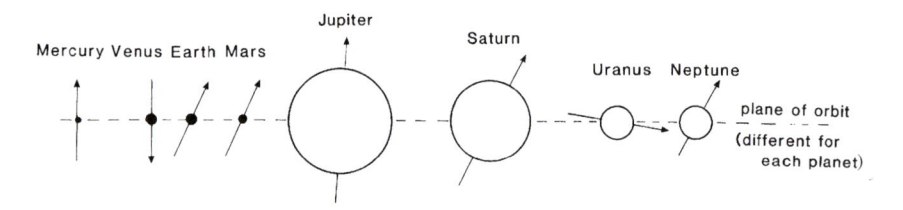

Fig. 10.6 The tilts of the planetary spin axes relative to their orbital planes.

an upward-pointing arrow. If the spin axis is perpendicular to the orbital plane but *retrograde* (opposite sense to prograde), then the tilt is 180° and the arrow would point directly downwards.

Several interesting features can be seen in Fig. 10.6. Working from the inside, Mercury has zero tilt but of more interest is the fact that its axial spin is coupled to the period of its orbit; for every two orbits, each lasting 88 days, it spins on its axis exactly three times. Venus has an axial tilt of 180° but is also distinguished by its very slow retrograde spin, with a period of 243 days. The Earth's tilt of 23.5° is responsible for seasonable climate variations. In the northern summer, the northern hemisphere is tilted towards the Sun and six months later it points away from the Sun to give the northern winter. The most notable feature of Mars' axial spin is its similarity to that of the Earth, with a tilt of 25.2° and a spin period of 24.6 h, slightly greater than one Earth day. Jupiter has a small axial tilt, 3°, but that of Saturn, 27°, is quite large. Next we come to Uranus which has an axial tilt of 98°, meaning that its spin is retrograde and the spin axis is very close to the orbital plane. This very strange relationship of the spin axis to the orbital plane must relate to some event in its evolution. Finally we have Neptune, which has a prograde spin with axial tilt 28°, very similar to that of Saturn.

The planets are the major bodies in the Solar System that accompany the Sun but there are several other types of body. The next we consider are satellites, the origins of which were considered in Chapter 9.

10.2. Satellites

In Chapter 9 a process was described by which satellites were produced from the material in a disk surrounding a newly formed planet. The disk would necessarily be in the equatorial plane of the planet and hence the satellites would be expected to orbit in that plane. In addition the orbit of a satellite formed in this way would be expected to be circular, or closely so. There are a number of substantial satellites orbiting the major planets that satisfy this condition of being in circular orbits in the equatorial plane; these are known as *regular satellites* and could have formed in the way suggested in Chapter 9. The number of detected satellites has increased enormously since the advent of the space age. For the most part, the satellites detected from spacecraft have been very small and many do not satisfy the conditions for being regular satellites. We first consider the satellites of the major planets, many of which are clearly of the regular kind.

10.2.1. *The satellites of Jupiter*

Jupiter has 63 known satellites, the greatest number among the major planets. Most of them are very tiny, just a few kilometres in mean diameter; satellites with a mean diameter greater than 10 km are listed in Table 10.1.

The satellites in the table fall into four groups. First there is the inner group, from Metis to Thebe, all satisfying the conditions for being natural satellites and of a size that suggests they may once have been either single satellitesimals or amalgamations of very few satellitesimals. Figure 10.7 shows two views of Amalthea, the largest of the group and the only one discovered from Earth-based observations, taken from the Galileo spacecraft. It has an irregular shape and is covered by what seem to be impact craters from bodies of smaller size. Astronomical bodies of mean diameter less than about 300 km tend to be irregular in shape; larger bodies are acted on by self-gravity that is strong enough to overcome the physical strength of the material and to pull material inwards as far as possible — the outcome being a spherical, or near-spherical shape.

The next group of satellites, discovered by Galileo in 1610 shortly after he constructed his telescope, are the Galilean satellites, shown in Fig. 10.8. They fall naturally into two pairs. The inner two, Io and Europa, are rocky bodies, as revealed by their densities, although Europa has a thin covering of ice. Before the Voyager 1 spacecraft reached Jupiter in 1979, a prediction was made by an American astronomer, S.J. Peale, that Io would show the presence of volcanoes. When the spacecraft arrived, that prediction was found to be true. A plume was seen from a volcano, now named Pele (Fig. 10.9), and subsequently a number of other volcanoes have been discovered on Io. The basis of the prediction was the 2:1 ratio of the periods of Europa and Io. Because of this, the two satellites are always closest together at the same points of their orbits and hence the perturbations due to their gravitational interaction are amplified. This causes the orbit of Io to be not quite circular so that the tidal stress on it due to Jupiter varies in a periodic way. Some of the energy involved in the stretching is converted into heat by a physical effect called *hysteresis* and it

Table 10.1 Jupiter satellites with average diameter greater than 10 km. i is the inclination of the orbit to the planetary equator, a the semi-major axis of the orbit and e its eccentricity.

Satellite	$i(°)$	a (10^3 km)	e	Mass (10^{22} kg)	Average diameter (km)	Density (10^3 kg m^{-3})
Metis	0.06	128	0.000		40	
Adrastea	0.03	129	0.002		20	
Amalthea	0.37	181	0.003		190	
Thebe	1.08	222	0.018		100	
Io	0.05	422	0.004	8.93	3,630	3.5
Europa	0.47	671	0.009	4.88	3,122	3.0
Ganymede	0.20	1,070	0.001	14.97	5,262	1.9
Callisto	0.20	1,880	0.007	10.68	4,821	1.8
Group of four	25 to 29	11,188 to 11,778	0.13 to 0.21		16 to 170	
Group of four	147 to 164	21,455 to 24,056	0.17 to 0.38		28 to 60	

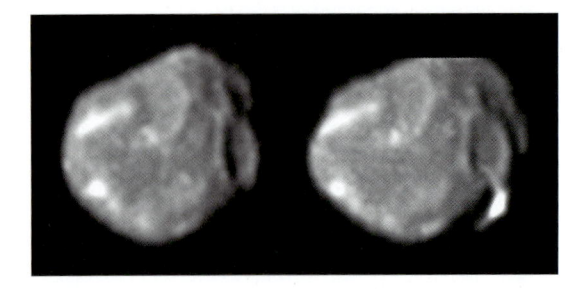

Fig. 10.7   Two views of the Jupiter satellite Amalthea (NASA).

Fig. 10.8   The Galilean satellites — Io, Europa, Ganymede and Callisto.

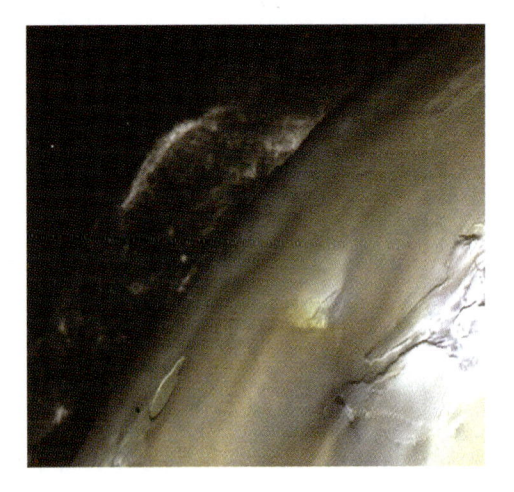

Fig. 10.9   The plume from the volcano Pele on Io.

is this heat that drives the volcanoes. The physics of this process, as it applies to Io, is described in Appendix AA. The surface of Io is an orange-yellow colour due to the sulphur and sulphur dioxide emissions from the volcanoes. Europa is also subjected to a periodic tidal stress that is thought to cause the cracking seen on its surface. There should also be some tidal heating although, because of its greater distance from Jupiter, the heating will be much less than on Io. Nevertheless, it could be enough to melt the ice below the surface so that surface ice may be floating on a vast ocean covering the whole satellite. Astrobiologists have suggested that the conditions in this ocean may have allowed some life forms to have developed within it.

The outer, and largest, satellites of the Galilean group, Ganymede and Callisto must both have a considerable ice content, as indicated by their densities. Ganymede is the most massive satellite in the Solar System, with more than twice the mass of the Moon. It is also larger than the planet Mercury. Ganymede has some regions showing old collision features, which indicates that it has a fairly stable surface. Other parts are younger and brighter in colour with the surface marked by bundles of parallel grooves. Callisto has a thick icy crust that is dark and shows the effect of many impacts, including a vast feature, Valhalla, consisting of a set of concentric rings, ripples from a major collision frozen into the surface.

The two outer groups of satellites, all quite small, should be considered together. The inner group are in prograde (direct) orbits but the outer group is, in contrast with most other bodies in the Solar System, in retrograde orbit around their primary body. A link between the two groups is suggested by the fact that a member of the inner group, Elara, with a mean diameter of 86 km and an orbit with semi-major axis of 1.17780×10^7 km and an eccentricity of 0.1948, has an apojovian distance (furthest distance from Jupiter) of 1.40723×10^7 km, which is just outside the perijovian distance (nearest distance to Jupiter) of 1.40637×10^7 km for Ananke, a member of the outer group with mean diameter 28 km, orbital semi-major axis 2.14550×10^7 km and eccentricity 0.3445. This apparent linkage between the two groups and the sense of their orbits suggests a common origin for both of them based on a collision of two bodies

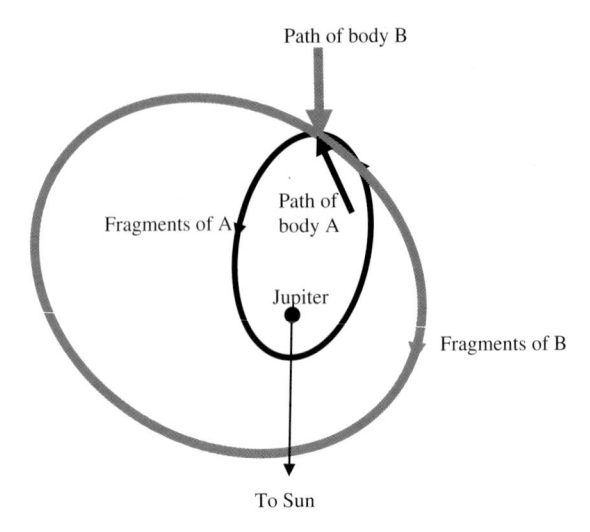

Fig. 10.10 A collision of two bodies near Jupiter giving rise to two outer families of satellites. All motions are shown relative to Jupiter.

in the vicinity of Jupiter. A proposed collision scenario is illustrated in Fig. 10.10. Two bodies, A and B, both in prograde orbits around the Sun, collide near Jupiter. The fragments of body A are deflected to follow a prograde orbit around Jupiter while those of body B are propelled into a larger retrograde orbit. Over time the orbits evolve due to perturbations by other bodies and by collisions between overlapping members of the two families of satellites.

A final, but not very prominent, feature of Jupiter is a system of faint rings. They consist of rather dark dust that reflects light poorly so that they are difficult to image clearly even from spacecraft.

10.2.2. *The satellites of Saturn*

There are 61 known satellites of Saturn, mostly very small and discovered by observations from spacecraft. Table 10.2 lists the 21 satellites with mean diameters greater than 20 km. Titan, which is second in mass only to Ganymede and is also larger than the planet Mercury, dominates the system. From its density, Titan clearly has a large ice component. It also has a very dense atmosphere, with a surface pressure 1.5 times that on Earth, which prevents its surface

Table 10.2 Saturn satellites with average diameter greater than 20 km. i is the inclination of the orbit to the planetary equator, a the semi-major axis of the orbit and e its eccentricity.

Satellite	$i(°)$	a (10^3 km)	e	Mass (10^{20} kg)	Average diameter (km)	Density (10^3 kg m^{-3})
Pan	0.00	134	0.000		30	
Atlas	0.00	138	0.001		31	
Prometheus	0.01	139	0.002		36	
Pandora	0.05	142	0.004		31	
Epimetheus	0.34	151	0.010		113	
Janus	0.17	151	0.007		179	
Mimas	1.57	185	0.020		397	
Enceladus	0.01	238	0.005		504	
Tethys	0.17	295	0.000	7.55	1,066	1.21
Telesto	1.16	295	0.000		24	
Calypso	1.47	295	0.000		31	
Dione	0.00	378	0.002	10.5	1,123	1.43
Helene	0.21	378	0.002		33	
Rhea	0.33	527	0.001	24.9	1,529	1.33
Titan	0.35	1,222	0.029	1345.20	5,151	1.88
Hyperion	0.57	1,481	0.123		292	
Iapetus	7.57	3,561	0.029	18.8	1,472	1.21
Phoebe	173.1	12,870	0.156		220	
Paaliaq	46.1	15,103	0.363		22	
Albiorix	38.0	16,267	0.477		32	
Siarnaq	45.8	17,777	0.250		40	

from being visible although details of the surface from infrared and radar images show features that might correspond to continents separated by oceans of some kind. There are four other satellites with diameters greater than 1,000 km, which would earn them the title of 'dwarf planets' (Section 10.3), were they in independent orbits around the Sun.

By virtue of their small orbital inclinations and close-to-circular orbits, all the satellites out to Titan satisfy the criterion for being regular satellites although many of them have dimensions that suggest they may have been either single satellitesimals or amalgamations of very few. This is particularly true for the tiny satellites not listed in the table.

There are some remarkable commensurabilities between the periods of some Saturn satellites, as noted in Section 9.5 for the three innermost Galileans. The periods of Titan and Hyperion are in the ratio 3:4. If there were no commensurability between the periods, then, once in a while, the aposaturnian (furthest distance from Saturn) of Titan and the perisaturnian (closest distance to Saturn) of Hyperion would happen at the same time and the satellites would be only 43,000 km apart, giving considerable disturbance of Hyperion's orbit. However, the commensurability acts in such a way that the two satellites never approach closer than about 400,000 km. Other commensurabilities are between the pairs Mimas–Tethys and Enceladus–Dione where the periods are in the ratio 1:2. From Table 10.2 it can be seen that the satellites Tethys, Telesto and Calypso have identical orbital radii, and hence periods, and the same is true for the trio Dione, Helene and Polydeuces; the last-named, having a diameter of about 3 km, is not included in the table. These are examples of a particular kind of stable dynamical system, the geometry of which is illustrated in Fig. 10.11. The body A is the central body, Saturn in the case we are considering, around which the other three bodies orbit. The body B (*e.g.* Tethys) has a mass considerably less than that of A, but substantially greater than those of bodies C_1 (Calypso) and C_2 (Telesto). The bodies C_1 and C_2

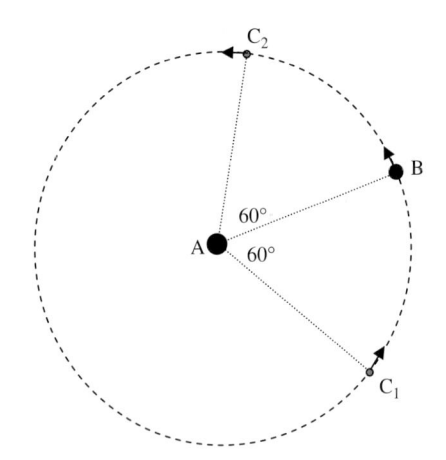

Fig. 10.11 Three satellites in the same orbit in a stable configuration.

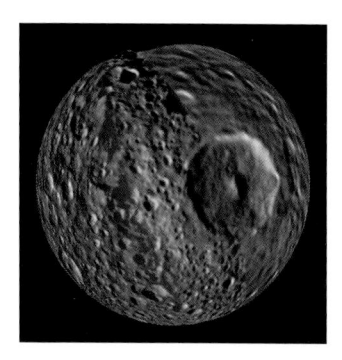

Fig. 10.12   Mimas showing the huge impact crater Herschel.

respectively follow and lead B by 60°. This situation makes the positions of C_1 and C_2 stable so that if for any reason they stray from the positions shown, then the resultant forces on them will return them to those positions (Appendix AB).

The image of Mimas, given in Fig. 10.12, indicates that it has had a charmed life. It shows an impact feature, Herschel, with more than one-quarter of the diameter of the satellite itself, which must have come close to completely destroying it.

The other large satellites of Saturn show well-cratered icy surfaces that are obviously quite old since any renewal of the surfaces would have removed all craters that previously existed. The other satellites of special interest are Iapetus and Phoebe. Iapetus has one dark hemisphere, the one facing its direction of motion, with the other quite bright; their *albedoes* — the proportion of incident light they reflect — are 0.05 and 0.5 respectively. This difference of hemispheres was an unsolved puzzle for many years but in October 2009, in a paper in the journal *Nature*, three American astronomers gave a fairly convincing explanation that involved the deposition of dark dust on the leading surface — dust that originated from the satellite Phoebe, which orbits Saturn well outside Iapetus. Phoebe's main claim to fame is that its orbit is retrograde so that it is almost certainly a captured body. The great majority of the satellites, mostly tiny, outside Phoebe have retrograde orbits (Fig. 10.13) and may be debris from a collision that left Phoebe in a retrograde orbit but retained only a few fragments of the other body in prograde orbits.

166   *On the Origin of Planets*

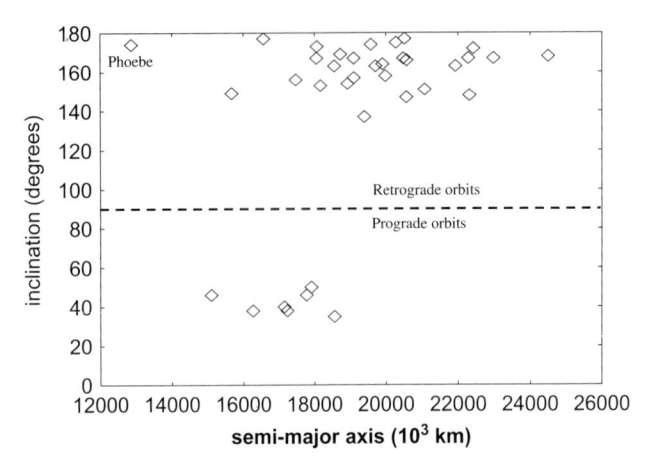

Fig. 10.13 The semi-major axes and inclinations of the small satellites of Saturn outside Phoebe.

The most spectacular feature of Saturn is its ring system that is easily seen with Earth-based telescopes and was even observed by Galileo although the resolution of his telescope did not enable him to see it for what it was. These rings contain much more material than the faint rings of Jupiter and they consist of icy particles that reflect light strongly. Gaps in the rings can be seen from Earth but, when viewed from a close distance by a spacecraft, the rings show a very complex structure (Fig. 10.14). Several of the gaps in the rings are at positions where the orbital period is related to that of the satellites.

Fig. 10.14 Saturn's rings (NASA/JPL).

For example, the very prominent Cassini division corresponds to one-half of the period of Mimas, one-third that of Enceladus and one-quarter that of Tethys. The other influence of satellites is that some of them stabilize portions of the ring; the F-ring lies between the satellites Prometheus and Pandora that act as *shepherd satellites*, preventing the particles in the ring from drifting either inwards or outwards.

10.2.3. *The satellites of Uranus*

The spin axis of Uranus has a tilt of 98° with respect to its orbital plane and the regular satellites of Uranus orbit the planet close to the resultant equator. The five largest satellites were known from Earth observation but a further 22 have been found from spacecraft observations. All the satellites with mean diameters greater than 60 km are listed in Table 10.3.

There are 13 small satellites within the orbit of Miranda with mean diameters between 18 and 162 km, seven representatives of which are included in the table. They all satisfy the conditions for

Table 10.3 Uranus satellites with average diameter greater than 60 km. i is the inclination of the orbit to the planetary equator, a the semi-major axis of the orbit and e its eccentricity.

Satellite	$i(°)$	a (10^3 km)	e	Mass (10^{18} kg)	Average diameter (km)	Density (10^3 kg m^{-3})
Cressida	0.006	61.8	0.000		80	
Desdemona	0.111	62.7	0.000		64	
Juliet	0.065	64.3	0.001		94	
Portia	0.059	66.1	0.000	1.7	135	
Rosalind	0.279	69.9	0.000		72	
Belinda	0.031	75.3	0.000		90	
Puck	0.319	86.0	0.000	2.9	162	
Miranda	4.232	129.4	0.001	66	472	1.5
Ariel	0.260	191.0	0.000	1,350	1,158	2.1
Umbriel	0.205	266.3	0.000	1,170	1,169	1.7
Titania	0.340	435.9	0.001	3,530	1,578	2.1
Oberon	0.058	583.5	0.001	3,010	1,523	2.0
Caliban	139.9	7,231.0	0.159		72	
Sycorax	152.5	12,179.0	0.522	2.3	150	

being regular satellites, having both small orbital inclinations and eccentricities. The next five satellites, from Miranda to Oberon, are those known from Earth observations. From their densities it can be inferred that they all have a large component of ice in their composition. Miranda's icy surface shows considerable evidence of disturbance with regions of parallel ridges and troughs that might have resulted from tidal action by Uranus when the satellite was less solid than it is now (Fig. 10.15a). Ariel shows signs of bombardment and valley systems that could indicate tension in the crust at some earlier time (Fig. 10.15b).

Outside Oberon there is a further group of eight satellites, with mean diameters between 18 and 150 km, with orbital eccentricities between 0.146 and 0.591 and in retrograde orbits with inclinations between 140 and 167°. The two largest members of this group are included in Table 10.3. These retrograde satellites seem almost certainly to be captured bodies as a result of a collision in the vicinity of Uranus. There is also a small satellite of diameter 20 km in addition to this outer group of eight with a prograde orbit of eccentricity 0.661 and inclination 51°.

In 1977, before any spacecraft visited the planet, ground-based observations had established the existence of a system of rings around

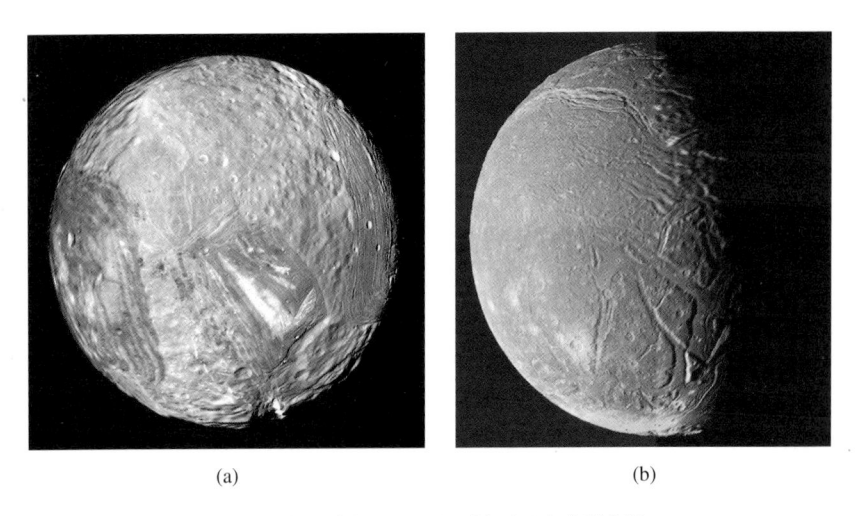

Fig. 10.15 (a) Miranda (b) Ariel (NASA).

Uranus. These could not be seen directly but were detected by stellar occultation, i.e. the blocking out of the light from a star by a body between the star and the Earth. The bodies giving the occultation were within the rings and on the basis of the observations it was concluded that there were five rings. Spacecraft have subsequently established the existence of 13 rings that contain little dust and mostly consist of bodies a few metres in diameter.

10.2.4. *The satellites of Neptune*

Prior to the space age there were two satellites observed around Neptune — Triton and Nereid — both rather remarkable in their relationship to the planet. Now 13 satellites have been detected and are listed in Table 10.4.

All the six inner satellites, out to Proteus, are regular in their characteristics. Proteus, which was discovered by the Voyager 2 spacecraft, is larger than Nereid, which was known from ground observation, but Proteus could not be detected from Earth because of its proximity to Neptune. Three of the five outermost satellites are

Table 10.4 All known Neptune satellites. i is inclination of orbit to planetary equator, a the semi-major axis of the orbit and e its eccentricity.

Satellite	$i(°)$	a (10^3 km)	e	Mass (10^{18} kg)	Average diameter (km)	Density (10^3 kg m^{-3})
Naiad	4.7	48.2	0.000		66	
Thalassa	0.2	50.1	0.000		82	
Despina	0.1	52.5	0.000		150	
Galatea	0.1	62.0	0.000		176	
Larissa	0.2	73.5	0.001		194	
Proteus	0.6	117.6	0.000	50	402	
Triton	156.8	354.8	0.000	21,400	2,707	2.05
Nereid	27.6	5,313	0.751	31	340	
Halimede	134.1	15,728	0.571		62	
Sao	48.5	22,422	0.293		44	
Laomedeia	34.7	23,571	0.424		42	
Psamathe	137.4	46,695	0.450		38	
Neso	132.6	48,387	0.495		60	

in retrograde orbits with similar inclinations and all five are probably the fragments from a collision in the vicinity of the planet.

Triton, the seventh largest satellite in the Solar System, has the odd feature that it is in a retrograde orbit so that, despite its almost circular orbit, it is certainly not regular. It is a feature of satellites in retrograde orbits that the effect of tidal forces between the satellite and planet reduces both the eccentricity and the semi-major axis. At some time in the future Triton will approach Neptune so closely that it will be torn apart by strong tidal forces. The effect of this is that a ring system might form rivalling that of Saturn in appearance. Triton possesses a very tenuous atmosphere of nitrogen with traces of methane. Its surface has few impact features, suggesting that it is young, but there are ridges and valleys all over its surface. Volcanic activity has been detected on Triton based on liquid nitrogen; one volcanic plume that was observed rose to 8 km above the surface. The image of Triton given in Fig. 10.16 shows an ice cap of frozen nitrogen and methane.

The final satellite of Neptune, Nereid, is also unique in that it has the highest eccentricity of any satellite. It could be a body captured as a result of a collision that also gave the other outer satellites; in Chapter 15 another possible explanation for its curious orbit will be suggested.

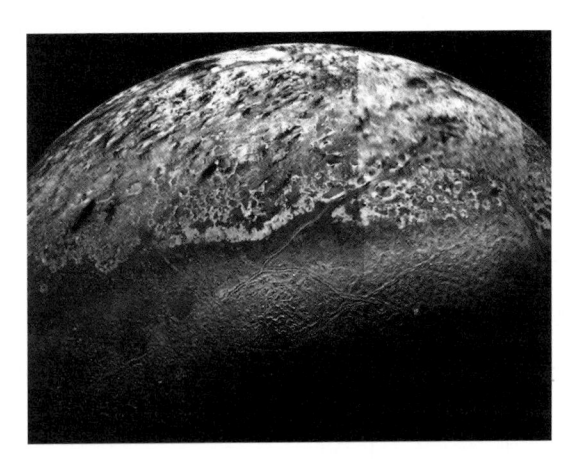

Fig. 10.16   Triton showing a frozen icecap (Voyager 2, NASA).

10.2.5. *Other satellites*

Of the terrestrial planets, only Mars and Earth have satellites. That of Earth, the Moon, has the distinction of being the only solar-system body outside Earth to have been visited by man. Our knowledge of the Moon transcends that of any other body in the Solar System, with the exception of Earth. It is the fifth biggest satellite in the Solar System, yet it is associated with a terrestrial planet. In addition, it has characteristics of its surface and internal structure that seems to demand explanations. This interesting body will be described in detail in Chapter 12 and explanations will be offered both for its structure and how it became associated with Earth.

The final satellites of the Solar System are two small companions of Mars — Phobos and Deimos, shown in Fig. 10.17. Both satellites are very small and almost certainly captured bodies. The larger satellite, Phobos, is roughly ellipsoidal with dimensions of $20 \times 23 \times 28$ km. It is heavily scarred with craters, the largest of which, Stickney, has a diameter of 10 km and can be clearly seen in Fig. 10.17(a). Phobos, the nearer to Mars, has an orbital period of 7 h 39.5 min, less than the Mars spin period, which has the effect that an observer on Mars would see Phobos rise in the west and set in the east. Deimos, with dimensions of $10 \times 12 \times 16$ km, has an orbital period of 30 h 24 min.

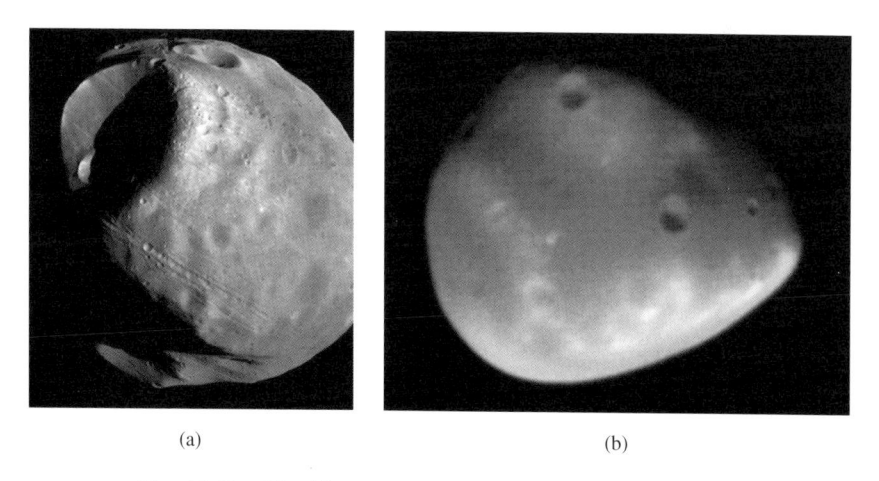

(a)   (b)

Fig. 10.17 The Martian satellites (a) Phobos (b) Deimos.

It also shows craters, but less distinctly than Phobos since it is covered by a thick layer of dust.

10.3. Dwarf Planets and the Kuiper Belt

A book written before 2006 describing the Solar System would have listed nine planets rather than just the eight I give here. The extra member of the planetary system would have been Pluto, orbiting for the most part outside the orbit of Neptune. At the end of the 18th century, knowledge of the planets extended out as far as Uranus, discovered by William Herschel in 1781. However, a few decades later, when the orbit of Uranus had been reasonably well established, it was found to be wandering slightly from its predicted path. Newton's laws of motion were used to take account of the gravitational effects of the other planets on Uranus but even so there was still a residual discrepancy. It was suggested that there was some other planet exterior to Uranus that was causing a drift in its predicted path. This problem came to the attention of a young mathematics student at Cambridge, John Couch Adams (1819–1892), who decided in 1841 that it could be solved by the application of Newton's laws but did not begin to work seriously on the solution until 1845, after he had graduated. Towards the end of 1845 he sent his solution, with the predicted position of the new planet, to the Astronomer Royal, George Biddell Airy (1801–1892), who asked for more information about how the solution had been reached, but communication between the two men seems to have broken down — for reasons that are not clear.

While this was happening in England, the problem was being independently tackled by the French mathematician, Urbain Le Verrier (1811–1877), who specialized in celestial mechanics. On 31 August 1846 he announced the predicted position of the new planet to the French Academy of Sciences. This predicted position was sent to Johann Galle (1812–1910) at the Berlin Observatory and five days later the new planet was discovered. By this time the Greenwich Observatory had begun a belated search using Adams' predicted position and after the Berlin discovery was announced they found that they had actually observed Neptune on two occasions in early August 1846 but had not recognized it for what it was. To his credit, Adams publicly gave priority to Le Verrier for both the prediction and discovery.

By 1906 there was a belief that Uranus and Neptune were both wandering off course and the explanation given was the existence of a ninth planet in the region beyond Neptune. A rich American businessman, Percival Lowell (1855–1916), had decided to devote himself to the study of astronomy and in 1894 he established the Lowell Observatory in Flagstaff, Arizona. Lowell searched for the new planet, generally referred to as Planet X, from 1906 until his death, but without success. Then in 1929 a young American astronomer, Clyde Tombaugh (1906–1997), was given the task of finding the new planet and after a year of painstaking work he was successful. The discovery of Planet X was announced in 1930 and it was eventually given the name Pluto, the first two letters being the initials of Percival Lowell.

The first estimate of Pluto's mass was six times that of Earth but over time this gradually fell and it is now known that its mass is 0.002 that of Earth — less than one-fifth that of the Moon. Pluto's status as a planet was always disputed but when, in 1978, it was discovered to have a satellite, given the name Charon, this seemed to reinforce its planetary status. However, in 1992 astronomers began to detect a myriad of small bodies orbiting the Sun just beyond the orbit of Neptune and in 2005 a body, Eris, which also had a satellite and was 27% more massive than Pluto, was found in that region. The choice was clear — either Eris had to be a tenth planet or Pluto had to be demoted. At a meeting of the International Astronomical Union in 2006 the latter course was chosen and a new category of bodies was defined — the *dwarf planets*, of which both Pluto and Eris were members. To be a dwarf planet a body has to be large enough to have adopted a spherical form and to be in an independent orbit around the Sun, the latter part of the definition ruling out any satellite body. At the time of writing there are five dwarf planets, which are listed in Table 10.5 in order of their discovery.

Table 10.5 Characteristics of the dwarf planets.

Dwarf planet	a (au)	e	i (°)	Diameter (km)	Mass (10^{21} kg)	Satellites
Ceres	2.77	0.080	10.6	975	0.95	0
Pluto	39.48	0.249	17.1	2,306	13.05	3
Haumea	43.34	0.189	28.2	1,150	4.2	2
Makemake	45.79	0.159	29.0	1,500	4?	0
Eris	67.67	0.442	44.2	2,400	16.7	1

Table 10.6 The progression of planetary radii out to Saturn compared with Bode's law.

Planet	Semi-major axis (au)	Bode's law
Mercury	0.387	0.4
Venus	0.723	$0.4 + 0.3 = 0.7$
Earth	1.000	$0.4 + 2 \times 0.3 = 1.0$
Mars	1.524	$0.4 + 4 \times 0.3 = 1.6$
?		$0.4 + 8 \times 0.3 = 2.8$
Jupiter	5.203	$0.4 + 16 \times 0.3 = 5.2$
Saturn	9.539	$0.4 + 32 \times 0.3 = 10.0$

Ceres

In 1772, before the discovery of Uranus, the German astronomer Johann Elert Bode (1747–1826) noticed that the orbital radii of the planets then known were close to a simple mathematical progression, as illustrated in Table 10.6.

When Uranus was discovered it was found that it fitted the Bode's law prediction quite well; the actual semi-major axis is 19.19 au and Bode's law gives $0.4 + 64 \times 0.3 = 19.6$. This agreement was taken by many to give Bode's law the status of a fundamental law and a search was pursued by many astronomers to find a planet to fill the gap in the progression of planets between Mars and Jupiter. On 1 January 1801, the first day of the 19th century, Giuseppe Piazzi (1746–1826) of the Palermo Observatory in Sicily found a body at the predicted distance. The body was named Ceres after the patron god of the island of Sicily. Its semi-major axis of 2.77 au

Features of the Solar System

was in satisfactory agreement with the Bode's law value of 2.8 and the pattern of planets seemed now to be complete. However, as we shall see later, Ceres was just the precursor of many other bodies to be found in the same region of the Solar System, although Ceres is the only one with dwarf-planet status.

It should be noted that when Neptune was discovered, with a semi-major axis of 30.07 au, this was a significant departure from Bode's law (prediction 38.8) and thereafter Bode's law was less highly regarded.

Pluto

The events leading to the discovery of Pluto, and the discovery itself, have already been described. Initially its image by ground-based telescopes was just as a small bright disk. But other than the fact that it had a bright surface, nothing more was known about it. In 1978 an image of Pluto taken by a terrestrial telescope showed a small bulge on one side of the disk that was identified as a satellite; later images showed this large satellite clearly and, later still, images showed two more very small satellites (Fig. 10.18). Once a satellite is found for a body, careful observation plus the laws of planetary (or satellite)

Fig. 10.18 A Hubble telescope image of Pluto, its large satellite Charon and two small satellites.

motion can be used to determine the mass of the central body and also of the satellite if this is a substantial fraction of the combined mass, which is certainly true for Charon. Its radius, 1,207 km, is more than one-half that of Pluto, and its mass about one-eighth. They are sometimes described as binary dwarf planets. Pluto has a retrograde spin with a period of 6.4 days and Charon orbits with the same period. The two bodies are tidally linked so that each always presents the same face to the other, much as the Moon always presents the same face to the Earth.

Haumea, Makemake and Eris

In 1951 the Dutch (later American) astronomer Gerard Kuiper (1905–1973) suggested that beyond Neptune there existed a region densely occupied by comet-like bodies moving close to the general plane of the Solar System. In 1992 his prediction was confirmed when the first such body was found, and subsequently many more have been discovered in what is now known as the Kuiper Belt, which stretches from 30 to 55 au from the Sun. Some of them are very substantial bodies and in the period up to 2009 three of them have merited the status of dwarf planets. About 1,000 Kuiper-belt objects have been observed and it has been estimated that there are about 70,000 bodies in the Kuiper Belt with diameters greater than 100 km.

The first of the three dwarf planets to be found in the Kuiper Belt was Haumea, the mass of which could be estimated reasonably well because it possesses two small satellites. The next to be discovered, Eris, also possessed a satellite that enabled its mass to be determined, which precipitated the discussion about the status of Pluto. An image of Eris and its satellite Discordia is shown in Fig. 10.19.

The final dwarf planet in this region is Makemake for which, since it has no satellite, only an approximate mass is available — based on an estimate of size made by comparing its image with that of Pluto.

There may be other potential dwarf planets in the region outside Neptune, perhaps at distances even beyond the reach of space telescopes. However, of one thing we can be reasonably certain: Pluto

Fig. 10.19 The dwarf planet Eris and its satellite, Discordia.

is not a full member of the family of solar-system planets but just the innermost member of a class of objects in the region outside Neptune.

10.4. Asteroids

10.4.1. *Types of asteroids and their orbits*

The discovery of the dwarf planet Ceres in 1801 seemed to satisfy the requirement of filling the gap in Bode's law between Mars and Jupiter. However, the situation soon became muddied by the discovery of other bodies in the same region of the Solar System. These bodies became known as asteroids and they were also discovered in regions of the Solar System within the orbit of Mars and outside the orbit of Jupiter. Table 10.7 lists a number of these bodies, chosen to illustrate the range of different characteristics they can have.

The first three in the list — Pallas, Juno and Vesta — are large bodies found shortly after Ceres that, until 2006, was classified as the largest asteroid. They, and the two that follow — Hygeia and Undina — occupy a region spanning the original gap in Bode's law.

Most asteroid orbits occupy the region between Mars and Jupiter, have eccentricities of less than 0.3 and inclinations of less than 25°. The table shows several exceptions to these general characteristics. The last four listed asteroids have large eccentricities and Pallas and

Table 10.7 Characteristics of some representative asteroids.

Name	Year of discovery	a (au)	e	i (°)	Diameter (km)
Pallas	1802	2.77	0.237	34.9	608
Juno	1804	2.67	0.257	13.0	250
Vesta	1807	2.58	0.089	7.1	538
Hygeia	1849	3.15	0.100	3.8	450
Undina	1867	3.20	0.072	9.9	250
Eros	1898	1.46	0.223	10.8	16
Hidalgo	1920	5.81	0.657	42.5	38
Apollo	1932	1.47	0.566	6.4	1.7
Icarus	1949	1.08	0.827	22.9	1.4
Chiron	1877	15.50	0.378	6.9	180

Hidalgo have inclinations well above 25°. There are some asteroids with inclinations even higher than Hidalgo's 42.5° but, significantly, there are no known asteroids with retrograde orbits.

The high eccentricity of Icarus, the largest for any known asteroid, in combination with its small semi-major axis, takes it within 0.19 au of the Sun — closer than any other asteroid. It is an example of an Earth-crossing asteroid, one whose orbit spans a distance of 1 au from the Sun. The first such asteroid discovered was Apollo; in the year of its discovery, 1932, it came within 3 million kilometres of the Earth, which counts as a near-miss in astronomical terms! Since 1932 many other asteroids, mostly very small, have been found in Earth-crossing orbits and these are known collectively as the *Apollo asteroids*. Another class of asteroids, the *Aten group*, orbit mostly within the orbit of the Earth and some of these are also in the Earth-crossing category. There are other asteroids, of which Eros is one, that are Mars-crossing but whose perihelia are outside the Earth's orbit and so are not Earth-crossing.

The Apollo and Aten asteroids are departures from the rule that asteroids just occupy the region between Mars and Jupiter and another such departure is shown by Chiron which moves mainly between Saturn and Uranus. It is possible that there are many other smaller asteroids, and hence more difficult to observe, further out in the Solar System. There are two other interesting groups of asteroids,

Features of the Solar System 179

not represented in the table — the so-called *Trojan asteroids*. These lead and trail Jupiter in its orbit, one group 60° ahead and the other 60° behind the planet. This arrangement is illustrated in Fig. 10.11 where now A is the Sun, B is Jupiter and C_1 and C_2 are the two groups of Trojan asteroids.

When the numbers of asteroids within narrow ranges of semi-major axis are plotted, the result is depicted in Fig. 10.20. The period of an orbit around any central body is determined by its semi-major axis. What is seen is that asteroid orbits avoid some values of semi-major axis and these dips are found to correspond to simple commensurabilities with the period of Jupiter. The gaps shown correspond to 1/3, 2/5, 3/7 and 1/2 of Jupiter's orbital period. The American astronomer Daniel Kirkwood (1814–1895) first noted this pattern in 1866. The gaps are due to perturbation by Jupiter; when its orbit is commensurate with that of Jupiter, an asteroid would repeatedly be closest to Jupiter at the same points of its orbit and the gravitational effects would build up until it is nudged into

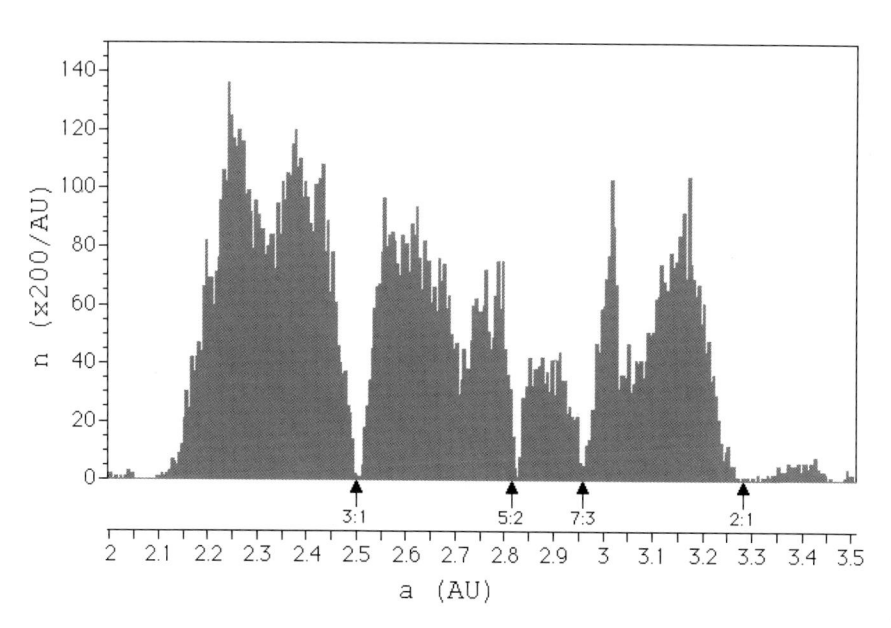

Fig. 10.20 An illustration of Kirkwood gaps (NASA).

a nearby non-commensurate orbit. This mechanism is also the one that produces the gaps in Saturn's rings.

10.4.2. *The composition of asteroids*

Smaller asteroids tend to be irregularly shaped objects and usually show signs of collision damage. Figure 10.21 shows the fairly large asteroid Mathilde, some 50 km in average dimension.

Asteroids vary greatly in their reflectivity, some being quite bright while others reflect very little light and, for that reason, are more difficult to detect. The key to understanding their compositions is the samples we have of asteroids on Earth — meteorites, fragments of asteroids caused by occasional collisions which happen to land on Earth. Using spectroscopes that analyse the reflectivity of meteorites over the visible and infrared range of the electromagnetic spectrum, it is possible to relate the pattern of the reflectivity of a meteorite to its composition. The reflected sunlight from asteroids can also be subjected to spectroscopic analysis and then, by comparison with the data from meteorites, the composition of the asteroid can be inferred. In this way it is found that asteroids, like meteorites, fall into three main classes — *stones*, consisting of different types of silicates, *irons*, consisting mainly of iron but with some nickel, and *stony-irons*, which are intimate mixtures of stone and iron. Within the stony type there is an important sub-type, the *carbonaceous* asteroids that are

Fig. 10.21 The asteroid Mathilde.

dark in colour and contain some volatile materials. We shall consider these types of material more fully when meteorites are dealt with in Chapter 16.

10.5. Comets

Even those who have not seen a comet will be aware of their general appearance, a bright luminous head with a long tail. Actually, there are normally two tails, as seen in Fig. 10.22. The longer tail consists of plasma — a mixture of ionized atoms and electrons — while the shorter one consists of fine dust particles. The bright head is called the *coma*, which is a ball of gas of radius 10^5 to 10^6 km. At the centre of the coma is the solid part of the comet, the *nucleus*, an intimate mixture of silicates and ices of various kinds. When the comet approaches the Sun, the ices vaporize to give the material for both the coma and the plasma tail. The jets of vapour prise dust off the nucleus so providing the material for the dust tail. Both the coma and the plasma tail fluoresce due to excitation by ultraviolet radiation from the Sun. Fluorescence is the phenomenon where material absorbs radiation over a range of wavelengths and emits it at wavelengths characteristic of its composition. Hence, from the wavelengths of the light coming from the coma and plasma tail, their chemical nature can be determined.

Fig. 10.22 Comet Mrkos, photographed in 1957, showing a long plasma tail and short stubby dust tail (Mount Palomar Observatory).

There is a popular misconception that the tails of a comet point in the opposite direction to its motion, much like the loose end of a scarf of someone in a moving open-top car. However, this is not so. The gas and dust leaving the nucleus is acted on by the *solar wind* — charged particles, mostly protons and electrons, pouring out of the Sun at very a high speed of order $400\,\mathrm{km\,s^{-1}}$. For this reason the tails point away from the Sun, regardless of the direction of motion of the comet.

The fact that comets were bodies in highly eccentric orbits around the Sun was first realised by Isaac Newton's friend, Edmund Halley (1656–1742), who postulated that the comet seen in 1682 was the one previously seen in 1607, 1531 and 1456 and would return in 1758. The prediction turned out to be correct although Halley did not live to see his prediction confirmed. Halley's comet is a *short-period comet*, one with a period less than 200 years and spends most of its time in the region occupied by the planets. Those with periods less than 20 years have prograde orbits but other short-period comets have more or less random inclinations. Halley's comet has an inclination of 162° (retrograde orbit) and because of planetary perturbations its period varies between 74 and 78 years. Comets with periods greater than 200 years are referred to as *long-period comets*.

There are about 70 short-period comets, with periods of between 3 and 10 years, modest inclinations of up to about 30° and aphelia of about 5 au, that seem to be linked to Jupiter. This *Jupiter family* were originally long-period comets, the orbits of which were perturbed by Jupiter either in a series of small perturbations or one very large one and swung into orbits with aphelia at about the distance of Jupiter from the Sun.

Another class of comets is that whose orbits have periods of tens of thousands to several million years corresponding to aphelia from thousands to tens of thousands of au. Since comets can only be seen when they are close to the Sun, this implies that they have orbital eccentricities just less than unity, the value at which an orbit ceases to be closed. Based on the fact that about one such comet is observed each year, the Dutch astronomer Jan Oort (1900–1992) estimated that there is a cloud of about 2×10^{11} comets, the *Oort cloud*, surrounding the Sun at a distances of tens of thousands of au. Normally the perihelia of such comets are too far from the Sun for them to be seen but occasional passages of stars close to the Sun perturb some of them into orbits with perihelia which enable them to be seen. The Oort cloud will be discussed further in Chapter 14.

10.6. Summary

There are two main types of planet, the terrestrial planets which consist of silicates with an iron core and the major planets in which most of the mass is in the form of gas, mostly hydrogen and helium, or volatile materials such as the ices of water, ammonia and methane. The terrestrial planets divide into two pairs, Venus and Earth which are similar in size and mass and are in near-circular orbits, and the much smaller Mars and Mercury with much more eccentric orbits. Similarly the major planets divide into two pairs — the more massive Jupiter and Saturn with some similarities in their surface features and compositions, and the less massive Uranus and Neptune, again with similar features linking the pair. Uranus is distinguished by its extreme axial tilt where the axis is close to the orbital plane and the spin is retrograde. Of the other planets, Venus also has a retrograde spin, but the axis is normal to the orbit and the spin is very slow.

All the major planets have substantial satellite families with the largest members being the four Galilean satellites of Jupiter, Saturn's Titan and Neptune's large satellite in a retrograde orbit, Triton,. There are many very tiny satellites, best understood as residual satellitesimals, and a few satellites, both in prograde and retrograde orbits, that can best be interpreted as the residues of collisions in the vicinity of the parent planet. Another large satellite, the Moon, is clearly anomalous in its linkage to Earth and will be considered later in more detail. The satellites of Mars, Phobos and Deimos, are both very small and are likely to be captured objects.

Somewhat smaller than the planets are the dwarf planets, of which five members are known. Four of them are associated with the Kuiper Belt, a region just beyond the orbit of Neptune, containing a large number of substantial bodies of diameter 100 km or more.

There are many small bodies within the Solar System. Asteroids orbit mainly in the region between Mars and Jupiter but there are others, such as the Apollo and Aten asteroids, that orbit much closer in, many of them Earth-crossing, and yet others, such as Chiron, much further out. Because of observational difficulties we cannot be sure how many there are in the outer regions of the Solar System. Asteroids can be of stone, iron or stony-iron composition, and within the stones there are carbonaceous asteroids, dark in colour and containing volatile materials.

Comets are stony bodies containing a considerable component of volatile material. When they move close to the Sun the volatiles in the nucleus become gaseous and form the coma and the plasma tail. Dust that is blown off the nucleus forms a separate dust tail. There are short-period comets, with orbital periods less than 200 years. Those with periods less than about 20 years are always prograde; the other short-period comets have random inclinations. The Oort cloud is a vast system of comets, estimated to be 2×10^{11} in number, situated at thousands to tens of thousands of au from the Sun. Perturbation by passing stars cause some to be deflected, giving visible comets that are seen at a rate of about one per year.

Chapter 11

Interactions Between Planets

11.1. The Precession of Planetary Orbits

The initial orbits of the planets are very extended, stretching out to 2,000 au or more from the Sun, with high eccentricity. If the only force acting on the planet were that of the Sun, then the orbit would never change and the planet would repeatedly pass through the same points of space. However, as we found in Chapter 7, the medium acts as a brake on the motion of the planet, causing the orbit to decay and, usually, round off over a period of one million years or so.

An aspect of the orbit that we have not previously considered is its inclination, which is the plane of the orbit relative to some reference plane. The obvious reference plane to use in this case is the mean plane of the resisting medium that has a disk-like form. In the SPH simulations of the capture-theory mechanism, as seen in Figs. 6.1 and 6.4, all motions are taken to be parallel to the plane of the figures. However, in a more realistic situation there would be motions out of the plane due, for example, to the spin of the protostar about an axis not perpendicular to the plane. For this reason we should not expect the orbital inclinations of the planets to all be the same, although they will only differ by a few degrees at most. If we take the case of a spinning protostar, then, since the spin rate is limited by the requirement that the protostar is not rotationally disrupted, the component of velocity of any body perpendicular to the plane is limited and it can be shown that all orbital inclinations will be contained within 10° from the mean plane.[1]

Observational evidence, and the SPH simulations of the CT, suggests that the masses of the disks around young stars are within

---

[1] Dormand, J.R. and Woolfson, M.M., 1977, *Mon. Not. R. Astr. Soc.* **180**, 243.

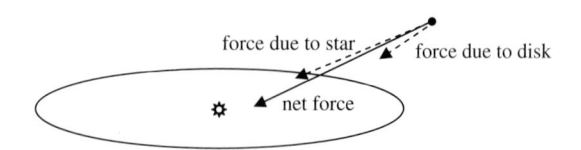

Fig. 11.1 The gravitational forces due to the star (pointing towards the star) and due to the disk (offset) giving a net offset force. For clarity, both the magnitude and offset of the disk force have been exaggerated.

the range of 0.01 to 0.1 times the solar mass. The medium not only offered resistance to the motions of planets but also influenced them gravitationally. An important aspect of this is that the net gravitational force of a disk on a point mass is not directed towards the centre of the disk but to some point off-centre. This means that the combination of the gravitational forces of the star and the disk is not directed towards the star and that as the planet moves, so does the point towards which the net gravitational force is directed (Fig. 11.1).

The overall effect of this combination of forces is that it gives *precession* of the planetary orbit. This motion is quite complicated and is described in Appendix AC but one component of it is illustrated in Fig. 11.2. The orbit is shown inclined to the mean plane, with A as the apastron point (that furthest from the star). As the planet moves around the orbit, so point A of the orbit moves around the dashed circle at some different rate. The main characteristic of precession is that it gives relative motions of the planets and, although quite complicated, this aspect of its effect is adequately illustrated in Fig. 11.2.

11.2. Close Interactions of Planets and the Tilts of Spin Axes

The rate of precession — the angular speed of point A round the dashed circle — depends on the characteristics of the planetary orbit and will vary from one planet to another. In general, at any instant, two orbits with different semi-major axes, eccentricities and inclinations will not intersect in space. In a projection onto the plane

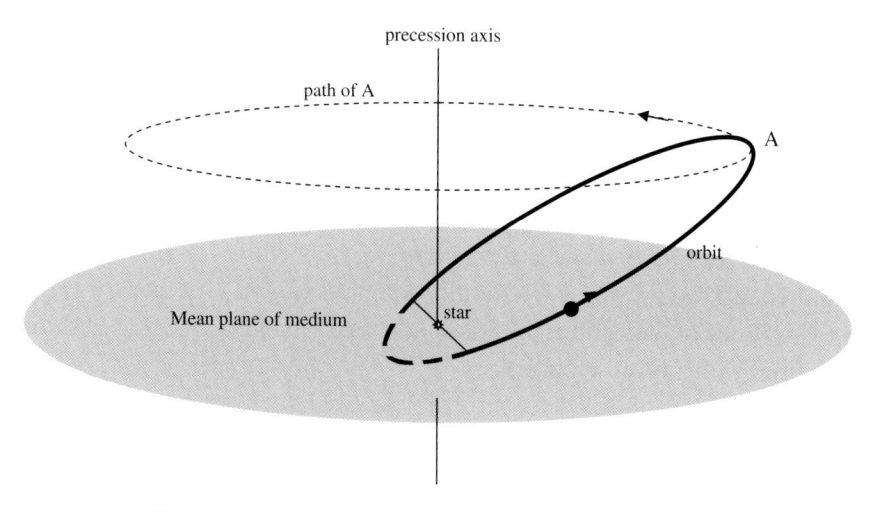

Fig. 11.2 The effect of precession on a planetary orbit.

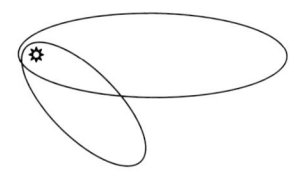

Fig. 11.3 Two orbits intersecting in projection but not intersecting in space.

of the medium, as seen in Fig. 11.3, they may appear to intersect but, because of their different inclinations, at the points of apparent intersection they will be separated in the direction perpendicular to the plane of the figure. However, because of the different rates of precession of the two orbits, the orbits *will* intersect from time to time. The periods of the precessions, *i.e.* the times for points such as A in Fig. 11.2 to go completely round the dashed path, are of the order of 50,000 years, so, in general, for each pair of planets there are several orbital intersections during the period of orbital decay and round-off.

The fact that orbits intersect, or nearly so, is a necessary condition for planets to closely approach each other but it is not a sufficient condition. It is also necessary for the planets to be in the vicinity of the intersection at the same time — a condition that will occur rarely but will happen occasionally. With notional starting orbits for the major planets, and their likely patterns of orbital evolution, calculations show that the probability of a close approach of any individual planet to some other during the period to be quite high — actually much more likely to happen than not. Here we define 'close approach' as having a centre-to-centre distance within five times the radius of the larger planet, within which distance major tidal effects would occur. It should be noted that the planetary radii at the times of close interactions might be considerably greater than the final radii we now observe.

Figure 10.6 shows the tilts of the planetary spin axes and it may be wondered how these came about. For planets produced as condensations in a filament, the most likely direction of a spin axis is perpendicular to the planet's orbital plane. This would come about because of tidal interactions between the star and a planet and between the planets themselves. Figure 11.4 shows how the tidal effect on a planet operates. As the planet moves past the tide-raising body, so its configuration changes, with the peak of the tide at three positions indicated as points A, B and C. The peak does

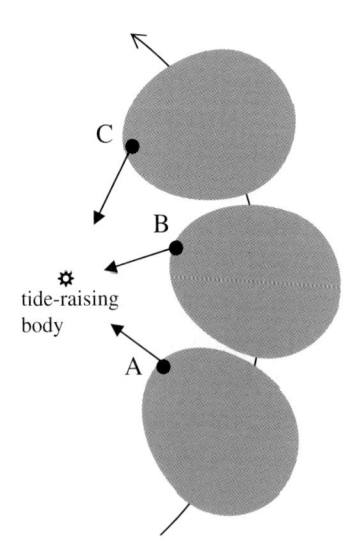

Fig. 11.4 Tides on a planet (grey body) at successive times with tidal bulges at A, B and C.

not point directly at the tide-raising body but lags behind because of inertia in the system; the material cannot react immediately to the ever-changing forces that are acting on it. The arrows show the gravitational force acting on the tidal bulge and in each case it is pulling the bulge material into rotation around the centre of the planet in an anticlockwise sense. When the planet leaves the vicinity of the tide-raising body, the body returns to a near-spherical shape and the rotational motion in the tidal bulge is distributed throughout the body. The imparted rotation will be about an axis perpendicular to the orbital plane of the motion and the final rotation will be a combination of the original spin of the body with what was added. Since all the planets in a filament move closely in one plane, it might be thought that the spin axis due to tidal effects between the bodies would be almost perpendicular to that plane.

How then can the observed tilts of the planetary axes be explained — in particular the extreme tilt of Uranus? The answer lies in the fact that orbits can intersect, or nearly so, and that they are not *precisely* coplanar. Figure 11.5 shows an early Uranus on an eccentric orbit with another more massive planet, taken as Jupiter, passing over it above its orbital plane. The tidal forces give a spin to Uranus about an axis almost in the plane of its orbit. Since Jupiter is much more massive than Uranus, its axis of spin is little affected; it is tilted at only 3° to its orbital plane. I have carried out a simulation

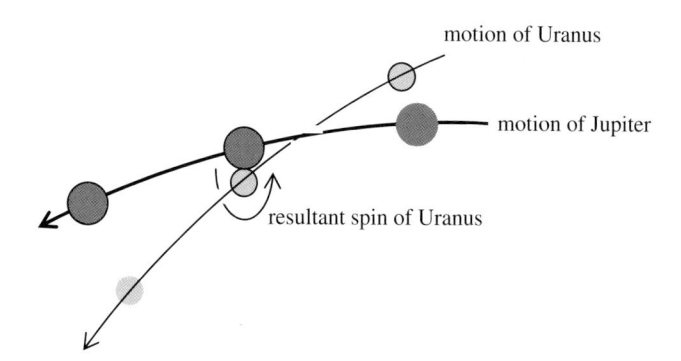

Fig. 11.5 The relative motions of Jupiter and Uranus giving the spin axis of Uranus close to its orbital plane.

of this mechanism using SPH and reproduced the axial tilt of Uranus almost perfectly. However, for reasons explained in Section 13.4, it may have been an interaction with another planet that gave Uranus its extreme tilt.

The smaller tilts of the other major planets can be explained in a similar way although it is not possible to define a specific scenario that would explain them all precisely. I shall deal with the tilts of the terrestrial planets later.

11.3. The Problem of the Terrestrial Planets

In the capture-theory model the planets originated as dusty gaseous condensations within an unstable filament. They eventually collapsed to form major planets with iron-silicate cores — the aggregated dust that settled in the centre of the condensation — and a gaseous envelope. The reason that a planet is able to form in this way is that the gaseous condensation has a mass greater than the Jeans critical mass. If it were of less than the critical mass, then it would disperse in its entirety — both the gas and the dust.

This scenario for planet formation raises the question of how the terrestrial planets formed. There are two possible ways this could happen. The first is that the SNT mechanism for producing planets was viable in the inner part of the Solar System and that solid bodies were produced that did not subsequently acquire a gaseous envelope. It is true that all the processes of the SNT — from the settling of dust to produce a dust carpet, the instability of the dust carpet to produce planetesimals and the collection of planetesimals to produce larger solid bodies — would be faster close to the Sun than at the distances of the major planets. Bodies with the mass of any of the terrestrial planets could not maintain a low-mass hydrogen-plus-helium atmosphere at their present distances from the Sun, so it is not unreasonable to suppose that they could not have started to acquire one if they were produced in this way. There are some marginal problems with this scenario, but not so strong that they exclude the possibility that this is the way terrestrial planets were formed.

The second scenario is that the Solar System, produced by the capture-theory mechanism, initially contained only major planets and that the terrestrial planets were a result of some evolutionary process. This is the possibility that I shall describe here. When dealing with a complex system such as the Solar System, with many and diverse characteristics, explanatory theories are regarded more favourably if a single proposed event explained many features of the system and linked them in some way. Alternatively, if every feature of the system were explained by a different *ad hoc* explanation, and these explanations were unrelated in any way, then this would normally be regarded less favourably. The great advantage of the explanation I shall give here for the origin of the terrestrial planets is that it simultaneously provides the background for explaining many other diverse, and apparently unrelated, features of the Solar System. However, before I embark on this explanation it is first necessary to describe an important property of solar-system material at the stage that planets were being formed.

11.4. Deuterium and the Major Planets

The standard model for the origin of the Universe is the Big Bang theory that proposes that the Universe began as a huge release of energy concentrated at a point. Before the Big Bang event there was neither space nor time. Time, which only has meaning in relation to change, came into existence as a result of the changes that occurred as the Universe expanded and space was created by that expansion. The Universe does not expand into anything because nothing exists outside the Universe. These are difficult philosophical concepts that challenge human comprehension.

As the Universe expanded, so some of the energy, in the form of radiation, was converted into matter; Einstein's equation linking energy, E, and mass, m, *i.e.* $E = mc^2$, in which c is the speed of light, governed this transfer from energy to material mass. In the early stages of the growth of the Universe, when temperature was extremely high, there were strange exotic particles produced, of a kind not encountered in everyday life, and sometimes energetic particles came together and disappeared, their mass being reconverted into energy. As the Universe expanded and became cooler, the processes of converting energy into mass and mass into energy slowed down and finally stopped altogether after which time radiation and matter had independent existences. At this stage the matter in the Universe was predominantly in the form of protons, which eventually picked up electrons to become hydrogen atoms (Fig. 11.6(a)). The electron, with a mass just 1/1836 that of the proton and with a negative charge that just balances the positive charge of the proton, gives an atom that is electrically neutral. Accompanying the hydrogen in this early material were α-particles, stable entities consisting of two protons and two neutrons, a neutron being an uncharged particle with a mass close to that of the proton. When α-particles acquired two electrons they became helium atoms (Fig. 11.6(b)) that account for about 20% of the total mass of the universe. In addition a very small amount of lithium (Fig. 11.6(c)), the third lightest element, was also produced at this early stage.

The above description of the material content of the early Universe is not quite complete. One thing it lacks is the presence of an isotope of hydrogen — deuterium (Fig. 11.6(d)). This has a nucleus consisting of a proton and a neutron. The number of protons in the nucleus solely dictates the chemical identity of an atom, so, although deuterium has its own chemical symbol, D, it is nevertheless just a form of hydrogen. Water is usually described as H_2O but HDO

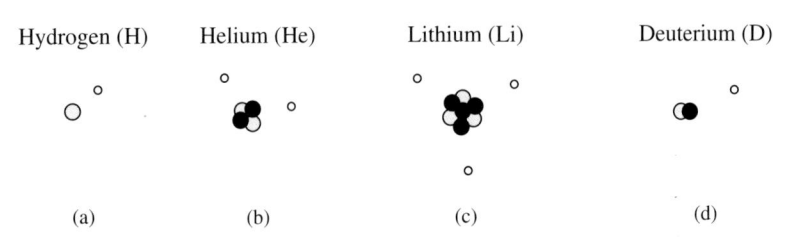

Hydrogen (H)　　Helium (He)　　Lithium (Li)　　Deuterium (D)

(a)　　(b)　　(c)　　(d)

Fig. 11.6 The dominant material content of the early Universe: (a) hydrogen (b) helium (c) lithium and (d) deuterium. Grey circle = proton, black circle = neutron, white circle = electron.

is also water and D₂O, referred to as *heavy water*, is used in the nuclear power industry.

The proportion of hydrogen in the form of deuterium in the Universe is very variable from one location to another. For the Universe as a whole, the ratio D/H, *i.e.* the relative numbers of the two kinds of atom, is about 2×10^{-5} and this is also the ratio in Jupiter, which has retained typical Universe material in a completely unmodified way since its formation. However the ratio D/H is very different in other kinds of solar-system bodies, as indicated in Table 11.1.

The outstandingly high ratio in the table is that for Venus, which is due to a special evolutionary factor for that body, which also led to it being very arid. Because it is quite close to the Sun it had a considerable amount of water vapour in its atmosphere that was dissociated by solar radiation by the reaction

$$H_2O \xrightarrow{\text{radiation}} OH + H$$

or, if the water contained deuterium,

$$HDO \xrightarrow{\text{radiation}} OH + D.$$

Deuterium has twice the mass of hydrogen and so its average speed when released into the atmosphere was smaller than that of hydrogen

Table 11.1 Deuterium/Hydrogen ratios for the Universe and various solar-system bodies.

Body	D/H
Universe	2×10^{-5}
Jupiter	2×10^{-5}
Saturn	2×10^{-5}
Uranus	6×10^{-5}
Neptune	6×10^{-5}
Earth	1.5×10^{-4}
Venus	1.6×10^{-2}
Comets	3×10^{-4}
Meteorites	up to 10^{-3}

by a factor of $\sqrt{2}$. For this reason, hydrogen was able to escape more easily from the gravitational pull of Venus and deuterium was gradually concentrated on the planet. Two OH groups would eventually combine to give water plus oxygen by

$$OH + OH \rightarrow H_2O + O$$

and the oxygen would combine with surface materials on Venus — for example, with sulphur from volcanic activity that would give sulphur dioxide and eventually the sulphuric acid that is present in Venus's atmosphere. The net effect of all these reactions would be a gradual loss of water and an increase in the D/H ratio.

There are parts of the galaxy where D/H ratios are even higher than on Venus and these are in various hydrogen-containing molecules that form the icy grains in cold molecular clouds and in the protostars that form within such clouds. Thus in the cold dense cloud L134N, the ratio of doubly deuterated ammonia, NHD_2, to normal ammonia, NH_3, is 0.05, meaning that 3.4% of the hydrogen in all the ammonia in that cloud is deuterium.[2] In the protostar 16293E, the D/H ratio in the ammonia it contains is more than 0.02 and the amount of methanol, CH_3OH, which is either singly or doubly deuterated, *i.e.* CH_2DOH or CHD_2OH, actually exceeds that of normal methanol.[3] The D/H ratios in the *whole* molecular clouds and protostars are the value for the Universe but there is a concentration of deuterium in the small fraction of those bodies that is solid and consists of hydrogen-containing ices. The reason for this is a phenomenon known as *grain-surface chemistry*. Atoms of gas coming into contact with grains dwell on the surface for a while. If the atom is deuterium, then it may change places with a hydrogen atom in one of the molecules of the grain. The change from hydrogen to deuterium in the molecule lowers its energy and, since physical systems are more stable with lower energy, the deuterium transfer

---

[2] Roberts, H., Herbst, E. and Millar, T.J., 2003, *Astrophys. J.*, **591**, L41.
[3] Parise, B., Ceccarelli, C., Tielens, A.G.G.M., Herbst, E., Lefloch, B., Caux, E., Castets, A., Mukhopadhyay, I., Pagana, L. and Loinard, L., 2002, *Astronomy & Astrophys.*, **393**, L49.

will be permanent. Over the course of time, the icy materials become increasingly deuterated; the consequent fall in the D/H ratio in the gas will be small and undetectable since gas accounts for the great majority of the mass and a very small amount of its deuterium can heavily deuterate the icy grains.

A protoplanet, formed from protostar material by the CT process, is initially at a very low temperature and it remains low for much of the protoplanet's collapse. During this stage the solid grains of iron, silicate and ices sink towards the centre of the collapsing planet, eventually forming an iron-silicate core with a surrounding mantle of ices, many of which contain deuterium-enriched hydrogen. The planet as a whole will be mostly hydrogen with no deuterium enrichment, plus helium. When it has collapsed to its final state, the centre of the planet will be at a high temperature and the ices will have vaporized and become part of the atmosphere, although mostly concentrated at the base of the atmosphere because they consist of heavier molecules. The upper atmosphere of Jupiter consists of 86% molecular hydrogen (H_2), 14% helium (He), with slight traces of water (H_2O), ammonia (NH_3) and methane (CH_4). The amount of materials that were originally ices is so small in the atmosphere that they do not substantially affect the measured overall D/H ratio. The same is true for Saturn.

The upper atmosphere of Uranus is quite different from that of Jupiter, mainly in that it contains 2.3% of methane — the remainder being 82.5% molecular hydrogen and 15.2% helium. If the D/H ratio in the methane is 8×10^{-4}, a very modest ratio in the light of measurements in protostars and molecular clouds, then the overall D/H ratio in the atmosphere would be 6×10^{-5}, as observed. There is a slightly different, but generally similar, explanation for the observed D/H ratio on Neptune. The comparatively low masses of Uranus and Neptune plus the much higher proportions of heavier elements in their upper atmospheres suggest that the tidal events that affected the direction of their spin axes may also have stripped away substantial parts of their outer atmospheres, thus denuding them of much of their hydrogen and helium while little affecting the quantities of heavier materials concentrated in their interiors.

11.5. Earth and Venus

We have seen that the tilts of the planetary spin axes can be explained in terms of near passages between pairs of planets, made possible by the differential precession of the planetary orbits while they were rounding off and decaying. We shall now look at another possible outcome of differential precession, an actual collision of major planets. Could such an event occur? I can offer two reasons why it might. The first is the calculation that a colleague, John Dormand, and I made in 1977, in which we showed that in a system of six major solar-system planets with evolving orbits, the probability that a collision would occur between one pair of them before they reached their final orbits is about 0.1 — small but by no means negligible.[4] The second reason is that in August 2009 NASA reported that its Spitzer Space Telescope had picked up traces of the remnants of a collision between two planets orbiting a young star, an event that had taken place within the last few thousand years. If collisions between planets were extremely rare, then the chance of observing the consequences of such a recent event within the brief time — a few decades — that we have been able to make such observations would be negligibly small. There is actually a third reason that I am reluctant to call on for support — although many would say it is relevant. This is the so-called *anthropic principle* that asserts that since life, and in particular *homo sapiens*, exists, then any event in the Universe that is necessary for life to occur must have happened. I shall invoke a planetary collision to explain the existence of Earth, which existence is a necessary precondition for the existence of humankind. Personally I do not think that this argument gives much support to the occurrence of a planetary collision but I offer it as a kind of argument sometimes put forward to support various postulated scenarios. Indeed, it was used to oppose the objection put forward by Harold Jeffreys to the Jeans tidal theory — that it was an extremely unlikely occurrence.

---

[4]Dormand, J.R. and Woolfson, M.M., 1977, *Mon. Not. R. Astr. Soc.* **180**, 243.

To set the scene for the planetary collision, we shall take an initial Solar System containing six major planets, as John Dormand and I postulated in 1977. These are the four major planets that exist today, plus two others that we shall call Bellona, the Roman goddess of war, and Enyo, her Greek counterpart. The characteristics of the two extra planets are given in Table 11.2.

The larger of the two additional planets, Bellona, has a mass about twice that of Jupiter, but comfortably within the range of masses observed for exoplanets. Enyo has a mass some 22% greater than that of Saturn. The two planets were taken to have collided in the inner part of the Solar System at about 1 au from the Sun, the distance of the Earth. An impression of the orbits is given in Fig. 11.7(a). For a planet on a very eccentric orbit at 1 au from the Sun, the orbital speed is of order $40\,\mathrm{km\,s^{-1}}$ and the approach speed of the planets when they are at some distance from each other, taking into account their directions of travel, is taken as $49\,\mathrm{km\,s^{-1}}$. Since the planets attract each other gravitationally, a calculation shows that their impact speed is $80\,\mathrm{km\,s^{-1}}$. Seen from the point of view of the centre of mass of the two planets (Appendix H), the planets approach on parallel paths with an offset (Fig. 11.7(b)), which is taken as $7 \times 10^4\,\mathrm{km}$ in the collision simulation to be described.

The progress of the collision is displayed as a SPH simulation in Fig. 11.8 where the planetary configurations are shown at approximately 500-second intervals. The main interest in this simulation is to follow the progress of the iron and silicate cores and hence, in modelling the planets, particles were distributed much more densely

Table 11.2 The characteristics of the planets Bellona and Enyo.

	Bellona	Enyo
Mass (Earth units)	617.55	116.35
Radius (km)	8.582×10^4	6.050×10^4
Central density (kg m^{-3})	162,000	98,000
Central temperature (K)	76,000	48,000
Mass of iron (Earth units)	2.75	1.875
Mass of silicate (Earth units)	11.00	7.50

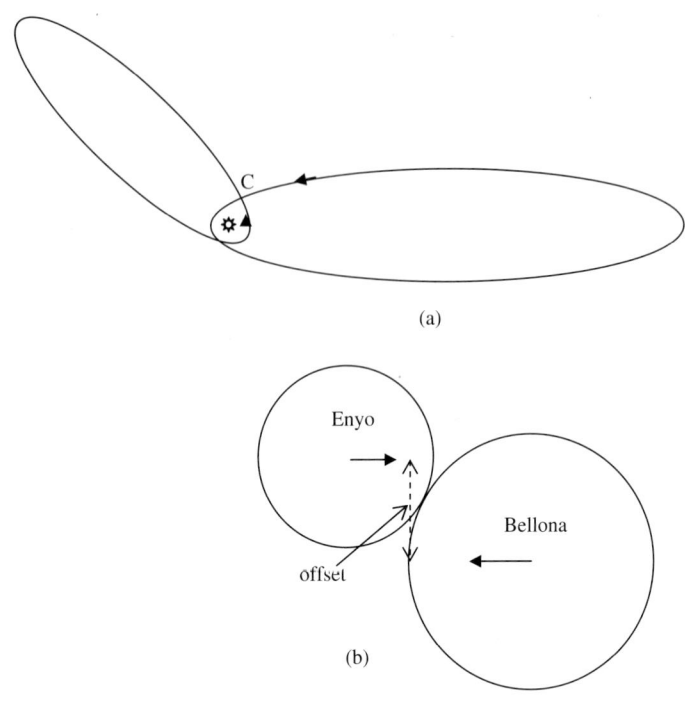

Fig. 11.7 (a) Planets approaching a collision point C. (b) The collision seen from the centre of mass.

in the planetary interiors, which enabled their behaviour to be followed in greater detail.

Frame (a) of the figure shows the planets when they first made contact. Frames (b) and (c) show heavy distortion of the planetary atmospheres with some material being sprayed outwards from the collision interface. At frame (d) the silicate region of Enyo becomes compressed and heated and at this stage something important happens. The upper part of the silicate region will be heavily impregnated with ices that have a high D/H ratio, which we take as 0.02, and in the simulation the temperature of the silicate region is found to rise to over 3×10^6 K (Appendix AD).

Nuclear reactions can occur when matter is raised to a high temperature and will occur more readily when the density of the matter is high. Rates of nuclear reactions are extremely sensitive to temperature and are proportional to the density of the involved materials. For most nuclear reactions the temperatures required are in the region of several hundred million degrees but there is a notable exception: reactions involving a pair of deuterium atoms, as given in (3.1), can occur at a few million degrees, in fact, at temperatures reached in the collision simulation. A product of this reaction is the radioactive isotope of hydrogen, tritium, and this too interacts with deuterium by the reaction

$$^2_1D + ^3_1T \rightarrow ^4_2He + ^1_0n \qquad (11.1)$$

in which the products on the right-hand side are the most common isotope of helium and a neutron. The important thing about these reactions is that they release considerable amounts of energy. If one-third of the atoms in the ice-impregnated silicates are hydrogen and if 2% of that hydrogen is deuterium, then the temperature attained can be over 100 million K (Appendix AE), at which temperature other reactions involving heavier elements can occur. The consequences of these further reactions will be considered in Chapter 16.

From frame (e) onwards, matter is thrown out violently from the region of the nuclear explosion, involving all the gas and also considerable amounts of the silicate and iron. The residues of the solid cores of the planets quickly move apart into independent orbits around the Sun and, since the medium around the Sun is still in place, these orbits will eventually round off and decay.

The parameters used in the simulation of this planetary collision are purely speculative and the aim of the exercise is to show that a scenario of this general type is feasible. The core residue of Bellona, the larger colliding planet, is more massive than that of Enyo so it is natural to identify these cores as Earth and Venus respectively. By placing the collision at 2 au and orienting the collision in a particular way, the orbits of the residues can be such that they round off approximately to where Earth and Venus are today.

The sideswipe collision of Bellona and Enyo would impart angular momentum to both of them, affecting the spin of the Earth and Venus residues. In addition, if one or both of the colliding planets had previously been involved in a close interaction, they would have

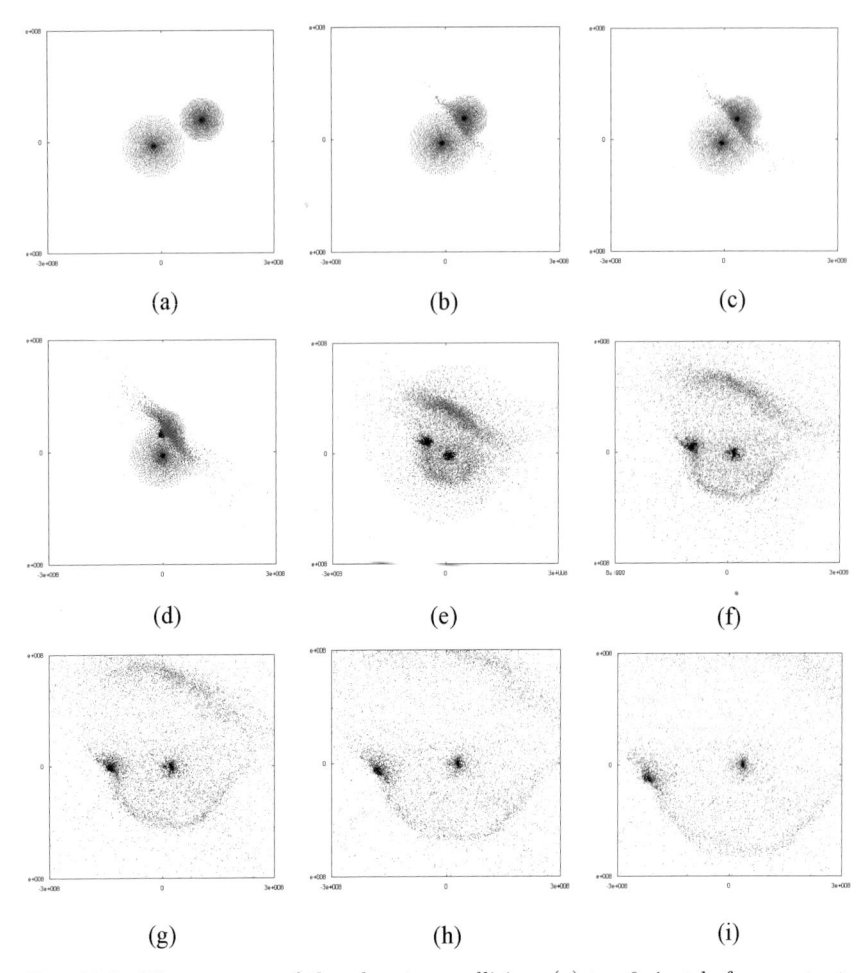

Fig. 11.8 The progress of the planetary collision. (a) $t = 0$, just before contact, (b) $t = 501$ s, (c) $t = 1001$ s, (d) $t = 1501$ s, (e) $t = 2003$ s, (f) $t = 2511$ s, (g) $t = 3004$ s, (h) $t = 3502$ s, (i) $t = 4005$ s.

collided with pre-existing spin-axis tilts. From Fig. 11.7b it can be seen that the offset collision gives retrograde spin to both Bellona and Enyo. A combination of an initial prograde spin for Enyo, plus the retrograde addition due to the collision, could account for the slow retrograde final spin of Venus. For the more massive Bellona that, like existing gas giants, could have had a spin period of about

10 hours, the retrograde impulse would just have slowed down the spin, so giving the slower spin of the residue that formed Earth.

If the postulate of a planetary collision explained the presence of Earth and Venus and nothing else, it would still be interesting but, because of its *ad hoc* nature, would carry little conviction. But, as we shall see, it explains much more.

11.6. Summary

Because of the gravitational effect of the resisting medium surrounding the Sun, the evolving orbits of the planets underwent precession, all at different rates. The eccentric and slightly inclined orbits intersected from time to time and gave the possibility of close interactions between pairs of planets. The tidal effects of such interactions explain the tilts of the planetary spin axes, in particular the curious axial tilt of Uranus.

Within cold molecular clouds, the birthplaces of stars, the ratio of D/H can be greatly enhanced in icy grains due to the influence of grain surface chemistry. This enhanced D/H ratio is present in the grains in protostars and in compressed regions within the cloud, both of which can be sources of planets due to capture-theory interactions. The loss of a large part of the original atmospheres of Uranus and Neptune can explain the observed enhanced D/H ratios of these planets compared with what is found for Saturn and Jupiter.

A collision is described between two postulated initial planets of the Solar System — Bellona, with twice the mass of Jupiter, and Enyo, with mass some 22% greater than that of Saturn. When the high temperature region of the collision reaches icy material close to the core of Enyo, in which the D/H ratio is taken to be 0.02, a nuclear explosion takes place, initially involving D-D reactions, that raises the temperature at the heart of the collision above 10^8 K.

The nuclear explosion disperses most material of the planets but residues of the solid cores survive and are identified as Earth and Venus. The collision takes place at about 1 au from the Sun and the final orbits of the core residues, after decay and round-off, are in the terrestrial region of the Solar System.

Chapter 12

The Moon

12.1. The Earth–Moon Relationship

In the hierarchy of solar-system satellites, the Moon is the fifth in terms of mass, with just under one-half that of the most massive satellite, Ganymede. The majority of the large satellites accompany Jupiter and Saturn. The satellites of the other major planets have masses considerably less than that of the Moon; Triton, the largest satellite of Neptune, has about one-third of the Moon's mass. Planetary scientists have often commented on the Earth–Moon relationship as being anomalous. Whatever models have been proposed concerning the formation of planets and satellites, it has been generally accepted that the formation processes of the two kinds of body are linked in some way and that it is probable that the most massive planets should have the most massive satellites. That would certainly be the conclusion one would draw from Fig. 6.2, which shows the formation of a disk, the source of satellites, around the collapsing protoplanet. It seems natural that the more massive the protoplanet, the greater the mass of the disk and the satellites that form from it. The Moon clearly does not satisfy the criteria for being a regular satellite so the question that planetary scientists have asked is: 'How did a small terrestrial planet like Earth acquire such a large satellite?'

In 1878 George Darwin (1845–1912), one of Charles Darwin's sons, suggested that, due to the Sun, the early Earth was subjected to tidal fluctuations that had a period equal to the Earth's natural frequency of oscillation. Because of this the amplitude of the Earth's oscillation built up by a resonance effect until instability set in and a chunk of the Earth, from the region that is now the Pacific Ocean, broke away and moved around the Earth in a close orbit. The original period of the orbit was about $5\frac{1}{2}$ hours but the tidal effects between the newly formed Moon and the Earth have caused the Moon to retreat slowly from Earth to where it is today.

The mechanics of the retreating Moon is sound and, even now, the Moon is slowly retreating from the Earth. Figure 12.1 shows an idealized representation of the tides drawn up on Earth by the Moon. These tides occur both on the side of the Earth facing the Moon and also on the far side. The tides do not precisely point towards and away from the Moon because the Earth's period of spin is less than the Moon's orbital period so, through friction, the peaks of the tides are dragged forward in the direction of the Earth's spin.

The main bulk of Earth gives a centrally-directed force on the Moon and the tidal bulges, both near and far, also make a contribution to the centrally directed force. However, the forces due to the tidal bulges also give force components perpendicular to the line connecting the centres of the two bodies and, since the force due to the near bulge is the larger, the net perpendicular force component due to both bulges is in the direction of motion of the Moon. Hence there is a torque that acts to increase the angular momentum of the Moon's orbit. For an orbit of semi-major axis a and eccentricity e, the angular momentum is given by

$$A = M_m M_\oplus \sqrt{\frac{Ga(1-e^2)}{M_m + M_\oplus}} \qquad (12.1)$$

in which M_m is the mass of the Moon and $M_\oplus$ that of Earth. It is clear that for a circular orbit ($e = 0$), an increase in A gives an increase in a, i.e. the Moon retreats from Earth.

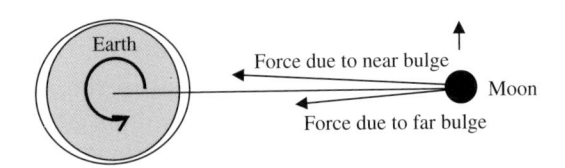

Fig. 12.1 Forces on the Moon due to tidal bulges on Earth.

For a non-circular orbit, Eq. (12.1) offers another way of increasing A, which is by reducing e to make the orbit more circular. This does not actually happen. As when we dealt with the effect of a resisting medium in Chapter 7, we can find the way in which the orbit is being modified by just considering what is happening at perigee and apogee. At perigee the force in the direction of motion increases the semi-major axis and also increases the eccentricity. At apogee the force in the direction of motion again increases the semi-major axis but this time reduces the eccentricity. At both extremes the semi-major axis is increased but the effects on eccentricity are in opposite directions. However, the influence at perigee is stronger; not only does the closer distance increase the size of the tidal bulge, it also increases the gravitational pull of the bulge on the Moon. For that reason, the effect on the Moon is that it retreats from the Earth with a slow increase in the eccentricity of its orbit.

Many difficulties with Darwin's theory were raised. A notable one was that periodic tidal stretching would give internal dissipation of energy within the Earth, converting mechanical energy into heat, as we now know happens within Io, and this would have prevented the build-up of oscillations to the point of instability.

Another idea that has been suggested is that the Moon was a body in independent motion around the Sun and was captured when it happened to closely approach the Earth. Conservation of energy is one of the basic laws of physics from which it follows that two bodies, isolated from all other bodies, approaching each other from a large distance, must depart from each other to a similar large distance unless some of the energy of the two-body system is converted into another form. Two bodies in independent orbit around the Sun are not isolated — the Sun is there — and it is also possible that a tidal interaction between the bodies could convert some mechanical energy into heat. These factors make capture just about feasible but the conditions for it to happen are so restrictive that there is a reluctance to accept the idea if anything more plausible presents itself.

Darwin's solution to the anomalous Earth–Moon relationship was a proposal that the Moon was formed in a different way from other satellites. A recent idea, which has received wide support from the astronomical community, is similar in that respect. It proposes that the Moon was formed as a result of a collision between Earth and a body with about the mass of Mars. The American planetary scientists W.K. Hartmann and D.R. Davis first suggested this idea in 1975 and it was successfully modelled by smoothed-particle hydrodynamics by W. Benz, W.L. Slattery and A.G.W. Cameron in 1986.[1] In their simulation the body struck the Earth obliquely, shearing off some of the Earth's mantle that mingled with the debris from the body and orbited the Earth. The Earth recaptured some of the orbiting material but other parts of it collected together to form an Earth satellite (Fig. 12.2).

The satellite produced by this mechanism contained little iron and was initially molten, characteristics that agree with what is expected for an early Moon. There are a few problems with this model, but not sufficiently severe to affect its basic plausibility. Although the Moon is iron-poor it does have an iron core just less than 400 km in radius.[2] The impact model, as presented, gives a Moon with too little iron but there is the possibility that different impact parameters might give more iron. As previously mentioned, it implies a mode of formation for the Moon different from that of the other large satellites although the Moon is similar to other large satellites in its general characteristics — for example, in being intermediate in both mass and density between Io and Europa.

Now I am going to show another way that the Earth–Moon relationship could have come into existence — one that follows from the description of the capture theory and the subsequent planetary collision, and that gives the same origin for the Moon as for other major satellites.

12.2. Satellites of the Colliding Planets

The two colliding planets, Bellona and Enyo, would, like the existing major planets of the Solar System, have been richly endowed with

---

[1] Benz, W., Slattery, W.L. and Cameron, A.G.W., 1986, *Icarus*, **66**, 515.
[2] Stock, J.D.R. and Woolfson, M.M., 1983, *Mon. Not. R. Astron. Soc.*, **202**, 287.

The Moon 207

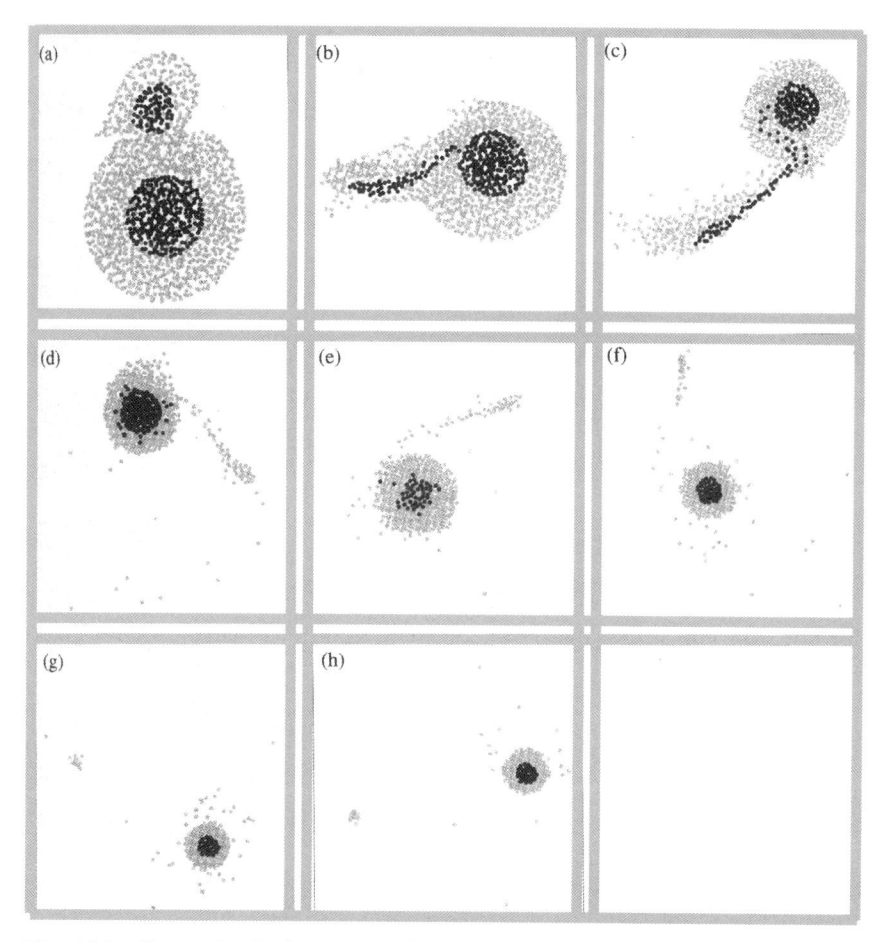

(a) (b) (c) (d) (e) (f) (g) (h)

Fig. 12.2 Stages in the formation of the Moon by the collision of a Mars-mass body with Earth. The scale of successive frames is reduced to show the formation of the Moon.

satellites. In particular Bellona, with twice the mass of Jupiter, would be expected to have had satellites that exceeded those of Jupiter both in number and mass. Logically there are five possible outcomes for the fate of these satellites following the collision:

(i) Released from the gravitational attraction of its planet, the satellite went into an independent orbit around the Sun.

(ii) As in (i), except that the satellite left the Solar System completely.
(iii) The satellite remained in orbit around the residue of its parent planet.
(iv) The satellite went into orbit around the residue of the other planet.
(v) The satellite was disrupted by the collision with the debris having one or other of the outcomes (i) to (iv).

Inserting a satellite in various situations into the smoothed-particle simulation that gave Fig. 11.8 shows that all outcomes (i) to (iv) are possible. We shall see that the satellites of Bellona and Enyo play an important role in giving the Solar System we see today.

12.3. Features of the Moon

The Moon, our nearest neighbour in space, has been closely studied for the last 400 years, ever since Galileo acquired his telescope. Even before the space age, its near surface had been imaged through powerful land-based telescopes and the lunar landscape was as familiar to astronomers as the surface of the Earth itself. The Moon presents one face towards the Earth although, due to a phenomenon known as *libration*, about 59% of its surface is visible from Earth at one time or another. The Moon's orbit around the Earth is elliptical with an eccentricity of 0.055 and for this reason its angular speed around the Earth varies, being greater when it is at *perigee* (closest to Earth) than at *apogee* (furthest from Earth). However, because of its large inertia the spin rate of the Moon about its axis is constant, so, as seen from Earth, during a lunar month the Moon will appear to rock slightly to-and-fro, exposing more than one-half of its surface during that period.

The model that we have for the formation of the Moon is by the gradual accumulation of satellitesimals (Section 9.5). As the satellite grew, so did its gravitational attraction and consequently satellitesimals struck the surface with ever-increasing speed. The energy of the impacting satellitesimals was transformed into heat and hence the temperature of the layers of the satellite increased outwards as it grew. Eventually the energy of impact was sufficient to melt the material and so a body the size of the Moon would have had a layer of molten material close to its surface. Over time, heat flows within and outwards from the Moon would have given the present temperature profile with temperature increasing inwards and the temperature at the centre making the material there partly molten (Appendix AF). For a few hundred million years, molten material was close enough to the surface to give the episodes of volcanism that created the mare but these gradually subsided. During the period of thermal transport within the Moon the fluid region would have moved towards the centre, so assisting the transport of denser iron inwards to form a core. The differentiation of material by density gave the Moon a composition of a crust with an average thickness of about 100 km consisting of low-density silicates, a mantle of heavier silicates, and a core of radius about 400 km consisting mostly of iron but with some admixture of nickel and sulphur contained in the mineral *troilite*, FeS.

The surface of the Moon, as shown in Fig. 12.3 shows several different kinds of terrain. The bright regions are the *highlands*, the original crust heavily covered with craters from an extended period

Fig. 12.3 The near side of the Moon (NASA).

of heavy bombardment when the Solar System was full of debris from the planetary collision. The darker, approximately circular, regions are *mare basins*, places where very large projectiles excavated what were essentially huge craters that became filled with molten magma coming from the lunar mantle, seeping through cracks in their floors. The mare basins did not fill completely with magma because the mantle material has a density of $3,300 \, \text{kg} \, \text{m}^{-3}$, higher than that of the crustal material, $2,900 \, \text{kg} \, \text{m}^{-3}$. The height of the magma within a basin is, in general, governed by the principle of *isostacy*. This principle, illustrated by the schematic basin with magma fill shown in Fig. 12.4, states that at some lower level in the body, known as the *compensation level*, the pressures should be the same everywhere. Hence it is clear that the surface of the dense magma must be below the level of the rim of the basin.

When the motions of spacecraft orbiting the Moon were analysed, it was found that they accelerated when approaching a mare and decelerated when departing from a mare region. This was not expected. If isostacy were in operation, a spacecraft moving well above the surface would not experience any significant acceleration or deceleration as it passed over a mare basin. It seemed that there was extra mass concentrated within the mare basins and these enhancements of the local gravitational field were referred to as *mascons*. The reason for a mascon is that within the mare basins the magma fill is higher than that required for isostatic equilibrium. In 1989 John Dormand and I[3] explained mascons as follows. From

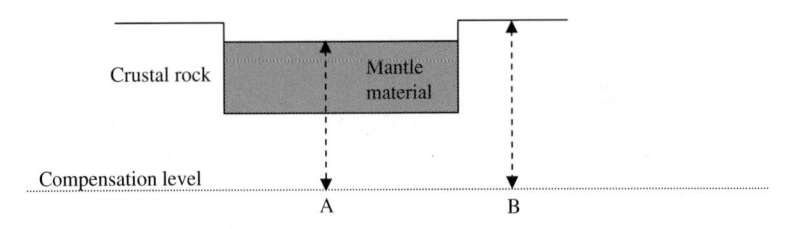

Fig. 12.4 The principle of isostacy requires the pressure at A and B to be the same.

---

[3]Dormand, J.R. and Woolfson, M.M., 1989, *The Origin of the Solar System: The Capture Theory* (Chichester: Ellis Horwood), pp. 272–276.

thermal evolution results, described in Appendix AF, it appeared that a solid thick crust appeared on the Moon quite quickly. As the crust cooled, it shrank and so exerted a pressure on the material below that would itself be heating up due to radioactivity within the crust and mantle, thus increasing the pressure further. The cracks through the crust, produced by the projectiles that created the mare basins, provided a route for material to exude from the interior, so reducing the pressure there. The consequent filling of the basins thus was higher than that for giving isostatic equilibrium. The magma came through the cracks like toothpaste being squeezed out of a tube, the shrinkage of the crust and the expansion of the matter contained within the crust providing the squeezing force. This model was tested numerically by Andrew Mullis, who found good agreement between the theoretical and observational values for mascon anomalies.[4]

Seen towards the bottom of Fig. 12.3 is the prominent crater *Copernicus* and emanating from it are several bright *rays* consisting of finely divided material thrown out of the crater. Less obvious at the scale of the figure are *rills*, long and deep cracks in the surface, some as large as the Grand Canyon.

When only ground-based images of the Moon were available, it was always assumed that the lunar surface seen from the Earth was a fair sample of the whole surface. Then in 1959 the Soviet Union began launching a series of spacecraft aimed at the Moon and the third of these, Luna 3, returned photographs of the rear side that had never been seen before. The pictures were of poor quality, but startling for all that since they showed that the rear face of the Moon is very different from the side we see from Earth. Figure 12.5 shows both the near side and the far side, from which the difference of the two sides is apparent. The far side is virtually all highlands with just one mare feature of any size, *Mare Moscoviense*. This then raised the important question of why the two sides of the Moon were so different. The Moon has hemispherical asymmetry — but why? That was a perplexing question and the impact model of Moon formation (Fig. 12.2) offers no obvious explanation for this feature.

---

[4]Mullis, A.M., 1991, *Geophys. J. Int.*, **105**, 778; 1992, *Geophys. J. Int.*, **109**, 223.

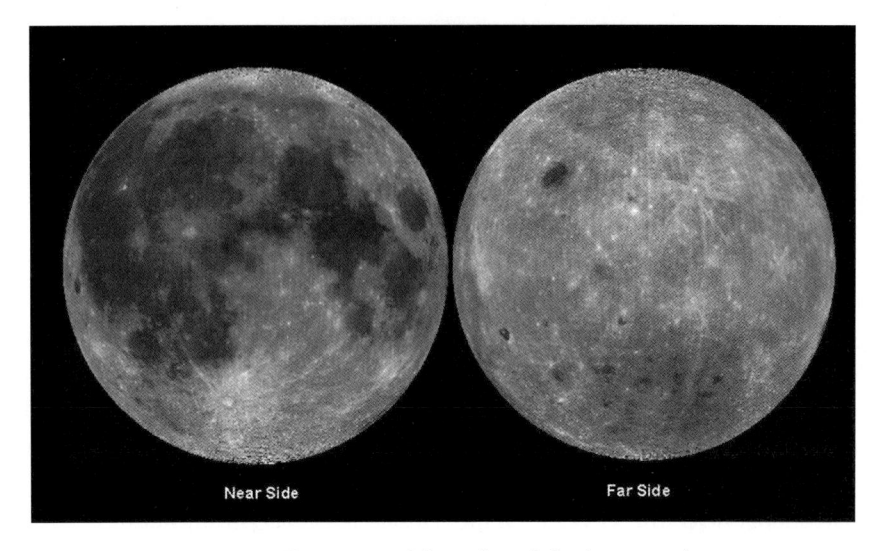

Fig. 12.5 The near and far sides of the lunar surface.

12.4. The Hemispherical Asymmetry of the Moon

Since mare basins are impact features, the most obvious explanation for the difference of the two sides is that, for some reason to be determined, the far side had not been bombarded to the same extent. One early theoretical idea was that the Earth behaved like a gravitational lens so that projectiles coming towards the Moon from the direction of the Earth were focused onto its surface. Mathematical analysis did not support this theory. It was further discredited when a lunar orbiter, carrying an altimeter with which it produced a profile of the Moon's surface, showed that there were as many large basins on the far side as on the near side — they just had not filled up with magma from below.

Since there *were* basins on the far side and they had not filled with magma, then it followed that, with the exception of Mare Moscoviense, the lava had been too far below the bottom of the basins to percolate through to the surface. Hence, *ipso facto*, if the basins were of the same depth on both sides of the Moon, then the crust must have been thicker on the far side. The Apollo missions to the Moon left seismometers on its surface, very sensitive instruments that could detect moonquakes, albeit mostly very feeble by terrestrial standards. However, some very shallow moonquakes have been recorded with a magnitude of 5.5 on the Richter scale, which is quite powerful and would cause structural damage as an earthquake. Analysis of the data from the lunar seismometers has enabled the internal structure of the Moon to be deduced, and this has confirmed that the crust on the near side, with a thickness of roughly 60 km, is indeed thinner than on the far side where the thickness is about 100 km. This provides the answer to the problem of the hemispherical asymmetry of the Moon. However, the difference in crustal thickness on the two sides of the Moon provides not just an answer to a question but also a question to be answered. Why is the crust thinner on the near side? This question is made more pertinent by the fact that a fluid or plastic satellite formed in synchronous orbit around a planet, so that one face always points towards the planet, should have a *thicker* low-density crust on the near side — not thinner. The appearance of the crust as it would form initially is shown in Fig. 12.6.

The planetary collision scenario offers a solution to this conundrum. The Moon as a synchronous satellite of a colliding planet — probably Bellona — would present one face towards the scene of the collision. With a planetary impact speed of $80\,\mathrm{km\,s^{-1}}$, the debris from the collision would fly out at a speed greater than that; experiments at NASA's Ames Research Center in California indicate that the debris from hypervelocity impacts can travel at up to three times the impact speed of the colliding bodies. To get an idea of the progress of this event, if the debris were moving at,

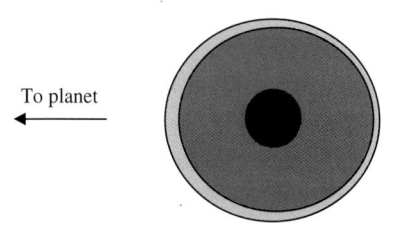

Fig. 12.6 The initial structure of a satellite formed in synchronous orbit around a planet, showing core (black), mantle (dark grey) and crust (light grey).

say, $150\,\mathrm{km\,s^{-1}}$ and the Moon's orbital radius were 10^6 km, then the debris would have arrived at the satellite in less than 2 hours after the impact took place, during which time the Moon would have rotated by about $6°$. Hence the face being bombarded would be, for the most part, just that facing Bellona.

Now we must consider the effect of such a bombardment on the lunar surface. For any astronomical body there is an *escape speed*; any material leaving the surface of the body at the escape speed or greater will depart from it and never return. If it leaves the surface at less than the escape speed, it will rise to a certain distance above the surface and then fall back. The escape speed from the Moon is $2.4\,\mathrm{km\,s^{-1}}$. Another aspect of the escape speed is that any projectile striking the body and coming from a large (theoretically infinite) distance, no matter what its initial speed was, would, by virtue of being attracted by the body, strike it with at least the escape speed. Figure 12.7 shows the effects of both slow and fast projectiles striking the surface of a solid astronomical body. The energy of the projectile is converted into various forms. Some of it is converted into heat

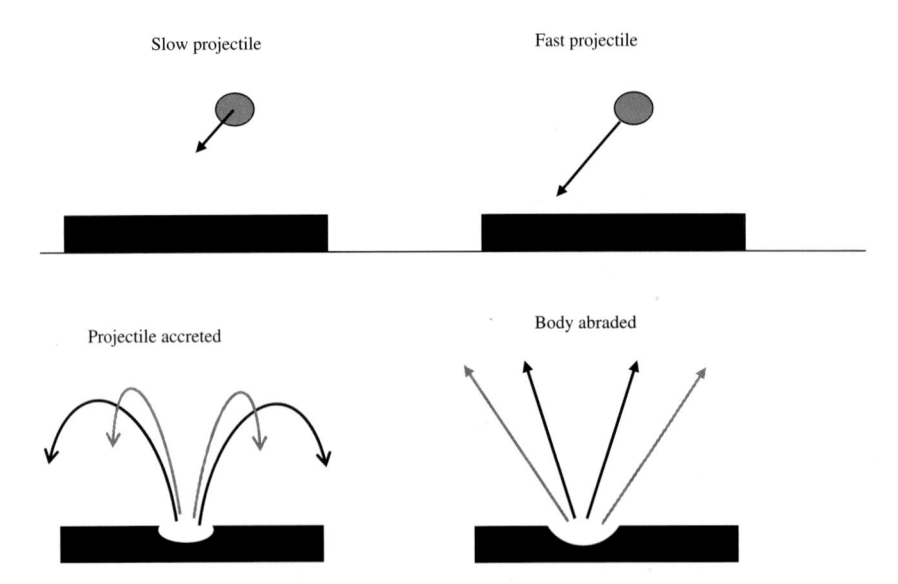

Fig. 12.7 The effect of slow and fast projectiles falling onto a solid surface.

and may even melt part of the surface. Some will be expended in breaking up surface material and, finally, some will be imparted as kinetic energy to fragments of surface material. If the impact speed is little more than the escape speed, the speed of the fragments will all be less than the escape speed and they will all fall back onto the surface. The projectile will have been *accreted*. Alternatively, if the impact speed were much greater than the escape speed, many of the fragments would themselves be given speeds greater than the escape speed and so depart from the surface never to return. The net result would be a loss of material from the surface, which would be *abraded*.

In the case of a projectile moving at $150 \, \text{km} \, \text{s}^{-1}$ striking the Moon's surface, if all the energy were transformed into kinetic energy of surface material, the projectile could abrade up to 4,000 times its own mass from the surface. In practice it would be much less but still there would be considerable abrasion. Given that the crust facing Bellona would have been thicker than that on the far side, in order for it to finish some 40 km thinner might have required the removal of, say, 60 km of crust material by abrasion. In Appendix AG it is shown that for the model of the planetary collision we are considering, this is a reasonable outcome.

12.5. The Evolution of the Moon's Orbit

I have already mentioned the regression of the Moon's orbit away from Earth due to the mechanism illustrated in Fig. 12.1. The semi-major axis of the Moon's orbit is increasing by about 3.8 cm per year and as it retreats, so the eccentricity of the orbit slowly increases; the present eccentricity is 0.055. This increase of the semi-major axis increases the angular momentum associated with the Moon's orbit, and to conserve angular momentum in the Earth–Moon system, angular momentum has to be lost from elsewhere. This comes from slowing the spin of the Earth, which lengthens the day. This process in which the month lengthens and the day lengthens would, in the absence of the Sun that is also a tide-raising body on Earth, eventually make both the day and month each equal to about 50 days.

At this stage the Moon and Earth would be synchronously locked together, each always presenting the same face to the other, and the Moon's orbit would thereafter cease to evolve further. Of course, it will not happen precisely like that because of the Sun's influence.

When Bellona and Enyo collided, they both suddenly changed their velocities and so our knowledge of the resultant velocity of the Moon relative to the Earth fragment is only constrained by the necessity that the Moon is retained in orbit around the Earth. It is very likely that the initial orbit, once the collision was over and the Earth and Venus residues well separated, would have been highly eccentric; a near-circular orbit would require a rather special relationship between distance and velocity that would be unlikely to occur just by chance. This being so, the same considerations would apply as for Io's eccentric orbit around Jupiter — tidal energy would have been generated in the Moon. Although the mass of the Earth is tiny compared to that of Jupiter, this is compensated for by the much larger expected eccentricity of the Moon's initial orbit so that the tidal energy generated in the Moon would have been quite significant.

It is shown in Appendix AH that the effect of tidal energy production in the Moon is to cause its orbit both to round off and decay. The energy of the Moon in orbit around the Earth is

$$E = -\frac{GM_m M_\oplus}{2a}, \qquad (12.2)$$

the negative sign indicating that the bodies are bound together. If energy is lost from the system by being tidally generated within the Moon, then, since that energy has to come from somewhere, it comes from the orbital energy. For this reason E decreases in value but increases in magnitude so that a must fall, representing decay of the orbit. For this mechanism there is no change in the angular momentum of the system. Hence, it will be seen from Eq. (12.1) that if A is to remain constant while a falls in value, then e must also become less — *i.e.* the orbit rounds off. As it does so, the rate of tidal energy production gradually falls and eventually reaches the level at which it equals the rate that energy is being pumped into the orbit by the process illustrated in Fig. 12.1. Thereafter, the Moon gradually retreats from the Earth with slowly increasing eccentricity.

This description of the Moon's orbital evolution is, of necessity, a simplified one since there are other factors that have not been taken into account — for example, the influence of the Sun and continental drift that will change Earth's moment of inertia. Nevertheless, it gives a good general account of the changing relationship between Earth and the Moon following the collision.

The tidal effect of Bellona on the Moon, giving it a thicker crust on the near side, has previously been mentioned. It also made the Moon slightly pear-shaped, a shape it retains despite abrasion, with the pointed end towards the Earth. Another feature of the mass distribution in the Moon, due to the removal of crust, is the offset of its centre of mass relative to the *centre of figure* — the position the centre of mass would have if the Moon had its present shape but a uniform density. This shift is about 2.5 km towards the Earth. The distribution of mass in the Moon, in particular its overall shape, is a major factor in ensuring that when the orbit became synchronous it gave the Earth–Moon configuration we have today.

At the beginning of this chapter I presented the question that planetary scientists asked about the Earth–Moon system: 'How did a small terrestrial planet like the Earth acquire such a large satellite?' The scenario described in this chapter is the answer to a somewhat different question: 'How did a large, normal satellite like the Moon become associated with such a small planet?'

12.6. Summary

There have been many proposed solutions to the problem of how the Earth–Moon relationship was established. One early suggestion involved large resonance oscillations of the Earth, caused by periodic tidal effects from the Sun, leading to instability and a portion of the Earth breaking off to form the Moon. Another suggestion was that the Moon, in an independent heliocentric orbit, closely approached the Earth and was captured. There are difficulties associated with both ideas. A more promising model, which has been investigated by numerical modelling, is that the Moon formed from debris created around the Earth when a Mars-mass body struck it obliquely.

A notable characteristic of the Moon is its hemispherical asymmetry. The near side has large numbers of mare features occupying a high proportion of the area seen, while the far side is predominantly highlands. The maria were produced by large projectiles that excavated basins which then filled with magma from below the surface. Due to the forces produced by the shrinkage of the Moon as it cooled, the maria overfilled with magma, above the level required by the principle of isostacy, to produce the mascon phenomenon. Altimeter determination of the lunar surface profile shows that large basins exist on the far side but did not fill with magma, the inference from which is that the lunar crust is thinner on the near side than on the far side. This conclusion is verified by the analysis of lunar seismic data that enables the internal structure to be determined. This situation with respect to crustal thickness is made more curious by the fact that a satellite in synchronous orbit should have formed a *thicker* crust facing its planet.

The colliding planets would, like the existing major planets, have had numerous large satellites. The Moon, as a satellite of, say, Bellona, would have received debris from the collision within two hours of the event and the bombarded surface would have been that facing Bellona. The debris would arrive at the lunar surface with a speed much greater than the escape speed from the Moon and this would have given abrasion of the surface. This abrasion removed sufficient crust from the lunar near side to explain the observed thinning of the crust there and hence the hemispherical asymmetry.

The orbit of the Moon around the residue of Bellona that constituted Earth would at first have decayed and rounded off and then later, through tidal interaction with Earth, slowly moved further away with a small increase of eccentricity.

Chapter 13
Mars and Mercury

13.1. Larger Solid Bodies of the Solar System

Earth and Venus are by far the largest and most massive solid bodies of the Solar System, bodies for which gaseous atmospheres form a very small part of the total mass. The other solid bodies in the category that we shall designate as 'larger' are the planets Mars and Mercury, the seven most massive satellites — Ganymede, Titan, Callisto, Io, Moon, Europa and Triton — and the five dwarf planets listed in Table 10.5. A plot of the masses and densities of these bodies is given in Fig. 13.1 and appears to show distinct differences between the three types. Planets are distinguished by having both densities higher than $3,800 \, \text{kg m}^{-3}$ and masses greater than $2 \times 10^{23} \, \text{kg}$. However, the apparent mass distinction between satellites and dwarf planets is a false one since only the most massive satellites have been selected. Including satellites such as Rhea, Iapetus, Titania and Oberon in the figure would have embedded the masses of dwarf planets among those of satellites.

Although it has been possible from the information in Fig. 13.1 to make a clear distinction between planets and the other kinds of bodies, another kind of relationship can be seen in Fig. 13.2 that just shows the densities of the bodies. When displayed in this way a new relationship becomes apparent — that Mars is more akin to satellites and dwarf planets in its density than it is to the other three planets. Is there any significance in this density relationship? The densities of the larger terrestrial bodies are affected by the effect of compression due to self-gravitational forces that increase their densities. The uncompressed mean density of Venus is

220 *On the Origin of Planets*

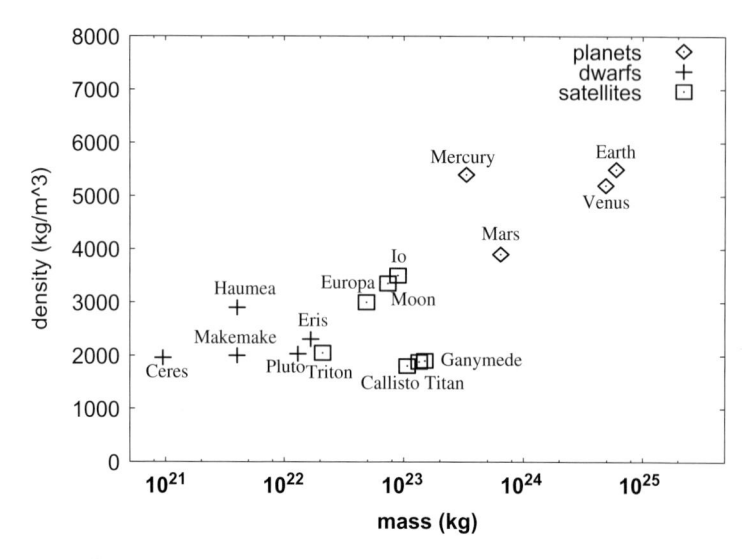

Fig. 13.1 The masses and densities of the 16 most massive solid bodies in the Solar System.

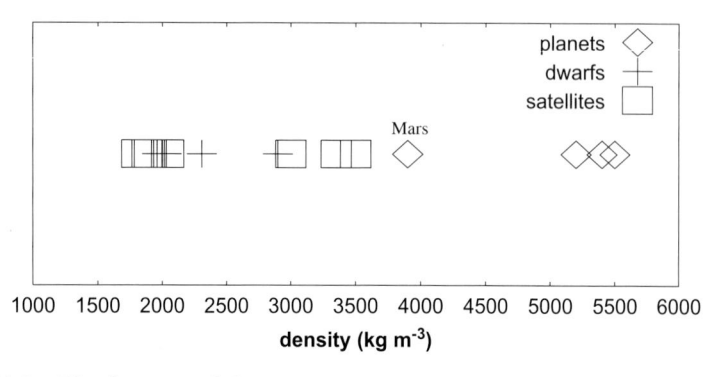

Fig. 13.2 The densities of the 16 most massive solid bodies in the Solar System.

4,400 kg m^{-3} compared to its compressed value of 5,200 kg m^{-3}, and the uncompressed density of Mars, 3,740 kg m^{-3}, is also less than its compressed value of 3,900 kg m^{-3}. Taking the uncompressed density values greatly reduces the gap in the progression of densities seen in Fig. 13.2 but, even so, the uncompressed density of Mars is much closer to that of Io, 3,500 kg m^{-3}, than it is to the uncompressed density of Venus. Mars also has a mass closer to that of Ganymede, a factor of 4.3 greater, than to that of Venus, a factor of 7.6 less. There is just a suggestion here — no more — that Mars may have more in common with the satellites of the Solar System than with the two largest terrestrial planets.

When we consider Mercury the message from its mass and density is a mixed one. Its uncompressed density, $5,300\,\text{kg}\,\text{m}^{-3}$, is considerably greater than that of either Venus or Earth. If high density is taken as a basic characteristic of a solid *planet*, as distinct from other types of solid body, then it is more planet-like than either of the large terrestrial planets. By contrast, its mass, just 2.2 times that of Ganymede and only half that of Mars, is more akin to that of satellites.

These comparisons are not conclusive but do suggest the possibility that both Mars and Mercury are sufficiently different from the larger terrestrial planets to indicate that they might have a different origin. I am now going to describe different origins for both these bodies — origins that have the desirable characteristic that they explain how these bodies acquired their distinctive properties.

13.2. Mars as a Satellite

The larger of the colliding planets, Bellona, is postulated to have had twice the mass of Jupiter, well within the range of observed exoplanet masses and by no means towards the upper end of the range. Now we consider the possibility that this planet, more massive than any existing solar-system planet, had satellites, more massive than any existing satellite. If Mars had been such a satellite, the possibility that it could have escaped into an independent heliocentric orbit has been confirmed by numerical calculations. Are there any indications that this could actually have happened?

13.2.1. *The hemispherical asymmetry of Mars*

A notable feature of Mars is that, just like the Moon, it has hemispherical asymmetry. The hemisphere mostly in the north of the planet is largely a plain covered with volcanic magma. It contains comparatively few craters, indicating that the volcanism continued until well after the period of greatest bombardment of solar-system bodies, so that early craters were covered. This hemisphere is divided by a scarp, a steep slope between 2 and 3 km in height, from the southern highlands that, like the lunar highlands, are heavily cratered. Information from the Mars Global Surveyor, a NASA spacecraft launched in 1996, indicates that the thickness of the crust in the northern plains area is 35 km compared with 80 km in the region of the southern highlands. The similarity with the lunar terrain is clear and a similar explanation, debris bombardment, can explain what is observed. Mars, being a much larger body than the Moon, would have had a much thicker molten layer extending downwards from the surface more than halfway towards its centre. Consequently it would have been much more volcanically active at the time it was bombarded. If it had been closer to Bellona, and hence had a greater thickness of crust removed — perhaps removing most of it on one hemisphere — then a large plain covered with lava, rather than a series of distinct mare basins, would have formed. Figure 13.3 shows the topography of the Martian surface. Blue represents the

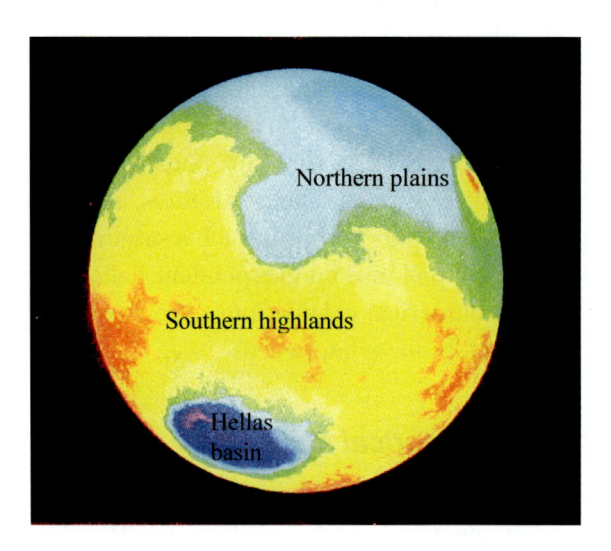

Fig. 13.3 The topography of the Martian surface.

lowest regions and red the highest. On average, the northern plains are about 4 km below the level of the southern highlands.

The southern highland region contains one very deep feature, the Hellas basin 1,800 km in diameter and about 4 km deep, produced by a projectile so massive and energetic that it penetrated deeply into the crust. There is a similarity here with the event that produced Mare Moscoviense in the lunar highlands on the far side of the Moon. When what is essentially a large and deep crater is formed in solid material, such as happened in forming the lunar mare basins, the pressure from below forces up the floor of the basin, so making the basin much shallower. This effect is shown in Fig. 13.4(a) and (b). Then, over time, magma rises from the mantle through cracks and the mare basin gets filled, first to satisfy isostatic equilibrium and then, through pressure caused by contraction of the solid surface layers, giving a slight overfill to give mascons. The depth of the surface of the magma below the level of the surrounding crust is proportional to the depth of the basin at the time it fills (Fig. 13.4(c)). However, if the basin produced is so deep that it penetrates through to a highly liquid magma, then it will very quickly fill with magma before the floor has risen to any extent — if there was a solid floor to rise. The basin that is filled in this case is a very deep one and consequently the final surface of the magma will be a long way below the level of the surrounding crust (Fig. 13.4(d)). This is the situation with respect to the Hellas basin.

13.2.2. Mars — now and in the past

The atmosphere of Mars is very tenuous by terrestrial standards, with an average surface pressure of about 0.006 atmosphere, rising

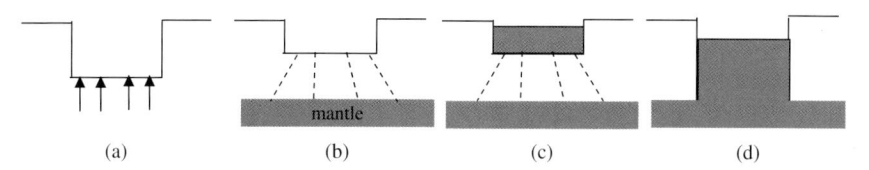

Fig. 13.4 (a) Basin with upward pressure on floor (b) Floor risen with cracks down into mantle (c) Basin gradually filled with magma (d) Deep basin quickly filled with magma.

Fig. 13.5 An image of Mars from the Hubble Space Telescope (NASA).

to nearly twice that level in the depths of the Hellas basin. Its composition is 95% carbon dioxide with the remainder mostly nitrogen and the inert gas argon. Notable features of Mars, which have seasonable variations, are the polar caps (Fig. 13.5). These have a permanent component of water ice and a seasonal component of solid carbon dioxide. With an axial tilt similar to that of Earth, Mars has similar seasons. In the northern summer the solid carbon dioxide at the northern pole sublimates and re-enters the atmosphere and a similar amount of atmospheric carbon dioxide is deposited in solid form at the southern pole. One-half a Mars orbit later, the process is reversed.

The surface of Mars is a reddish colour due to it being covered by a layer of iron oxide in the form of a fine dust. Occasionally there are enormous dust storms caused by energetic winds picking up the dust and moving it over large distances. When these events occur, large areas of the surface are hidden from view.

Mars has two spectacular surface features that are the largest of their kind in the Solar System. Valles Marineris (Fig. 13.6(a)) is a huge canyon that stretches 4,000 km in length along the equator with a width of 200 km and depth up to 7 km. For comparison, it is eight times the length, eight times the width and four times the depth

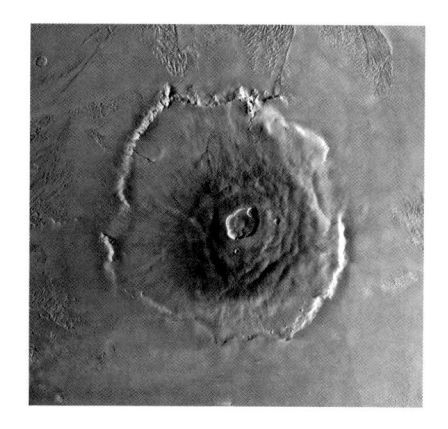

Fig. 13.6 Two giant Martian features (a) Valles Marineris (b) Olympus Mons.

of the Grand Canyon in the USA. The other feature is the extinct volcano Olympus Mons (Fig. 13.6(b)). It towers to a height of 27 km, more than three times the height of Mount Everest, above the mean surface level of Mars. There are many other extinct volcanoes on Mars that are thought to have become inactive about three billion years ago.

There is evidence that early in its history Mars had a far more substantial atmosphere than it does now, with a consequent greenhouse effect that would have given a higher temperature and liquid water on its surface. Many images of the Martian surface show features that suggest the one-time presence of flowing water, such as that seen in Fig. 13.7 which seems to show a system of dried-up river beds. It is likely that there was once a completely different climate with large amounts of water vapour in the atmosphere, the occurrence of rain and consequently features such as rivers, lakes and possibly even small oceans.[1] Such water as is now present on the planet is locked up as ice at the poles and also occurs as permafrost below the surface — a feature detected by spacecraft observations.

With its low escape speed of $5\,\mathrm{km\,s^{-1}}$, less than one-half that of Earth, and a high temperature, early Mars would have been

---

[1]Connell, A.J. and Woolfson, M.M., 1983, *Mon. Not. R. Astr. Soc.*, **204**, 1221.

Fig. 13.7 Possible dried-up river beds on Mars.

unable to retain permanently a carbon dioxide atmosphere. The loss would have been slow but as the atmosphere was lost, so the greenhouse effect would have lessened, which would have led to cooling of the planet, thus increasing its ability to retain the residual atmosphere. This process continued until the temperature fell to its present value, where the existing atmosphere gives a greenhouse effect estimated to increase the temperature by 10 K more than it would be otherwise.

13.2.3. *The Martian spin axis and hemispherical asymmetry*

The reason that we see just the abraded hemisphere of the Moon from Earth is because the spin axis is contained within the plane that separates the different hemispheres. The Moon had a shape and distribution of mass which ensured that one face was always facing Bellona and that the same face is towards Earth.

The arrangement of the spin axis to the hemispherical symmetry of Mars is very different with the spin axis making an angle of 55° to the plane dividing the hemispheres (Fig. 13.8).

Unlike the Moon, Mars was released into an independent heliocentric orbit and thus was not subjected to tidal forces due to

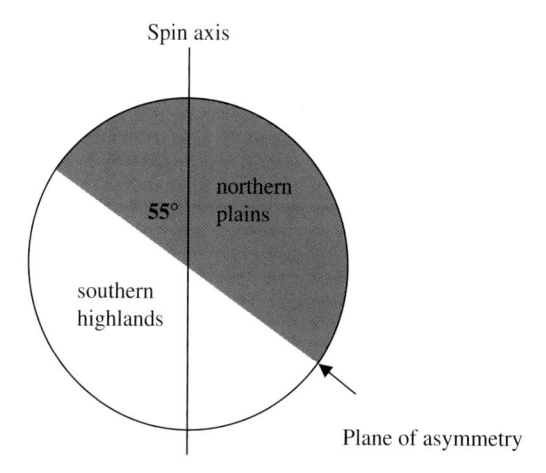

Fig. 13.8 The spin axis of Mars in relation to the plane of asymmetry.

being bound to a planet. However, like the Moon, it has a centre of mass displacement from the centre of figure, about 3.5 km, in the direction towards the centre of the northern plains. Because Mars is a considerably larger body than the Moon, it would have had a much more extensive fluid region below the crust and, in addition, that fluid would have been at a higher temperature and hence of lower viscosity. These are the conditions for something to happen that we know has taken place on Earth — continental drift. The movement of surface features relative to the spin axis is one way of interpreting what we see in Fig. 13.8.

There is a good reason why the crust should have moved over the molten mantle. In 1972 two planetary scientists, P.L. Lamy and J.A. Burns, showed that, as a consequence of internal energy dissipation within a body with a molten mantle, the surface features would drift into positions such that material moved as far as possible from the spin axis. There are many features on the surface of Mars that represent either extra local concentrations of mass, such as Olympus Mons and various plateau features, or represent local or larger deficiencies of mass, such as the whole of the northern plains and the Hellas basin. In 1983 Tony Connell and I showed that the orientation of the spin axis relative to the surface features of Mars is very close to the requirements of the theory given by Lamy and Burns (Appendix AI).[2]

To summarize, the proposed origin of Mars as a one-time satellite explains its hemispherical asymmetry and the arrangement of the spin axis to those features is a consequence of crustal motion, in conformity with the Lamy and Burns theory.

13.3. Mercury as a Satellite

In terms of mass there is no difficulty in accepting the idea that Mercury was a one-time satellite of one or other of the colliding planets. However, its density, higher than the uncompressed density of any other substantial body in the Solar System, demands an explanation. The reason for this high density — an obvious one — is that it has a particularly large iron core in relation to its overall size. The iron core has a radius that is 75% of that of the whole planet, meaning that more than 40% of the volume of the body is iron. A comparison of the structures of Mercury and Mars is shown in Fig. 13.9.

It can be seen from the figure that the iron cores of Mercury and Mars are similar in size. An interpretation of the structure of Mercury that has often been made in the past is that it was originally similar to Mars but suffered a collision with some large

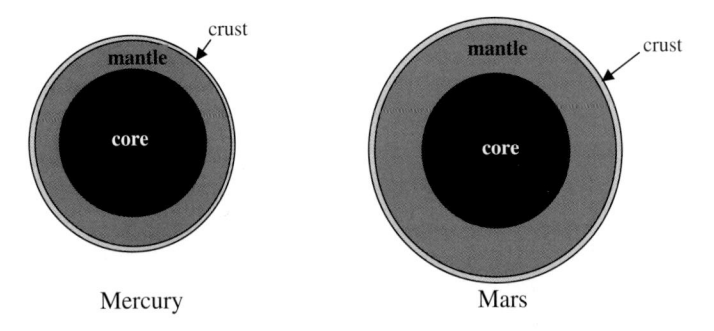

Fig. 13.9 The internal structures of Mercury and Mars.

---

[2]Connell, A.J. and Woolfson, M.M., 1983, *Mon. Not. R. Astr. Soc.*, **204**, 1221.

projectile that stripped away a large part of its mantle. The planetary collision gives another scenario that links features of Mercury to features of both the Moon and Mars. If Mercury had been both close to the colliding planets and in a position that exposed it to the maximum concentration of debris (Appendix AG), then it could have suffered abrasion to a much greater extent than either the Moon or Mars. With most of the mantle stripped from one hemisphere, the remaining mantle material would have rearranged itself to form a more or less uniform layer around the core. The stages of this process are indicated in Fig. 13.10 and are:

(a) Debris from the collision approaches a satellite similar to Mars.
(b) A large part of the mantle in one hemisphere is lost by abrasion. Mantle material from the sheltered hemisphere moves round to fill the vacated region.
(c) The surge of material flowing from the sheltered hemisphere overfills the abraded hemisphere and creates a small column where material coming from different directions collides at the centre of the abraded hemisphere.
(d) The surplus of mantle material flows back towards the sheltered hemisphere.
(e) There is a slight overfill of the sheltered hemisphere with a very small column forming at its centre.

The oscillations of mantle material between the two hemispheres would have quickly subsided and then, once the crust had reformed by lighter silicates rising to the top of the newly formed mantle, Mercury would appear as shown in Fig. 13.9.

The surface of Mercury does not show hemispherical asymmetry and, indeed, the origin we have described would give a uniform appearance over the whole surface. The surface is heavily cratered, with volcanic plains between craters and within the larger craters. A prominent feature on the surface is the *Caloris Basin* (Fig. 13.11) about 1,550 km across. There are a series of circular features within the crater, which is filled with magma, and a ring of mountains some 2 km in height surrounds it. The profile is rather similar to what

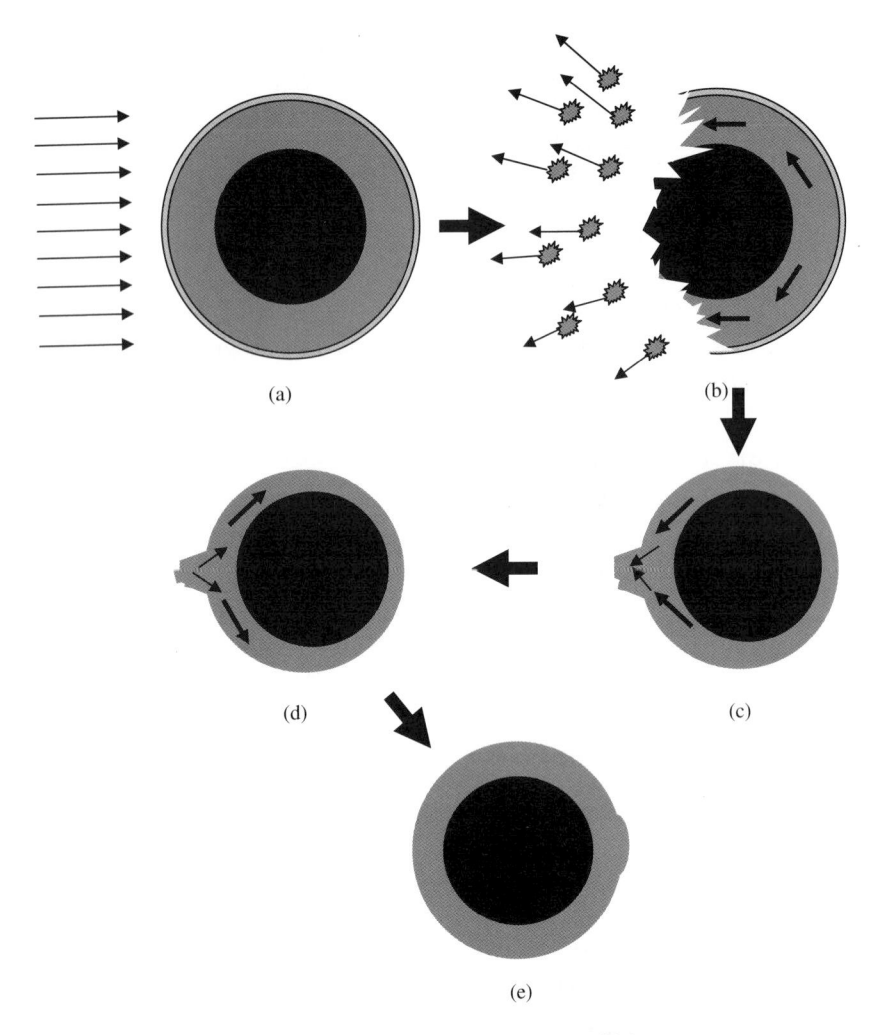

Fig. 13.10   Stages in the development of Mercury.

happens when a stone is thrown into a pond and ripples extend outwards from the point of impact; the Caloris Basin is normally described as an impact feature. In the antipodal region, diametrically opposite the Caloris Basin, there is a region of rather disturbed surface material known as the *Chaotic Terrain*. This is thought to be due to shock waves emanating from the impact, travelling around the surface by different paths and meeting in the antipodal region.

Fig. 13.11 One half of the Caloris Basin, centred at a point on the left-hand edge.

The spin period of Mercury, 58.64 days, is exactly two-thirds of its orbital period with the consequence that Mercury presents the same face to the Sun at perihelion every alternate orbit. This kind of tidal linkage is known to be related to distributions of mass within the secondary body — in this case Mercury. The Caloris Basin and the region of the Chaotic Terrain are the sub-solar points at alternate perihelion passages, which is the explanation for the name of the Caloris Basin (*Calor* in Latin is *heat*). There are a number of possible explanations for this arrangement. The first is that it is just a coincidence that these prominent features of the surface are linked to perihelion passages in the way that they are. This is possible but seems unlikely. The second is that the event that created the Caloris Basin also influenced the mass distribution within Mercury so that it became coupled in this way. This is also possible but the precise mechanism for this to have happened is not obvious. The third possibility is that its tidal linkage as a satellite with one of the colliding planets produced the mass distribution in Mercury, giving an elongation along the line of centres of the two bodies. In this interpretation the Caloris Basin is at the centre of the bombarded face, which was the site of the column produced at the centre of the face seen in Fig. 13.10(c). When this column collapsed it would have produced disturbance in its surroundings similar to that produced by a projectile. Similarly the region of the Chaotic Terrain is at the antipodal region where material spattered together due to the returning surge of mantle material. If these explanations for the nature of the Caloris Basin and of the Chaotic Terrain are tenable, they link together the origin of Mercury, the event that gave it its present density and features of its surface in a coherent and self-consistent way.

13.4. The Orbits, Spins and Tilts of Mercury and Mars

The velocity of a satellite relative to the Sun is a combination of the velocity of the planet relative to the Sun and that of the satellite relative to the planet. If the planet were suddenly to cease to exist or, as would have happened due to the collision, its gravitational influence were greatly reduced within a short time period, then this combination of velocities could yield any of the outcomes (i) to (iv) listed in Section 12.2.

At the time of the planetary collision, the resisting medium around the Sun was reduced from its original mass but still effective; the colliding planets were still in elliptical orbits that enabled a collision to occur. A satellite thrown into an independent heliocentric orbit would be subjected to the influence of the resisting medium and hence its orbit would round off and decay. However, if the initial orbit were highly eccentric, then complete round-off may not have occurred within the remaining lifetime of the disk and the eccentricity of the final orbit would then not be close to zero. The eccentricities of Mercury and Mars, 0.206 and 0.093 respectively, are significantly higher than those of any other planet, may well derive from such a cause.

The spin of Mercury, both in its tilt and period — the latter commensurate with its orbital period — is completely governed by the tidal influence of the Sun, which means that nothing can be inferred about its original spin parameters before it became linked to the Sun. By contrast Mars became completely isolated so that its spin, both in tilt and period, probably indicates the corresponding values when it was in orbit around Bellona. Since Mars as a satellite would have been in synchronous orbit around Bellona, its present spin period, 24 hours 37 minutes, was also the period of its orbit. For the mass we have taken for Bellona, this implies an orbital radius of 375,600 km, a not unreasonable value. As a regular satellite, Mars would have been in orbit in the equatorial plane of Bellona with its spin axis perpendicular to its orbital plane and hence its axial tilt of 25.2° indicates the tilt of Bellona's spin axis. To obtain such a tilt for a planet of Bellona's mass would have required a very strong tidal interaction with another planet and for that reason it is identified as a very likely candidate for the planet that caused the extreme change in the axial tilt of Uranus. Since the original mass, spin period and other physical characteristics of Uranus are unknown, it is not possible to explore this postulate numerically.

13.5. Summary

The masses and densities of Mars and Mercury indicate that their origin may differ from that of the larger terrestrial planets Venus and Earth. The colliding planet, Bellona, with twice the mass of Jupiter, could have had satellites more massive than those presently in the Solar System. With Mars as a satellite of Bellona, its hemispherical asymmetry can be explained in the same way as that of the Moon, by debris bombardment of the hemisphere facing the collision. Because it is more massive than the Moon, Mars would have had greater volcanism so that the abraded hemisphere would have been completely covered with magma to give the northern plains. The southern highlands region, similar to the lunar highlands, has one feature similar to a lunar mare basin, the Hellas basin, caused by a very energetic projectile that managed to penetrate the unabraded crust. The relationship of the spin axis of Mars to its hemispherical asymmetry is due to the dissipation of energy in its interior requiring continental drift to move mass as far as possible from the spin axis.

Mercury was initially similar to Mars in its overall structure but due to very intense bombardment by debris from the collision, it was stripped of most of its crust in one hemisphere. The remaining crust rearranged itself to cover the core fairly uniformly; in this process it surged to and fro from one hemisphere to the other, first giving rise to a temporary column of magma forming at the centre of the bombarded hemisphere and then a smaller column in the antipodal region. Prominent features on the surface of Mercury are the Caloris Basin and the Chaotic Terrain at the diametrically opposite position. Because of the 2:3 ratio between the spin and orbital periods of Mercury, the Caloris Basin and the Chaotic Terrain are at the sub-solar point at alternate perihelia. It is suggested that the figure (shape) of Mercury was formed when it was a satellite and, since the figure governs the present tidal linkage with the Sun, the sub-solar points at perihelion would be the centre of the bombarded region and its antipodal point. This gives the possibility that the Caloris Basin is not an impact feature but was caused by the collapse of the column formed by colliding streams of mantle material.

Chapter 14
Neptune, Triton and Pluto

14.1. The Neptune–Pluto Relationship

Figure 14.1 shows the orbits of Neptune and Pluto projected onto the ecliptic, the plane of the Earth's orbit. Neptune's orbit has an inclination of under 2° while that of Pluto is 17°. The relationship between the two orbits is evident. Neptune's orbit is close to circular while that of Pluto has an eccentricity of 0.249. But at Pluto's perihelion, which last occurred in 1989, it is slightly closer to the Sun than Neptune is. The inclination of Pluto's orbit and the 3:2 relationship of its orbital period to that of Neptune ensures there will be no collision between the two bodies; rather curiously, Pluto can approach Uranus more closely than it can approach Neptune. Despite this separation of the two bodies, it has often been speculated that in the distant past some event had occurred that left its mark in their orbital relationship.

14.2. The Strange Satellites of Neptune

Before the space age, two satellites had been detected orbiting Neptune, both with rather remarkable orbits. The larger of the two, Triton, is the seventh largest in the Solar System with a mass about one-third that of the Moon, and 63% more massive than Pluto. It orbits Neptune in a circular orbit with a radius of 355,000 km but the orbit is *retrograde*, which immediately rules it out as a regular satellite produced by the process described in Chapter 9 or, indeed, by any other process that has ever been suggested for satellite formation. This raises the problem of how Triton became associated with Neptune. Another curious feature of Triton is that, like Io, the

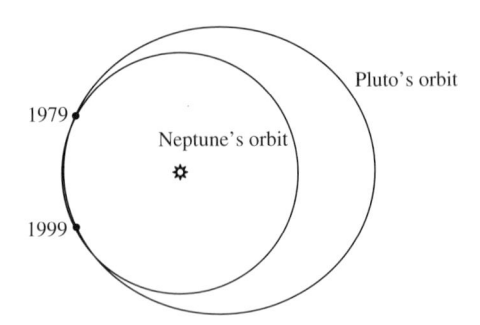

Fig. 14.1 The orbits of Neptune and Pluto showing the position of Pluto in 1979 and 1999.

innermost Galilean satellite of Jupiter, it has volcanoes. These are probably driven by nitrogen; the spacecraft image of part of Triton given in Fig. 10.16 shows a polar cap consisting of solid nitrogen.

The other satellite of Neptune known before the space age is Nereid. This has the distinction of having the largest eccentricity of any satellite in the Solar System, 0.749, and a very large semi-major axis, 5,513,400 km. This means that its distance from Neptune varies from 1.38 to 9.64 million km. It too does not have the characteristics of a regular satellite, although it has a direct orbit. Spacecraft observations have now increased the number of known satellites of Neptune to 13 and one of the newly discovered satellites, Proteus, is larger than Nereid with a diameter of 420 km against Nereid's 340 km. However, with a comparatively small orbital radius, 117,647 km, it was too close to the planet to be detected by Earth-based telescopes.

14.3. The Neptune–Triton–Pluto Relationship Explained

In 1999 I gave an explanation, involving all three bodies, for the relationships between Neptune and Pluto and between Neptune and Triton.[1] The starting point for this explanation had Pluto as a regular satellite of Neptune, orbiting in a direct sense, and Triton as an

---
[1] Woolfson, M.M., 1999, *Mon. Not. R. Astr. Soc.*, **304**, 195.

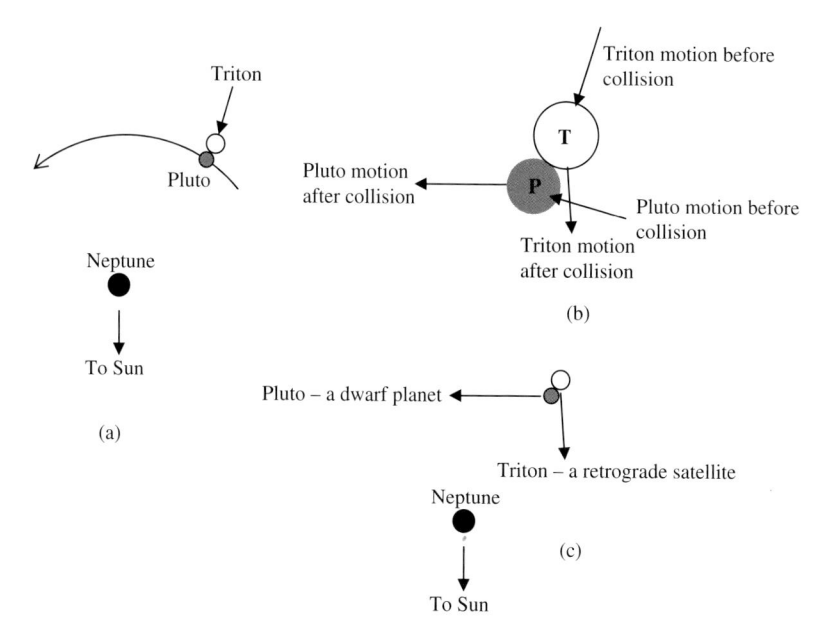

Fig. 14.2 The Triton–Pluto collision. (a) Triton, travelling towards the Sun strikes Pluto. (b) Motions before and after the collision. (c) The final outcome.

escaped satellite from the collision with an orbit that ranged out far beyond the orbit of Neptune. The process that then gave rise to what we have today is shown in Fig. 14.2. It has been subjected to mathematical analysis with the numerical results that now follow.

In Fig. 14.2(a) Pluto is in a circular orbit of radius 544,800 km around Neptune, and Triton is on a Sun-bound path with a semi-major axis of 29.09 au and an eccentricity of 0.9122. Figure 14.2(b) shows the detail of a Triton–Pluto collision in which Triton sideswipes Pluto, propelling it at a faster speed in a direction that deviates slightly from its original path. This speed is greater than the escape speed from Neptune and so Pluto moves into a heliocentric orbit of semi-major axis 39.49 au and eccentricity 0.253 — very close to the present orbital parameters, 39.55 au and 0.249. The sideswipe shears off part of Pluto that goes to form Charon, the large satellite of Pluto, while two smaller fragments become the satellites Nix and Hydra (Fig. 10.18). Pluto is set spinning, and the satellites go into orbit, in a retrograde sense — which is what is observed.

The final motions of Pluto and Triton immediately after the collision are shown in Fig. 14.2(c). The form of the collision is such as to decrease Triton's speed, which is reduced to below the escape speed from Neptune in the vicinity of the collision. The semi-major axis of the initial orbit of Triton is 436,500 km and its eccentricity is 0.881. The corresponding present orbital parameters are 355,000 km and 0.000. The direction of its motion puts it into a retrograde orbit. In Section 12.1 the mechanism was described by which, due to tidal effects, the Moon is slowly retreating from the Earth with a gradual increase in the eccentricity of its orbit. The process depends on the tide raised by the Moon on the Earth and the spin period of the Earth (1 day) being less than the orbital period of the Moon (a lunar month). In the case of a satellite in a retrograde orbit, the effect is exactly the reverse and the satellite slowly approaches the planet with reducing eccentricity.[2] The present orbital parameters of Triton are completely consistent with the original parameters given above.

Because of the gradual approach of Triton to Neptune at some time in the future, the satellite will be torn apart by the tidal effects of the planet. When this happens Neptune will acquire a substantial ring system.

We have mentioned the peculiarity of Nereid's orbit and this does not seem to be related directly to the Triton–Pluto collision. There are two obvious possibilities but there may be more. If Triton, either in the incursion that produced the collision with Pluto or in a previous incursion, had passed close to Nereid, a regular satellite, then it could have been perturbed into an extreme orbit. In view of Nereid's average distance from Neptune, this does not seem to be very probable. Alternatively, and with higher probability, Nereid could be a captured body, a large asteroid or a small escaped satellite that underwent a collision in the vicinity of Neptune, lost energy and was captured by the planet.

---

[2] McCord, T.B., 1966, *Astron. J.*, **71**, 585.

14.4. Summary

The relationship of Pluto's orbit to that of Neptune and the retrograde orbit of Triton, Neptune's largest satellite, indicate that the three bodies may be related by some event in the early Solar System. A model is described in which Pluto is initially formed as a satellite of Neptune in a direct, near-circular orbit. Triton is postulated as an escaped satellite from the planetary collision in a highly eccentric and extended heliocentric orbit. An impact by Triton on Pluto when Triton is travelling inwards towards the Sun can both propel Pluto into its present orbit and slow down Triton so that it is captured by Neptune into a retrograde orbit, initially highly eccentric but now rounded-off due to tidal effects. The sideswipe nature of the collision explains both the retrograde spin of Pluto and how material is sheared off it to form its satellites.

Chapter 15
Dwarf Planets, Asteroids, Comets and the Kuiper Belt

So far, the model I have described accounts satisfactorily for the formation of the major planets, the terrestrial planets and satellites, both regular and irregular. However, the Solar System contains many other bodies, mostly very small. Although they account for only a tiny part of the total content of the Solar System in terms of mass, they are a significant component and any plausible theory for solar-system origin should explain their presence.

15.1. Dwarf Planets

The defining characteristic of a dwarf planet is that it must be a spherical, or near-spherical, body smaller than a terrestrial planet, in an independent heliocentric orbit. Five such bodies had been designated in that category by the end of 2009. They also all have the characteristic that there are regular satellites both of much larger and much smaller mass. This being so, a fairly obvious interpretation, in relation to the planetary collision hypothesis, is that they are escaped regular satellites of the colliding planets. What needs to be explained is how they attained their present orbits.

15.1.1. *Ceres*

Ceres, the smallest and least massive of the dwarf planets, has a low mean density and almost certainly consists of a rocky core covered by a thick mantle of ice — mostly water ice. In those characteristics it is similar to, although slightly smaller than, Dione, a regular satellite of Saturn. Its orbit, in the space between Mars and Jupiter, has an eccentricity of 0.079, which fluctuates somewhat with time due to the gravitational influence of Jupiter. If it had been released from its orbit around a colliding planet into a heliocentric orbit of somewhat higher eccentricity than now, but still keeping it within, say, 4 au from the Sun, then the effect of the resisting medium in the inner part of the system, where it operated most strongly, could have modified the orbit to its present characteristics.

The explanation for both the existence of Ceres and of its orbit is similar to that for both Mars and Mercury; the only reason for distinguishing between them is their relative masses. If the collision took place at about 2 au from the Sun and Mars and Mercury went into orbits taking them mostly further inwards while that of Ceres went mostly outwards, then the final characteristics of the orbits can be well explained. To end up where it is now, Mercury would initially have had to move on an orbit that went much closer to the Sun and hence was much more eccentric than that of Mars. If the medium had evaporated and became ineffective before the rounding-off process was complete, then the moderate eccentricities of the orbits of Ceres and Mars, 0.079 and 0.093 respectively, and the higher orbital eccentricity of the orbit of Mercury, 0.206, could have been the outcome.

It may have occurred to the reader that the explanation for the origin of Ceres given here offers an alternative definition of a dwarf planet. Defining a dwarf planet as a near-spherical ex-satellite of a major planet that is now in a heliocentric orbit would include both Mercury and Mars in that category and seem much less arbitrary than a distinction made just on the basis of mass.

15.1.2. *The outer dwarf planets*

The outer dwarf planets share the characteristic that they spend either most of their time, in the case of Pluto, or all their time, in the case of Haumea, Makemake and Eris, in the region just beyond the orbit of Neptune. In Chapter 14 the origin of Pluto as an ex-satellite of Neptune was described, an origin that explains well the relationship between the present orbits of Pluto and Neptune. The alternative definition of a dwarf planet, suggested at the conclusion of the previous section, still includes Pluto, although its origin is distinctly different from that of the other dwarf planets — Mercury and Mars included.

The orbit of Pluto is not problematic in any way. It began its heliocentric orbit at 30 au from the Sun (Neptune's orbital radius) and was propelled by its collision with Triton in such a way that 30 au was close to its perihelion distance. Since the orbits of both Neptune and Pluto were close to each other at the time the latter went into heliocentric orbit, it might be wondered why, after a sufficient passage of time, they had not approached closely once more. There are in fact several reasons for this. Due to perturbation by other major planets, the orbits of both Pluto and Neptune have undergone precession.[1] This, combined with the high inclination of Pluto's orbit, 17°, moved the closest points of the orbits apart. Another factor is that, through gravitational interactions between them, the periods of the orbits of Pluto and Neptune are now locked in the ratio 3:2 so that every three orbits of Neptune, about every 495 years, the pattern of their relative positions with time repeats itself. The mechanics of the Neptune–Pluto interaction makes the 3:2 orbital commensurability quite stable so that Pluto can never approaches Neptune closer than about 17 au — although it can sometimes be only 11 au from Uranus.

For the other three outer dwarf planets, the reason why their orbits should keep them so far from the Sun is not so obvious. When they escaped from the colliding planets, they would have gone into orbits that repeatedly brought them back to the region of the collision. Their initial paths may indeed have taken them far from the Sun at aphelion but orbital evolution in a resisting medium,

---

[1] Another source of precession comes about from Einstein's general theory of relativity, which is a significant contributor to the precession of Mercury's orbit. It also gives a much smaller rotation of the perihelion of Pluto's orbit, although quite appreciable over the lifetime of the Solar System. However, it is much smaller than that given by gravitational perturbation due to major planets.

on its own, could not have given the orbits we now observe. To resolve this problem we must consider what was happening to the surviving four major planets at the time the satellites escaped into heliocentric orbits. These planets started on orbits extending out to more than 2,000 au but, on a timescale of one million years or so, the orbits were modified to their present states. The planetary collision took place when the orbits of the colliding planets were still highly eccentric — indeed, that was a necessary condition for them to collide violently — and the orbits of the other planets would have been similarly extended at that time. The characteristic of a highly eccentric heliocentric orbit is that the planets spend most of their time near aphelion where they are moving most slowly. Thus the incipient dwarf planets travelled on extended orbits into regions where potentially they could interact with major planets. Since the dwarf planets have tiny masses compared with the major planets, an interaction between the two kinds of body would barely affect the major planet but could greatly affect the dwarf planet. A potential interaction is illustrated in Fig. 15.1 where the dwarf planet is perturbed from an orbit with a small perihelion distance to one of lesser eccentricity and larger perihelion distance.

An escaping satellite in an orbit that entered the region of the present major planets may have undergone more than one interaction before the whole Solar System settled down or it may not have undergone any interaction. In the latter case it could not possibly have survived in an orbit that repeated brought it into the inner Solar System; at some time it must either have interacted gravitationally

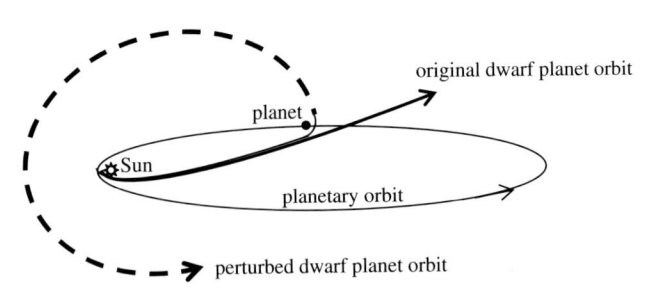

Fig. 15.1 A dwarf planet perturbed from an orbit with small perihelion to one with a large perihelion.

with a major planet and been thrown out of the Solar System, never to return, or it would have collided with and been swallowed up by a major planet and so ceased to exist.

This scenario explains the presence of Haumea, Makemake and Eris within the region of the Kuiper Belt just beyond Neptune — a 'safe' region for bodies of that size. It also suggests the possible existence of other ex-satellite dwarf planets that reacted with major planets further out and so are too distant to be easily observed, at least for the time being.

15.2. Asteroids and Comets

Now that Ceres has been promoted to the status of a dwarf planet, the remaining, as yet unconsidered, bodies of the Solar System are all smaller than a few hundred kilometres in mean diameter and range down in size to sub-micron particles. We recognize two general types of larger bodies in this category — asteroids and comets, distinguished by their compositions and locations, although the distinctions may sometimes be rather blurred.

15.2.1. *Asteroids*

Historically there have been two extreme views about the nature of asteroids. It was realised fairly early that they were of irregular shape; as they tumble in space, the amount of light they reflect towards Earth varies and so they fluctuate in brightness. An early view was that they represented the debris of a broken planet but since nobody could envisage any source of energy that could break up and then disperse a planet, this idea fell out of favour.[2] At the other extreme they were regarded as the building blocks of planets, which would fit in with the solar-nebula model in which smaller bodies — planetesimals — combined to form terrestrial planets or the cores of major planets. The capture-theory model, and the subsequent planetary collision, provides the mechanism by which

---

[2]Napier, W.McD. and Dodd, R.J., 1973, *Nature*, **224**, 250.

planets *can* break up and provide fragments and this is what will be considered here.

It is not possible to estimate the period of time between when the planets first condensed from a filament and then contracted, as illustrated in Figs. 6.1, 6.2 and 6.4, and when they collided. For the collision to have been sufficiently violent to generate the temperature at which nuclear reactions involving deuterium could have taken place, it was necessary for the orbits of the colliding planets to still be highly eccentric, although their eccentricities could have been much lower than the original values. From Fig. 7.7 it is clear that this time period could have been about a million years, which is important because there would have been considerable time for the planetary materials to segregate, with the densest material at the centre and the least dense at the outside. The separation may not have been complete. There may have been regions with various admixtures of metal and silicates, others where silicates were mixed with material consisting of water, ammonia and methane, and yet others at the lower part of the gaseous atmosphere with various mixtures of heavier and lighter gases. A schematic, simplified impression of a planet at the time of collision is shown in Fig. 15.2.

The calculation that gave rise to Fig. 11.8 showed that material from every level of the colliding planets, from core to atmosphere, was ejected from the collision region. The further out from the centre

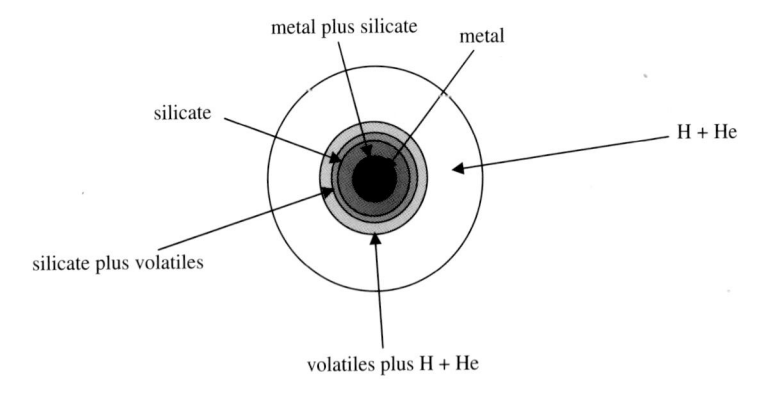

Fig. 15.2 A schematic partially segregated major planet.

the material was, the greater its likely speed of ejection, the actual speed depending on its initial position relative to the collision region. Of the material that could potentially form solids, the silicate plus volatiles tended to be thrown out furthest, with lesser average speeds successively for silicate, silicate plus metal and metallic core material. We shall see that this pattern explains in a general way what is observed.

Small bodies produced by the collision can only have survived until the present time if their final orbits kept them well clear of planets, in particular the major ones. We have already considered the origin of the dwarf planet Ceres that stayed well within the orbit of Jupiter when it was released, and some fraction of the silicate, silicate plus metal and metal debris would have done likewise. This would have formed the bulk of the material now in the asteroid belt. Excluding the largest body in the asteroid belt, Ceres, the total mass of all known asteroids is just about 3% of the mass of the Moon. Since the total mass of ejected solid material would have been hundreds or even thousands of times greater, this indicates that what we now observe is the vestigial remains of what once existed. This ties up with the observational evidence of the damaged surfaces of solid bodies in all regions of the Solar System that shows that, early in its existence, there was a period of intense bombardment. Most projectiles would have fallen on the major planets but, because of their nature, the evidence for bombardment of these bodies is no longer visible.

Asteroids are usually designated in three main categories as determined by analysis of the sunlight they reflect. The great majority, 75% of them, are C-type carbonaceous chondrite asteroids. They are very dark, typically reflecting from 3–10% of the light falling on them. From the laboratory examination of carbonaceous chondrite meteorites, which have a similar light-reflection pattern, it is known that these asteroids contain a great deal of volatile material, including water within hydrated minerals such as serpentine. This type of asteroid is mainly found in the outermost region of the asteroid belt and can be identified as having an origin in the region of silicate plus volatiles in the colliding planets.

The other two types of asteroid, both of which reflect light far better than the C-type, are known as the S-type and M-type. It was originally thought that S-types were mostly silicate with perhaps some small content of metal while M-types were mostly metal with perhaps some small content of silicate. However, there are several lines of evidence, based on their reflection characteristics, that many M-type asteroids may indeed be mostly of silicate composition so that the distinction between S-types and M-types is now somewhat uncertain. Anyway, these combined types can be confidently categorized as consisting of silicate and metal in various proportions, with little volatile content, if any, and, as expected for material from the deepest parts of the colliding planets, they are mostly found in the inner parts of the asteroid belt.

15.2.2. Comets and the Kuiper Belt

The volatile-rich regions of the colliding planets, which were thrown out to the largest distances, provided the source material for comets. Unlike the surviving asteroids, their orbits took them into the region of the major planets and also well beyond the present orbit of Neptune. If their orbits brought them repeatedly into the planetary region after the planets had settled down in their present positions, then they would have had a very low probability of survival. To have any possibility of long-term survival, they would need to have undergone the type of interaction shown in Fig. 15.1 that threw them into an orbit with perihelion outside the present planetary region. There they would be temporarily exposed to possible absorption by a major planet while the planetary orbits were still evolving but eventually, if they survived for a million years or so, they would be safe.

Just as for asteroids, the majority of debris material of the type that formed comets would have been lost by collisions with major planets. Additionally, much of it may have been thrown out so violently that it left the Solar System altogether. What survived was that which ended up outside the orbit of Neptune, which might have included some asteroid-type material thrown out unusually far, either directly or through a gravitational interaction with some larger body. The fraction of this material that underwent the kind of interactions shown in Fig. 15.1 constitutes the Kuiper Belt that stretches from the inner region just outside Neptune, within which bodies are now being observed, outwards to an indeterminate distance.

It is believed that short-period comets, those with orbital periods of 200 years or less, are produced by Neptune's perturbation of inner bodies of the Kuiper Belt. However, there are other types of visible comet that come from a completely different source by a completely different process.

15.2.3. *Long-period comets and the Oort cloud*

The existence of the Oort cloud of comets was described in Section 10.5. There is a vast family of comets, estimated to be 2×10^{11} in number, with aphelia at distances of tens of thousands of au and perihelia well outside the planetary region. They have a more or less permanent existence since they are extremely unlikely to be appreciably perturbed by any member of the Solar System. However, when they are at their greatest distances from the Sun, they are subjected to occasional perturbations, most frequently by passing stars, and may have their orbits modified so that they subsequently pass close enough to the Sun to become visible (Appendix AJ).

The energy associated with a heliocentric orbit consists of two parts — energy of motion, kinetic energy, which is positive, and gravitational potential energy, negative in sign, which depends on the distance of the body from the Sun at any time. As the body moves around its orbit, these energies both change but their sum, the total energy, remains constant. For a closed elliptical orbit, the energy must be negative. If it is positive, then the body is in a hyperbolic orbit, not bound to the Sun and will leave the Solar System. The intrinsic energy (energy per unit mass) E_T of a body of small mass orbiting the Sun is

$$E_T = -\frac{GM_\odot}{2a} \qquad (15.1)$$

in which $M_\odot$ is the solar mass and a is the semi-major axis of the orbit. Since all quantities other than a on the right-hand side of (15.1) are constant, it is convenient to express the intrinsic energy in terms of $1/a$ so an intrinsic energy expressed as $-10^{-4}\,\mathrm{au}^{-1}$ corresponds to an elliptical heliocentric orbit with a semi-major axis of 10^4 au. In the case of an Oort cloud comet, the two component types of energy are nearly in balance but the combined energy is slightly negative so that the comet is a bound member of the Solar System. When a star perturbs one of these comets, modifying its orbit and taking it into the planetary region, then the planets, particularly the major planets, can further perturb it. The expected magnitude of the energy change of the comet's intrinsic energy due to planetary perturbation will be small and depend on how far into the planetary region the comet penetrates (Appendix AK). If its perihelion is less than 5 au, so that it becomes visible, then the magnitude of the expected change of intrinsic energy will be of order $10^{-3}\,\mathrm{au}^{-1}$. For a comet with an initial semi-major axis of 20,000 au, the intrinsic total energy, in the units given above, is $-5 \times 10^{-5}\,\mathrm{au}^{-1}$ so the expected change of energy has a large magnitude compared to the magnitude of the comet's original intrinsic energy. If the planetary perturbation adds energy, then the new energy will probably be positive and after passing out of the planetary region, the comet will be on a hyperbolic path that takes it out of the Solar System. Alternatively, if the perturbation subtracts energy from the comet's orbit, then the new negative energy will be many times the original negative energy and the new orbit will be one with a much smaller semi-major axis — probably of the order of 1,000 au. The comet will then return to the neighbourhood of the Sun at intervals of a few tens of thousands of years and after a few tens of millions of years its inventory of volatile materials will be exhausted. It will then become a dark object, similar to a carbonaceous chondrite asteroid although not on a typical asteroid orbit. Eventually it will either be thrown out of the Solar System or absorbed by colliding with a major planet. Because comets coming from the Oort cloud are very unlikely to return to it and then re-enter the inner Solar System from the same region, such comets are referred to as *new comets*. Another sense in which they are *new* is that the periods of their modified orbits are normally much larger than the period of recorded history of mankind, so that no previous apparition will be known. New comets are seen at the rate of somewhat less than one per year, on the basis of which figure, by making a number of assumptions, Oort estimated that there are 2×10^{11} comets in the Oort cloud.

15.2.4. *The survival of the Oort cloud*

One of the great mysteries of the Oort cloud is how it has survived for the lifetime of the Solar System. We know the local stellar number density and the average speed of local stars relative to the Sun. From this it is possible to deduce the expected number of passing stars that have approached within any given distance of the Sun since the Solar System formed; it can be shown that more than 1,000 stars have approached the Sun to within 50,000 au during that period and about 50 have passed within 10,000 au. Each of these stars would have considerably perturbed the Oort cloud, to the extent that it must be wondered how so many of the comets in the cloud are still where they are. This problem of survival is exacerbated by another kind of perturbing influence that may be even stronger than that of stars.[3] Within the galaxy there are *Giant Molecular Clouds* (GMCs), large regions within which gas exists in a molecular form such as H_2 and carbon monoxide (CO). The mass of one of these clouds can be from ten thousand to a million times the solar mass and they have dimensions stretching over many tens of parsecs. From the observed number density of these clouds in the galaxy, it can be deduced that the Solar System should have passed through three or four of them during its lifetime. If GMCs were of uniform density, then they would not be very effective in perturbing the Oort cloud. However, the mass within them exists in clumps — they are starforming regions — and this greatly enhances their effectiveness as a perturbing agent.

---

[3] Bailey, M.E., Clube, S.V.M. and Napier, W. McD., 1990, *The Origin of Comets* (Oxford: Pergamon).

The solution to the problem of the survival of the Oort cloud was suggested by the British astronomer Mark Bailey in 1983.[4] He postulated that there is an inner reservoir of comets so that when outer comets were removed by a severe perturbation, the same perturbation moved inner comets outwards to replenish the Oort cloud. Indeed, this is what the planetary collision scenario would predict. Initially the cometary bodies would be thrown out over a wide range of distances with orbits ranging from those that were partially within the present planetary region to those leaving the Solar System completely. Any that were perturbed by evolving planets to keep them outside the present orbit of Neptune, by the mechanism shown in Fig. 15.1, would survive for a long time, if not indefinitely. Those that went out so far that they were readily perturbed by even very distant passing stars would quickly either have left the Solar System or become new comets and had their orbits substantially modified. The net result after a sufficient passage of time would be the present situation with a continuous distribution of cometary bodies stretching from the Kuiper Belt to the Oort cloud. This model, with the Kuiper Belt the inner boundary and the Oort cloud the outer region of a continuous distribution of cometary bodies, the latter being constantly replenished from within, differs from that assumed by Oort when he estimated the total number of comets in the cloud. For this reason the total number of potential comets in the Solar System may be very different from his estimation.

15.3. Summary

The dwarf planets did not originate in the same way although they are all ex-satellites of planets. Ceres, a satellite of a colliding planet, was released into an independent heliocentric orbit that kept it within the asteroid belt region, before evolving to its present state. Pluto is an ex-satellite of Neptune, projected into its present orbit by a collision with Triton, as described in Chapter 14. The other three dwarf planets, Eris, Makemake and Haumea, were all satellites

---

[4]Bailey, M.E., 1983, *Mon. Not. R. Astr. Soc.*, **204**, 603.

of colliding planets, released into extended orbits that took them beyond the present orbit of Neptune. By interacting with surviving major planets, also on extended orbits at that time, they were swung into orbits that kept them beyond Neptune in the inner reaches of the Kuiper Belt.

The debris from the collision was roughly segregated according to its volatile content; that with the highest content, which would form comets, was thrown out the furthest. The surviving inner debris, for the most part composed of silicate and iron, forms the asteroid belt. The cometary material that stayed too long in the region of the major planets would have collided with, and been absorbed by, one of them. Other material was thrown out at high speed and left the Solar System. There would have remained a distribution of potential comets stretching from the inner reaches of the Kuiper Belt out to the extremes of the Oort cloud. Short-period comets are mostly due to perturbations by Neptune on nearby bodies within the Kuiper Belt. New comets are produced by stellar perturbation of the Oort cloud and long-period comets are probably the result of perturbation by planets of new comets when they enter the planetary region.

The Oort cloud has survived the ravages of rare intrusions of stars very close to the Sun and of the passage of the Solar System through Giant Molecular Clouds. When outer Oort cloud comets are removed from the Solar System by these episodes, they are replenished by comets from further in that are perturbed outwards.

Chapter 16
Meteorites: Their Physical and Chemical Properties

Meteorites are Nature's gift to planetary scientists. Mostly they are small fragments of asteroids, chipped off by the occasional collisions they experience, but sometimes they have an even more exotic source. So, without the trouble and expense of travelling to asteroids to find out what they are made of, bits of them come to us. They have stories to tell about the early history of the Solar System and may indeed help in deciding on the relative plausibility of different theories of its origin.

The mass of meteorite material striking Earth per day is between 100 and 1,000 tonnes. If this rate has occurred over the age of the Solar System, the addition to the mass of Earth would have been of the order of 10^{-7} Earth mass, covering Earth with a 40 cm layer of material. In the early period of the Solar System the rate of bombardment would have been greater but most early material would have disappeared beneath the surface and been subsumed into Earth's substance by the process of plate tectonics.

Meteorites come in all shapes and sizes, from tiny fragments to great boulders with masses of several tonnes. Once in a while, an object will arrive of much greater size, giving rise to global catastrophes, but these can be thought of as asteroids rather than meteorites. All objects from space enter the atmosphere with a speed of at least $11\,\mathrm{km\,s^{-1}}$, the escape speed from Earth. A body of diameter 1 km would have a mass of about 1,500 million tonnes and the energy of impact on Earth would be the equivalent of about

20,000 average hydrogen bombs. The fall of a small asteroid, a few km in diameter some 65 million years ago, is thought to have wiped out the dinosaurs that dominated Earth at that time. The evidence for this is that marine clays from all over the world deposited at that time have high iridium content and iridium is a far more common element in meteorites than on Earth.

There are many types of meteorite and within each type there are sub-classifications. The very fine detail of their compositions is of great interest to meteoriticists, those who specialize in the structure of meteorites, but is confusing to those who, like us, wish to concentrate on the broader picture. In that vein I now offer an overview of the types of meteorite that occur.

16.1. The Broad Classes of Meteorites

The broadest classification that can be made of meteorites is that they are, for the most part, either *stones* or *irons*, *i.e.* they predominantly consist of either silicates or metallic iron. A small number are known as *stony-irons* because they contain an intimate mixture of appreciable proportions of the two basic types of material.

Meteorites can be either *falls* or *finds*. The former variety usually consists of those where a body enters the atmosphere from space, breaks up, becomes incandescent from atmospheric friction and then is seen to fall in a well-defined area that can be subsequently searched for the fragments. Rarely, a meteorite may fall within sight of an observer or even on a house, in which case finding the object is quite simple. Falls are important as a source of meteorites because, although they will be contaminated by travelling through the atmosphere and contact with the ground, they are comparatively pristine. Finds, by contrast, are meteorites that were not seen to fall and may have been on the Earth for thousands or, potentially, even millions of years. These will obviously have sustained the ravages of weathering over the years. The condition for discovering a 'find' meteorite is that it must be recognized for what it is. A rocky meteorite in a region where terrestrial rocks abound will probably pass unnoticed. However, an iron meteorite will certainly be recognized as an alien object, probably visually, because of its dark colour but certainly due to its high density. Meteoriticists search for meteorites in regions where they would stand out — in deserts or in the Antarctic. In arid regions, weathering is a slow process and meteorites retain their characteristic appearance for much longer. In the Antarctic, where land is covered by several kilometres of ice, one can be reasonably sure that a stony object on or near the surface has come from above rather than from below (Fig. 16.1).

By and large the proportion of stones, irons and stony-irons arriving on Earth reflects the proportions that exist in space and this would also be reflected in the proportions derived from falls. For finds, the proportions are biased towards irons, which are much easier to recognize. However, since finds are so easy to recognize in the Antarctic, the proportions from there are quite similar to the proportions for falls outside the Antarctic. This is shown in Table 16.1, based on a 2006 NASA report, which makes allowance for the fact that meteorites often fragment so that several specimens found in the same area may derive from the same meteorite.

Fig. 16.1 An Antarctic meteorite photographed with a catalogue numbering device (NASA).

Table 16.1 The percentages of finds and falls of various types of meteorite from outside Antarctica compared with the percentages from Antarctic finds.

Type of meteorite	Non-Antarctic falls	Non-Antarctic finds	Antarctic finds
Stone	93.6	53.4	94.9
Iron	5.1	43.0	4.3
Stony-iron	1.2	3.6	0.9

16.2. The Physical and Chemical Characteristics of Meteorites

16.2.1. *Stony meteorites*

There are two main types of stony meteorite — *chondrites* and *achondrites* — that differ both in their physical and chemical characteristics. Chondrites, which account for about 86% of stony meteorites, usually, but not always, contain *chondrules*, small, glassy, spherical silicate inclusions (Fig. 16.2). Achondrites, as their name suggests (the prefix *a* or *an* in Greek means *without*), never contain chondrules and they often resemble the igneous differentiated rocks found on Earth's surface (Fig. 16.3).

The chondrites can be divided into five categories on the basis of their chemical composition. These are: E (enstatite), H (high iron), L (low iron), LL (low iron and low metal) and C (carbonaceous). The amount of iron decreases as one goes from E to C and the degree

Fig. 16.2 Part of a chondritic meteorite showing many chondrules.

Fig. 16.3 A typical achondrite resembling a terrestrial rock.

Table 16.2 The iron content of ordinary chondrites.

	H	L	LL
Total iron (%)	27	23	20
Metallic iron (%)	20	8	2

of oxidation of the iron increases so that in enstatite chondrites the iron is either in metallic form, or as the sulphide, the mineral *troilite* (FeS), while in carbonaceous chondrites iron is mostly in the oxidized state — for example, in *magnetite* (Fe$_3$O$_4$). The proportions of total iron, and of that in the metallic state for the H, L and LL types — the so-called *ordinary chondrites* — are given in Table 16.2. The steady decrease in both the proportion of iron and of metallic iron as one goes from H to LL is evident.

Enstatite chondrites contain a considerable proportion of the mineral enstatite, Mg$_2$Si$_2$O$_6$. They are mainly distinguished by having a lower-than-average Mg:Si atomic ratio, less than 0.85, while the ratio for ordinary chondrites is about 0.95 and that for carbonaceous chondrites is greater than 1.05.

The accepted interpretation of chondrules is that they were formed when droplets of liquid silicates condensed out of a silicate vapour, forming into spheres under the influence of surface tension. After solidification they were then incorporated into masses of silicate fragments and the bodies that were subsequently formed were massive enough for gravitational forces to compress the material into rock. Some chondrules are broken up and others, which were in a plastic state, are often distorted but their origin as spherical droplets is evident. When individual chondrules are examined they are found to contain mixtures of minerals. In the original silicate vapour there would have been no recognizable minerals but rather stable components such as SiO_2, Na_2O, MgO or Al_2O_3 that could combine in various ways to form different minerals. Such combinations would occur when the liquid droplet was first formed but while the silicate was still a very hot liquid, or even a very hot solid, the components would be continuously changing partners with the tendency to form more stable minerals — ones with the lowest energy. Given enough time, the minerals formed would together have the lowest possible energy consistent with using all the components available; the mixture of minerals would then be *equilibrated*. However, when the mixtures of minerals in chondrules are examined, they are extremely *non-equilibrated* — that is, they correspond to a total energy far greater than the minimum possible. The reason for this is that the chondrules became solid and cooled so quickly that the non-equilibrated state became frozen in; individual stable groups did not have enough energy to jostle their way through the material to form more favourable associations. The message from this is that the chondrules cooled very quickly and estimates for the rate of cooling are in the range of 700–4,000 K hour^{-1}, suggesting that the total cooling time from the formation of a liquid chondrule to the chemically frozen state was of the order of one hour. Although a hot nebula was the *raison d'être* for the origin of the SNT, an idea that has long since been abandoned, it is difficult to see how such a nebula could have had such a fleeting high-temperature existence.

The most distinguishing feature of carbonaceous chondrites (Fig. 16.4) is that they contain volatile materials. They can contain up to 22% water although not in liquid form but as water of crystallization in minerals. Of great interest is their organic content, including many hydrocarbons such as benzene. They also include many amino acids, the building blocks of proteins, and three of

Fig. 16.4 Three fragments of carbonaceous chondrites (NASA).

the four nucleotide bases that are the components of DNA — thymine, guanine and adenine — with only cytosine missing. This has prompted some astrobiologists to speculate that life on Earth may have originated from materials brought in by carbonaceous chondrites and comets.

Rather curiously, for bodies that contain so many volatile materials, carbonaceous chondrites also contain some highly refractory substances, materials that are stable and only melt at very high temperatures. These are in the form of white inclusions of minerals; they are called CAI, standing for calcium-aluminium-rich inclusions.

There are several different kinds of achondrites but on the whole they are igneous rocks that appear to have formed on the surface of a planetary-size body. With few exceptions they all seem to have solidified about the same time — 4,500 million years ago. The exceptions relate to three kinds of meteorite, the shergottite, nakhlite and chassignite classes — collectively called SNC meteorites — where the solidification ages are much more recent, just 1.3 billion years ago in the case of the nakhlites. It has been suggested that these meteorites may be material that was blasted off Mars by an asteroid impact and there are several lines of evidence to support this. Perhaps the most compelling is the analysis of gas trapped within the meteorites, which is closely similar in characteristics to that of the Martian atmosphere.

16.2.2. Iron meteorites

Iron is a common material in planets, being the end product of exothermic nuclear reactions, ones that provide energy when they occur. To produce elements heavier than iron, a source of energy is needed and such elements are produced as a result of supernova events (Section 3.4). Consequently iron grains are present in the interstellar medium and are also present in planets, satellites and smaller derived bodies. Although we refer to some meteorites as being iron they are actually a mixture of iron and nickel with 5–25% of nickel. The metal exists in the form of two iron–nickel alloys, nickel-poor *kamacite* and nickel-rich *taenite*. In the liquid state the two metals are intimately mixed but when it solidifies, it first forms taenite. As it cools, so kamacite crystals occur and the amount of kamacite increases as the temperature falls. At about 800 K the atoms in the mixture have insufficient energy to jostle about to change places with each other and to reorganize themselves in different arrangements, so the arrangement at 800 K will be retained no matter how much further the temperature falls.

There are a number of iron meteorites with 5–6.5% nickel that are mostly kamacite, with a little taenite. Seen under a microscope they show a characteristic pattern related to the crystal structure of kamacite and they are called *hexahedrites*. At the other end of the scale, with about 20% or more nickel, the meteorite will be almost pure taenite and will show no structure under a microscope; they are called *ataxites*, which means *without form* in Greek.

An iron meteorite with between 6.5% and 13% nickel will contain a mixture of kamacite and taenite. As it cools and becomes solid, it will first be uniformly taenite. At a certain temperature tiny kamacite plates will appear and nickel will migrate out of the kamacite regions into the taenite regions. However, the nickel more easily leaves the kamacite than it enters the taenite so there is a build-up of nickel content in a rim around the kamacite plates. Once the material has cooled to a level where atoms are no longer able to migrate through it, then the pattern is congealed and can be seen in a polished surface of the meteorite. These patterns are called Widmanstätten figures (Fig. 16.5). From their appearance it is possible to estimate the rate

Fig. 16.5 Widmanstätten figures.

of cooling of the meteorite material. These rates of cooling are usually in the range of 1–10 K per million years. This is consistent with iron meteorites having been parts of bodies of small asteroid size or near the surface of larger asteroids.

16.2.3. *Stony-iron meteorites*

Most meteorites, possibly all, contain both iron and silicate but for the majority of them one component dominates to the extent where we say that they are either stones or irons. However, there are a small number of meteorites, about 1% of the total, in which both components are substantially present. These are designated as stony-irons. There are two main groups of stony-irons — *pallasites* and *mesosiderites*.

Pallasites show silicate materials, predominantly olivine, set in a metallic matrix (Fig. 16.6). One could imagine this being derived from a body that has formed a liquid iron core within which a considerable amount of silicate was trapped. Globules of silicate, of lower density, would then rise through the molten metal but at an ever-slower rate as the temperature fell until eventually the material froze with silicate globs trapped within a solid metal framework.

The appearance of mesosiderites is quite different. The rock is partially in fragments, mostly minerals consisting of plagioclase, a silicate mineral containing sodium, calcium and aluminium, plus

Fig. 16.6 A Pallasite with olivine crystals within a metal matrix (Oliver Schwartzbach).

calcium-bearing pyroxene, a mineral found in igneous rocks, but also globules of olivine, a major component of Earth's mantle. The minerals are in forms that are only stable at low pressures, below 3 kbar (3,000 atmospheres), so they could not have originated from deep within a massive body. The metal is partly in the form of globules and partly as veins occupying the space between minerals.

16.3. Interpreting the Physical Properties and Appearance of Meteorites

The planetary-collision hypothesis gave a scenario in which much of the material of the cores of two major planets was violently dissipated into space. The material in the region of the collision interface would have been heated to an extremely high temperature, one so high that nuclear reactions occurred. Here the material would be in a highly ionized state with no chemical associations of any sort between atoms. The parts of the cores remote from the collision interface would have been at a much lower temperature, perhaps not much more than their temperature in the undisturbed planet, but they would still be propelled into space. When the shock waves from the collision region reached the far side of the core, spallation of the material there took place. A similar process can be observed by gripping a brick or some other frangible object in a vice and hitting one end with a hammer. Fragments of the brick fly off at the other end.

Material that was not in the collision region would have undergone less dispersion than that near the collision region and should have more readily assembled into comet-size or asteroid-size bodies that would eventually cool to give what we have today. Some of these would be uncontaminated by chondrules and provide the asteroids from which achondritic meteorites would eventually be derived. The very hot material in the collision would have quickly cooled, reformed itself into stable mineral components, then formed mineral associations and finally cooled into liquid in the form of spherical droplets — chondrules. These would have formed an admixture with cooler silicate material and the asteroidal bodies from this combined material would have become the asteroids that are the source of chondritic meterotites.

Some ejected material, coming from close to the core, would be predominantly iron and so form metallic asteroids — of the M-type or possibly S-type (Section 15.2.1). The cooling rates of iron asteroids a few kilometres in diameter would be consistent with the cooling rates estimated from Widmanstätten figures. Some other asteroids, but comparatively few if segregation in the planets was well advanced, would be mixtures of silicate and iron. If such a mixture started rather quiescently, then, in the asteroid so formed, the silicate would tend to accumulate into globules that would slowly rise towards the surface in the weak gravitational field. This would be a slow process and probably be incomplete by the time the asteroid had solidified, at least in its outer regions. Fragments from such an asteroid would give pallasites. Mesosiderites, which show rather more chaotic mixtures of stone and iron, could be the result of iron and stony material coming together violently because of the collision rather than having been previously associated in the planets.

There are many subtle differences between meteorites even within the same general type. Thus there are many forms of chondritic meteorite with differing chemical compositions and different degrees of clarity in the appearance of the chondrules they contain. Indeed, some chondritic meteorites have no visible chondrules but are classified as chondritic because of their compositions; they were probably at a high temperature, about 1,500 K, for an extended period which caused the chondrules to merge with the surrounding material. For each type of meteorite it is possible to imagine some particular circumstance within the planetary collision environment, which contains a wide variety of physical conditions, that would give rise to it.

16.4. Summary

There are three main categories of meteorites — stones, irons and stony-irons — and within each of those categories there are different types. The general types within the stony category are chondrites and achondrites. Chondrites are distinguished by containing chondrules, small glassy spheres typically a millimetre in diameter, embedded within a matrix of silicate grains. There are five sub-categories within the chondrite group, distinguished by their chemical compositions. Enstatite chondrites have a high content of the mineral enstatite, $Mg_2Si_2O_6$, and three other kinds, constituting ordinary chondrites, are H (high iron), L (low iron) and LL (low iron, low metal). The final sub-category is carbonaceous chondrites, distinguished by having a high content of volatile materials such as water (as water of crystallization) and various organic compounds.

Achondrites are igneous rocks that are similar to terrestrial surface rocks. Some achondrites, designated as SNC, probably originated on the surface of Mars.

Iron meteorites, which contain 5–25% of nickel, contain two main iron–nickel alloys — nickel-rich taenite and nickel-poor kamacite. When iron meteorites containing 6.5–13% nickel cooled, kamacite plates formed, their size increasing with time while the solid metal was hot enough for nickel atoms to migrate through it. From the Widmanstätten patterns that form from this process, it is possible to estimate the cooling rate of the material. Cooling rates of 1–10 K per million years indicate that they originated in bodies of small asteroid size.

Stony-iron meteorites are of two types — pallasites and mesosiderites. Pallasites consist of silicate globules within an iron matrix and are the result of the cooling of a region in which the globules were rising under gravity within surrounding iron. By contrast, mesosiderites are a rather chaotic mixture of silicates and iron that look as though they splattered together and then quickly cooled.

The physical and chemical properties of meteorites can be interpreted as a consequence of a planetary collision. Vaporized material, quickly cooling, would have produced chondrules which then further cooled so quickly that the minerals within them were unequilibrated. These would have become incorporated in debris that had never melted to give asteroids, the source of chondrites. Achondrites are then interpreted as coming from material that was too remote from the collision region to become associated with chondrules. Iron with associated silicate material, thrown out from the collision in coherent globs, would have formed mixed-material asteroids that would then have become the source of pallasites. Iron and silicate material that violently came together in a less coherent form would have given mesosiderite source bodies.

Chapter 17
Isotopic Anomalies in Meteorites

17.1. Isotopes and Anomalies

When John Dalton (1766–1844) established the concept of elemental atoms, it was assumed that all the atoms of a particular element were identical. Later it was realised that atoms had a structure. At the centre of an atom is its nucleus, a highly compact entity consisting of protons, particles with a positive charge, and neutrons, of mass virtually equal to that of protons, but with no charge. Surrounding the nucleus are electrons, equal in number to that of protons and with a negative charge equal in magnitude to the positive proton charge. The electrons contribute very little to the total mass of atoms; the mass of an electron is just 1/1836 that of the proton.

The idea of the structure of an atom and the meaning of the term isotope was introduced in Section 11.3, which also showed the structures of hydrogen, deuterium, helium and lithium in Fig. 11.6. To further this description, in Fig. 17.1 the three stable isotopes of oxygen are illustrated. The chemical nature of an atom is related to the number of electrons it possesses, for chemical bonding depends on interactions of some sort between the electrons of neighbouring atoms. Since it is possible to strip electrons from an atom by a process known as ionization, it is better to identify the type of atom by the number of protons it has; a singly ionized oxygen atom has only seven electrons but its eight protons establish its identity as basically oxygen. To remind you again of the convention for describing atomic nuclei, the symbols under the isotopes in Fig. 17.1 are O, the chemical symbol for oxygen, the bottom prescript, 8, giving the number of

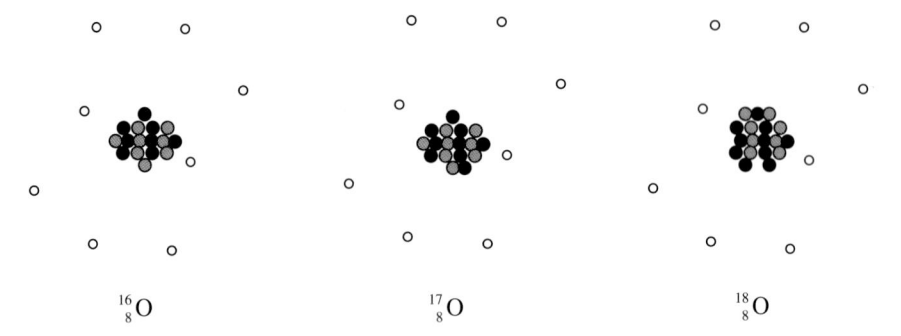

Fig. 17.1 The three stable isotopes of oxygen. Grey = proton, black = neutron, white = electron.

protons and the top prescripts 16, 17 and 18, which give the combined number of protons and neutrons, a measure of its mass. The 8 is really redundant since for oxygen there must be eight protons but I shall continue to use this form. It is also possible to refer to the three isotopes as oxygen-16, oxygen-17 and oxygen-18 and within figures, for brevity, I shall use a shorter form — O16, O17 and O18.

On Earth oxygen occurs in a great variety of materials — in water, rocks and in living matter. When terrestrial oxygen from different sources is analysed, it is found that the proportion of the three isotopes is always very nearly the same. The usual proportions quoted are:

$$^{16}_{8}O : ^{17}_{8}O : ^{18}_{8}O = 0.9527 : 0.0071 : 0.0401.$$

This composition is referred to as SMOW (Standard Mean Ocean Water). Samples of terrestrial oxygen do depart from the SMOW composition but only by a little and in a very systematic way. Physical or chemical processes, such as diffusion in a thermal gradient, or rates of chemical reactions may be linearly dependent on mass, giving *mass-dependent fractionation*, so that the deviation of the concentration of $^{18}_{8}O$ from $^{16}_{8}O$ may be just twice the deviation of that of $^{17}_{8}O$ from $^{16}_{8}O$. The difference in the ratio of, say, $^{17}_{8}O$ to the most common isotope $^{16}_{8}O$ in a meteorite sample, from the same ratio in SMOW is

$$\left\{\frac{n(^{17}_{8}O)}{n(^{16}_{8}O)}\right\}_{\text{sample}} - \left\{\frac{n(^{17}_{8}O)}{n(^{16}_{8}O)}\right\}_{\text{SMOW}}$$

where $n(Q)$ represents the concentration of Q in atoms per unit volume. Dividing this quantity by the ratio for SMOW gives the fractional change in the ratio and multiplying by 1,000 gives the fractional change *permille*. A change permille is like a percentage change except it is parts per thousand rather than parts per hundred, and the symbol used is ‰ rather than %. Thus the fractional change permille is given by

$$\delta^{17}_{8}O(\text{‰}) = \frac{\{n(^{17}_{8}O)/n(^{16}_{8}O)\}_{\text{sample}} - \{n(^{17}_{8}O)/n(^{16}_{8}O)\}_{\text{SMOW}}}{\{n(^{17}_{8}O)/n(^{16}_{8}O)\}_{\text{SMOW}}} \times 1000. \tag{17.1}$$

When there is mass-dependent fractionation, if $\delta^{17}_{8}O$ is plotted against $\delta^{18}_{8}O$, known as a *three-isotope plot*, then the result is a straight line of slope 0.5. The Earth–Moon line in Fig. 17.2 is that given by samples from both Earth and the Moon, indicating that they probably come from a common source. Samples from some achondrites also give a line with a slope of 0.5 but displaced from the Earth–Moon line. This suggests either a different source of oxygen or the same source as for Earth that became contaminated in some way.

Many meteorites show isotopic compositions, or variations of their isotopic compositions, that are very different from those on Earth. While there is no reason to regard isotopic compositions on Earth as some kind of cosmic standard, it is customary to regard large departures from what is observed on Earth as an *anomaly*. The various anomalies that have been observed in meteorites can be explained in terms of the products of the planetary collision.

17.2. The Planetary Collision and Nuclear Reactions

In Section 11.3 I described the effect of surface chemistry in concentrating deuterium in hydrogen-containing molecules constituting the grains contained in dark cool star-forming clouds and in protostars. This deuterium enrichment ends up in the inner parts of a newly formed planet and will give a localized D/H ratio of at least 0.01 and possibly considerably higher. The planetary collision gives an enhancement of density by the compression of material in the collision interface and the model shown in Fig. 11.8 indicates that the ignition temperature is reached for D-D reactions to take place — about 3×10^6 K. The mechanism by which this occurs when major planets collide is explained in Appendix AD. In Appendix AE it is shown that with D/H = 0.02, the temperature attained is about 2.6×10^8 K, at which temperature reactions involving other light elements can take place.

Formulae for the rates at which nuclear reactions occur, dependent on the concentrations of the reacting nuclei and particles and the temperature, have been given by the American astrophysicist William Fowler (1911–1995; Nobel Prize for Physics, 1985) and his colleagues.[1] In 1995 Paul Holden and I reported on a calculation we carried out in which a mix of materials, corresponding to the most abundant components of an silicate-impregnated icy region with some iron present, was at a density of just over 10^4 kg m^{-3} and was subjected to a triggering temperature of 3×10^6 K.[2] The D/H ratio used was 0.016, which happens to be the ratio for Venus. At the time that Holden and I published our results, there was no observational evidence for high D/H ratios in protostars, first obtained in 1999, and the only information about D/H ratios in the Solar System was that given in Table 11.1; we took the highest value in that table, that for Venus for the hydrogen in our calculation. In retrospect the total amount of hydrogen used in that calculation was too high for just taking account of hydrogen-containing molecules, so that using the D/H ratio we did for all the hydrogen present was unjustified. For that part of the hydrogen belonging to the atmosphere, the basic value of D/H = 2×10^{-5} was more appropriate. In addition the

---

[1] Fowler, W.A., Caughlan, G.R. and Zimmerman, B.A., 1967, *Ann. Rev. Astron. P.*, **5**, 525; 1975, *Ann. Rev. Astron. P.*, **13**, 69.
[2] Holden, P. and Woolfson, M.M., 1995, *Earth, Moon Planets*, **69**, 201.

analysis shown in Appendix AD had not been done at that time so the triggering temperature was justified by some rather handwaving arguments. In the event, observation and theory eventually caught up with hypothesis!

To put the results on a better footing, I have repeated the calculations with a firmer foundation in both theoretical and observational results. The computational program began with a mixture of stable isotopes of the elements, H, He, C, N, O, Ne, Na, Mg, Al, Si, S and Fe. The amounts taken were consistent with an icy region impregnated with about 24% silicates plus some atmospheric gas — although the term icy does not mean that the materials were at a low temperature, but rather that they were materials such as water, ammonia and methane and other hydrogen-containing materials that readily formed ices at low-enough temperatures. The mixture used, in terms of concentrations of elements, is given in Table 17.1, corresponding to a density of 10^4 kg m^{-3}, and the overall D/H was taken as 0.02. At the commencement of the calculation, the concentrations are calculated for each isotope of those elements present. In the program the reaction rates for 274 reactions were included, plus large numbers of reverse reactions since many reactions

Table 17.1 The concentrations of each element in the nuclear-reactions calculation.

Element	Number density (m^{-3})
H	8.308×10^{29}
He	4.000×10^{28}
C	7.331×10^{28}
N	3.665×10^{28}
O	1.625×10^{29}
Ne	2.000×10^{20}
Na	2.444×10^{27}
Mg	4.887×10^{27}
Al	4.887×10^{27}
Si	1.222×10^{28}
S	4.000×10^{27}
Fe	2.444×10^{27}

can also operate in a backwards direction. There were also 40 radioactive decays to take into account, for 38 of which the product of the decay quickly gave a new stable product. For the other two decays, those of $^{22}_{11}$Na and $^{26}_{12}$Al, sodium-22 and aluminium-26, the half-lives were much longer and affected the outcome in terms of final isotopic composition in a different way. All these reactions and decays are listed in Appendix AL.

It is always important in theoretical work to include all known features that would operate *against* the result being sought, which means in this case anything that lowers temperature and would thus inhibit nuclear reactions. The iron that was included in the mixture of materials is not involved in reactions under the conditions of the planetary collision, so it will provide extra particles to take up energy and act as a coolant. Similarly the ionization of material was taken into account and the degree of ionization was computed at every step of the calculation, using equations developed by the Indian astrophysicist, M.N. Saha (1893–1956), by which the ionization of a mixture of materials can be found as a function of temperature and density. An ionized atom produces more particles, *i.e.* electrons. Every kind of particle, independent of its mass, takes up the same expected energy so, for a system containing a certain amount of thermal energy, the more particles there are, the cooler it becomes. However, it is also necessary to find the number of electrons present for another reason, as they are involved in some reactions.

The temperature rise is slow at first but then accelerates. The period during which most of the reactions of interest are taking place lasts about 0.1 second and during that interval the temperature increases rapidly from under 3×10^8 K to 10^9 K. If the calculation were continued, it would increase further. Within the heart of the explosion the material is shielded against cooling by exterior material but it would be blown outwards by the sudden increase in pressure and would then quickly cool to a temperature at which most nuclear reactions would cease. Material towards the boundary of the reaction region would be thrown out soonest and hence reactions within it would have ceased at a lower temperature. The maximum temperature in the heart of the reacting region has been taken as 10^9 K but this is just an estimate that could be wrong by a considerable margin without affecting the general conclusions that follow.

17.3. Explanations of the Anomalies

Under the conditions of the planetary collision, there would have been a wide range of temperature regimes. In regions remote from the collision interface, material would be almost unaffected and would have been thrown outwards as gas, liquid or solid. Closer to the collision there would be regions where the material would have been considerably heated, but not to the extent that nuclear reactions would have occurred. Closer still, there would be regions where nuclear reactions did take place, but perhaps to a lesser extent than in the heart of the collision region. The material from regions with different nuclear processing would have mixed to some degree before the asteroids, first liquid and then solid, are formed that subsequently become the source bodies for meteorites. We should really take into account the variety of materials, processed to different extents, in interpreting the computational results in relation to observations. However, there are so many variables involved that it is impossible to deal with them all. For example, there are variations in the temperature regime, density, duration of containment and local composition, all of which would change the results of the calculation. For this reason all that can be expected is that an outcome from the collision, found by mixing material processed by the nuclear reaction with unprocessed material, gives results that are in broad agreement with what is observed.

17.4. Individual Isotopic Anomalies and How They Are Produced

Here we shall be dealing with anomalies that, one way or another, involve most of the elements in Table 17.1. There are many other anomalies that have been observed, several involving much heavier elements, and supernovae rather than the comparatively low-energy planetary collision would have produced some of these.

17.4.1. *The oxygen anomaly*

When three-isotope plots for oxygen were found for some meteorite samples, for example, from anhydrous carbonaceous chondrite material or the chondrules from some ordinary chondrites, it was discovered that the plot had a slope of 1.0, or nearly so. These plots are shown in Fig. 17.2 together with those from Earth and Moon material. This unit slope could be interpreted as the result of mixing different amounts of pure $^{16}_{8}O$ with some standard mixture of the three isotopes, although the standard mixture to which the pure $^{16}_{8}O$ was added was different for the two types of material — chondrules and anhydrous material. To explain how pure $^{16}_{8}O$ could be produced and then mixed with other oxygen in meteorites, it has, in the past, been suggested that reactions between the common carbon isotope, $^{12}_{6}C$, and helium, $^{4}_{2}He$, in distant stars produced $^{16}_{8}O$ that was then transported to the Solar System in grains. The oxygen in the grains then interchanged with some of the oxygen in meteorites that was of solar-system origin.

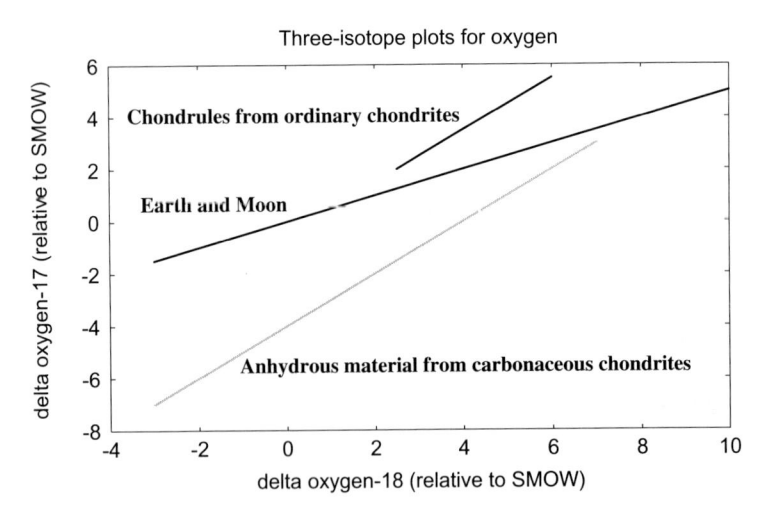

Fig. 17.2 Three-isotope plots for oxygen for various solar-system materials.

In considering the amount of three isotopes of oxygen produced by the nuclear reactions, we must take into account the production of two isotopes of fluorine, $^{17}_{9}F$ and $^{18}_{9}F$, which are radioactive with half-lives of 1.075 minutes and 1.83 hours and decay into $^{17}_{8}O$ and $^{18}_{8}O$ respectively. In the final material that creates asteroids, these isotopes appear as oxygen. In Fig. 17.3 the concentrations of $^{16}_{8}O$, $^{17}_{8}O$ (including $^{17}_{9}F$) and $^{18}_{8}O$ (including $^{18}_{9}F$) are shown plotted against temperature as the nuclear reactions progress.

What will be seen in Fig. 17.3 is that beyond a temperature of about 6×10^8 K, the concentrations of both $^{17}_{8}O$ and $^{18}_{8}O$ fall markedly (notice the logarithmic scale) while that of $^{16}_{8}O$ has hardly changed. That being so, the material at that stage is almost pure $^{16}_{8}O$ and, if admixed with unmodified oxygen in various proportions, it would give a three-isotope plot with a slope very close to 1.0. In fact the experimental slopes are not actually 1.0 but are about 0.94 and taking material from the nuclear reactions corresponding to various temperatures above 6×10^8 K and mixing it in various proportions with unmodified oxygen gives slopes between about 0.92 and 0.97.

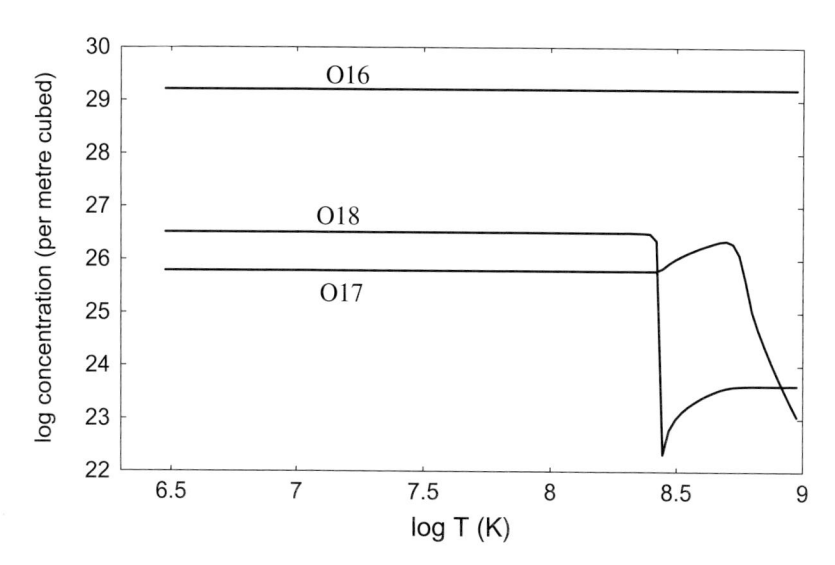

Fig. 17.3 The variation of the isotopes of oxygen (including radioactive fluorine) with temperature.

The basic assumption of almost any theory of formation of the Solar System is that planets and satellites are formed from the same initial pool of material — in the case of the Capture Theory, the protostar or dense region that was the source of the filament within which planetary condensations formed. That being so, in the absence of any nucleosynthetic event, the isotopic composition of all bodies, from the size of planets downwards, would necessarily be almost the same, differing only to the extent that mass-dependent fractionation would allow. If SMOW was that basic initial oxygen and pure $^{16}_{8}O$ was added to it in various proportions to give the unit-slope line, then all the $\delta^{17}_{8}O$ and $\delta^{18}_{8}O$ values for contributing samples would have to be negative. It is clear from Fig. 17.2 that they are not. The reason for this is that SMOW is not the basic initial oxygen that existed before the nuclear explosion and characterized all solar-system material at that time. The Earth and Moon material were contaminated by the pure $^{16}_{8}O$ and other matter from which the unit-slope lines were obtained had undergone some mass-dependent fractionation before having additions of $^{16}_{8}O$. The δ-values of the unit-slope lines roughly show the extent to which $^{16}_{8}O$ was added to the meteorite sample more-than or less-than it was added to terrestrial and lunar material. This kind of consideration indicates that any attempt to explain isotopic anomalies without allowing for the possibility of large-scale local nucleosynthetic activity, that could contaminate the surface material of whole planets, is bound to have problems. Indeed, some meteoriticists have proposed that there had to be two sources of contaminating oxygen from outside the Solar System, one with an $^{16}_{8}O$ surplus and the other with an $^{16}_{8}O$ deficit. This seems highly unlikely.

We have already noted that features of oxygen isotopic composition within grains in meteorites have previously been interpreted in terms of the grains having been of pre-solar origin, with the material processed by nuclear reactions within distant stars. The planetary collision hypothesis offers an alternative source of nuclear reactions, an event that can also account for many other features of the Solar System.

17.4.2. The magnesium anomaly

Magnesium has three stable isotopes, $^{24}_{12}$Mg, $^{25}_{12}$Mg and $^{26}_{12}$Mg, which occur in the approximate proportions of 0.790:0.100:0.110. For most meteorite samples a three-isotope plot shows the expected slope of 0.5, corresponding to mass-dependent fractionation. In 1976 three American meteoriticists, Lee, Papanastassiou and Wasserberg, found that CAI in carbonaceous chondrites contained an excess of $^{26}_{12}$Mg that was proportional to the amount of aluminium in the sample.[3] Different grains in a particular inclusion would contain different minerals and hence the proportions of magnesium and aluminium present would vary from one grain to another. The effect was easiest to measure in grains that contained a high ratio of aluminium to magnesium where the excess of $^{26}_{12}$Mg was very evident.

There is only one stable isotope of aluminium, $^{27}_{13}$Al, but there is an unstable isotope, $^{26}_{13}$Al, with a long half-life of 720,000 years. The decay product of $^{26}_{13}$Al is $^{26}_{12}$Mg by the reaction

$$^{26}_{13}\text{Al} \to {}^{26}_{12}\text{Mg} + {}^{0}_{1}\text{e} + \nu, \tag{17.2}$$

where the last two products on the right-hand side are a positron and a neutrino.

The interpretation of the Lee et al. measurements for magnesium is that, when the CAI was formed, a small proportion of the aluminium it contained, typically 10^{-5}, was $^{26}_{13}$Al. Each individual grain in the inclusion had an aluminium-to-magnesium ratio that depended on the mineral form of the grain and hence the proportional excess of $^{26}_{12}$Mg would vary from one grain to another. Plotting $^{26}_{12}$Mg/$^{24}_{12}$Mg against $^{27}_{13}$Al/$^{24}_{12}$Mg gives a straight-line relationship as shown in Fig. 17.4. The mineral anorthite (CaAl$_2$Si$_2$O$_8$) can have magnesium as an impurity and the two samples shown contain different amounts of magnesium.

It has generally been assumed that the radioactive aluminium was produced in a supernova which was the trigger for forming the Solar System. This assumption puts a time constraint of a few million

---

[3] Lee, T., Papanastassiou, D.A. and Wasserberg, G.T., 1976, *Geophys. Res. Lett.*, **3**, 109.

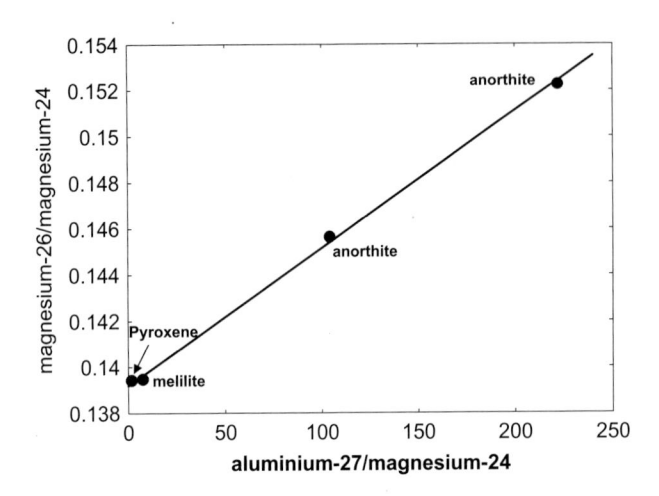

Fig. 17.4 Excess $^{26}_{12}$Mg plotted against the amount of $^{27}_{13}$Al in a CAI.

years on the period between a supernova event and the formation of the Solar System — but this is not a serious constraint.

When we considered the concentrations of the stable oxygen isotopes in the final processed material, we also had to consider the unstable fluorine isotopes that decayed to oxygen. So it is for magnesium where unstable $^{24}_{11}$Na and $^{24}_{13}$Al decay to $^{24}_{12}$Mg with half-lives of 15 hours and 2.1 seconds respectively and $^{25}_{11}$Na and $^{25}_{13}$Al decay to $^{25}_{12}$Mg with half-lives of 59 and 7.1 seconds respectively. Including these components, the variation in the magnesium isotopes, and also of $^{26}_{13}$Al and $^{27}_{13}$Al, with temperature are shown in Fig. 17.5.

At a temperature of about 2.5×10^8 K, $^{26}_{13}$Al begins to be produced in some quantity and builds up quickly thereafter. An admixture of a small amount of processed material with unmodified material gives an initial ratio of $^{26}_{13}$Al/$^{27}_{13}$Al, typically 10^{-5} that will be characteristic of the particular CAI in a carbonaceous chondrite. Since so little processed material is added, the initial magnesium isotope mixture will be very close to that of unprocessed material. Then, in a particular mineral grain, with its individual aluminium-to-magnesium ratio depending on the minerals it contained, $^{26}_{13}$Al would completely decay over the course of a few million years and add to

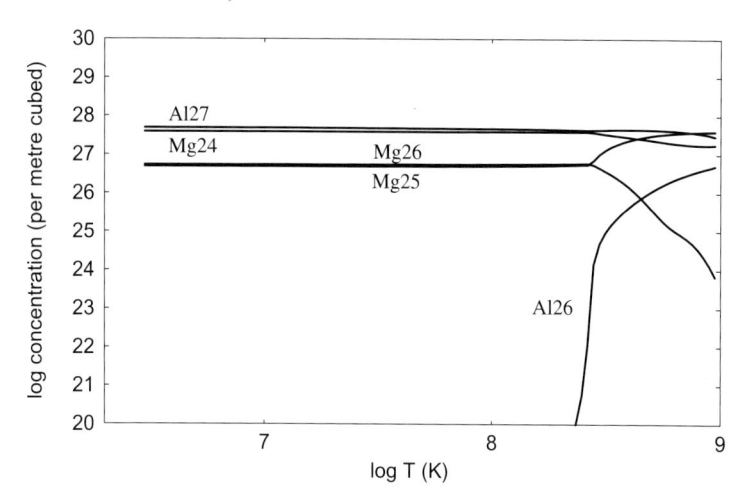

Fig. 17.5 The production of magnesium and aluminium isotopes during the nuclear explosion.

the complement of $^{26}_{13}$Mg to give what is seen in Fig. 17.3, which corresponds to an initial ratio of $^{26}_{13}$Al/$^{27}_{13}$Al, about 5×10^{-5}.

The $^{26}_{12}$Mg content of a CAI is not always related to the aluminium content. There are some samples where there are no $^{26}_{12}$Mg excesses even for large Al/Mg ratios, which would correspond to CAIs condensing from vaporized material that had little if any nuclear processing. Other samples show a $^{26}_{12}$Mg deficit which could be accounted for by the massive fall in concentration of $^{26}_{12}$Mg at higher temperatures as shown in Fig. 17.5. It is clear from the variety of observations of $^{26}_{12}$Mg and indeed of many other isotopic anomalies that the environments in which they were being established were not homogeneous. The range of conditions pertaining to the collision scenario, varying both with place and time, can explain the variations in observations made.

Measurements of $^{25}_{12}$Mg anomalies have also been made in CAI material. Mostly these are small — in the range of -12 to $+31‰$ — and they have been ascribed to mass-dependent fractionation, the evidence for this being that the value of δ^{25}_{12}Mg is proportional to δ^{30}_{14}Si. However, one sample gave δ^{25}_{12}Mg $= 350‰$, which is well outside any fractionation limit. From Fig. 17.5 it can be seen that at higher temperature there is a substantial increase in $^{25}_{12}$Mg and also a decrease in $^{24}_{12}$Mg. A somewhat larger-than-normal component of processed material, one part in 10^3, at 4.98×10^8 K gives δ^{25}_{12}Mg $= 263‰$ and a slightly larger component would give an even larger value.

17.4.3. *Neon in meteorites*

Neon is a very common element in the Universe at large, accounting for about 1/750 of its total mass, but it is rare on Earth, being a very minor component of the atmosphere. Since it is an inert gas, it does not form chemical compounds and so its presence in meteorites is in the form of pockets of gas trapped in interstices within the minerals they contain. There are three stable isotopes of neon, with the following proportions in the atmosphere:

$$^{20}_{10}\text{Ne} : ^{21}_{10}\text{Ne} : ^{22}_{10}\text{Ne} = 0.9051 : 0.0027 : 0.0922. \quad (17.3)$$

To determine the presence of neon or any other trapped gas in a meteorite, it must be heated to a temperature that will give the gas atoms sufficient energy to joggle their way through the solid material and escape. Neon from meteorites has a wide range of isotopic compositions, mostly falling within the triangle ABC in Fig. 17.6, which shows the ratio $^{20}_{10}$Ne/$^{22}_{10}$Ne plotted against $^{21}_{10}$Ne/$^{22}_{10}$Ne. There are samples that fall outside the triangle and the most interesting of these fall very close to the origin — they are almost pure $^{22}_{10}$Ne, which forms 9% of terrestrial neon. It is because of the rather special status of $^{22}_{10}$Ne that the ratios in Fig. 17.6 are given relative to that isotope instead of the dominant isotope, $^{20}_{10}$Ne.

There is an obvious explanation for the presence of this almost pure $^{22}_{10}$Ne that, in this context, is called neon-E. There is only one stable isotope of sodium, $^{23}_{11}$Na, but there is a radioactive isotope, $^{22}_{11}$Na, which decays by

$$^{22}_{11}\text{Na} \rightarrow ^{22}_{10}\text{Ne} + ^{0}_{1}\text{e} + \nu. \quad (17.4)$$

If, when the meteorite formed, radioactive sodium were present that became incorporated in sodium-containing minerals, then, as long as the meteorite was cold when the sodium decayed, the resultant $^{22}_{10}$Ne

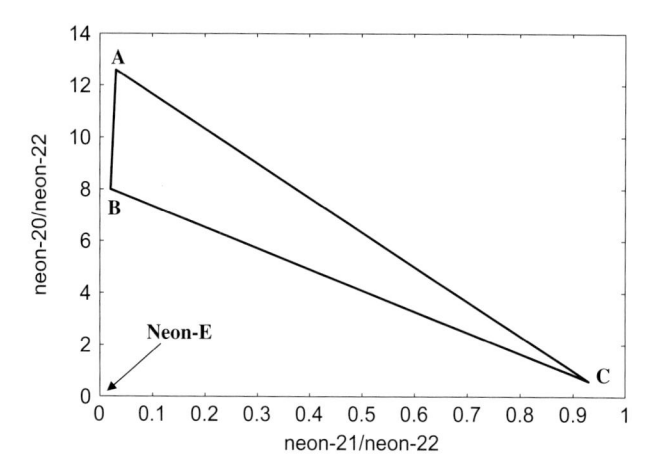

Fig. 17.6 The containing triangle for most meteorite neon samples and the region of neon-E.

would be trapped within it. There is, however, a problem with this scenario. The half-life of $^{22}_{11}$Na is just 2.6 years, which means that the radioactive sodium would have to be produced by a nuclear reaction and then be incorporated within a cool meteorite on a timescale of a few times 2.6 years. The usual supposition that has been made in the past is that the only kind of event that would produce $^{22}_{11}$Na is a supernova and it seems doubtful, but perhaps not impossible, that the time constraint could be satisfied with this form of production. To get round this problem it has been suggested that $^{22}_{11}$Na is produced by the action of energetic protons on $^{21}_{10}$Ne, that then was somehow incorporated into meteorites.

To explain the existence of neon-E, there are requirements other than having some $^{22}_{11}$Na produced which become incorporated in a sodium-containing mineral within the meteorite. Another requirement is that the vaporized material quickly forms grains and that these grains cool quickly to a temperature at which the neon produced by decay will be retained. In the case of the decay of $^{22}_{11}$Na, with a half-life of 2.6 years, the time for grain forming and cooling would need to be of that order or less, but in fact the times for these processes are hours, or days at the most, when the system is on as small a scale as a planetary collision (Appendix AM).

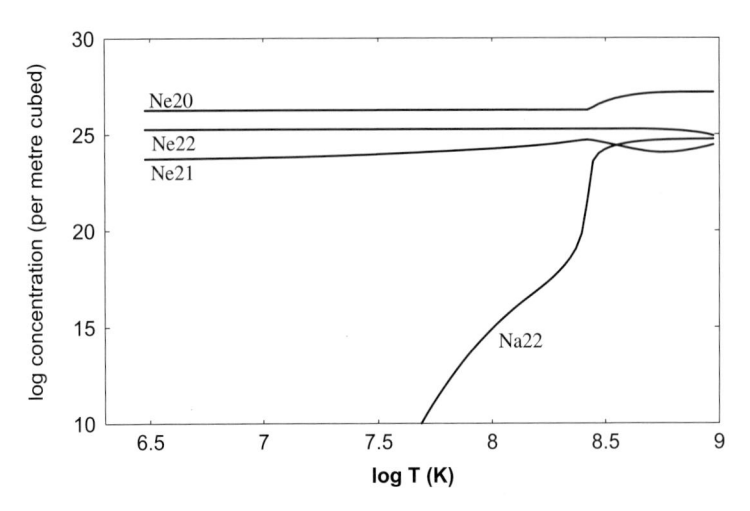

Fig. 17.7 The concentrations of stable neon isotopes and of sodium-22 during the nuclear explosion.

The concentrations of the stable isotopes of neon and of $^{22}_{11}$Na during the course of the nuclear reactions are shown in Fig. 17.7. It will be seen that at the top end of the temperature range, a large amount of $^{22}_{11}$Na is present — ample in quantity to explain the neon-E observations.

There is one other interesting neon anomaly that will be described and explained in the following section.

17.4.4. Anomalies associated with silicon carbide

17.4.4.1. Silicon in silicon carbide

Many chondritic meteorites contain grains of silicon carbide, SiC, which show anomalies in silicon, carbon and many other elements, including some present in just trace amounts. Silicon has three stable isotopes, $^{28}_{14}$Si, $^{29}_{14}$Si and $^{30}_{14}$Si, for which the terrestrial proportions are 0.9223 : 0.0467 : 0.0310. A three-isotope plot of silicon from different silicon carbide samples gives a line of slope about 1.3 with scatter around the mean line and with the δ^{29}_{14}Si and δ^{30}_{14}Si values all positive. Various admixtures of $^{28}_{14}$Si with normal silicon cannot explain this since, as for oxygen, that would give a slope of 1.0 with at least some negative δ values. Rather confusingly there are other meteorite samples that give large negative values for δ^{29}_{14}Si and δ^{30}_{14}Si and yet others giving negative values for δ^{29}_{14}Si and positive values for δ^{30}_{14}Si. These variations are probably indicative of the non-homogeneity of the material after undergoing nuclear reactions, such as would be expected across the full extent of the collision interface and in locations away from the interface. The formation of silicon carbide requires a high ratio of carbon to oxygen in the cooling vapour, indicating that the region from which the silicon carbide formed was gas-rich and silicate-poor, since silicates contain a great deal of oxygen. For nuclear processing of the mixture given in Table 17.1, the variation of the amounts of the three stable silicon isotopes with temperature is given in Fig. 17.8. The amount of $^{28}_{14}$Si takes into account the contribution of $^{28}_{13}$Al, with a half-life of 2.24 minutes, and also of $^{28}_{12}$Mg, that first decays into $^{28}_{13}$Al with a half-life of 12.9 hours. The amount of $^{29}_{14}$Si includes the contributions of $^{29}_{13}$Al, with a half-life of 6.5 minutes, and of phosphorus-29, $^{29}_{15}$P, with a half-life of 270 milliseconds, and that of δ^{30}_{14}Si includes the contributions of $^{30}_{13}$Al, with a half-life of 3.7 seconds, and of $^{30}_{15}$P, with a half-life of 2.5 minutes. It is clear from Fig. 17.8 that, for this particular processed material,

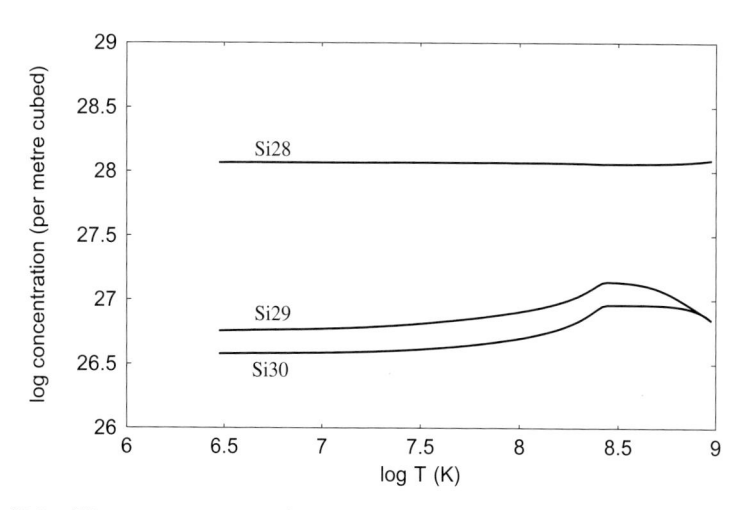

Fig. 17.8 The concentrations of stable silicon isotopes during the progress of the nuclear explosion.

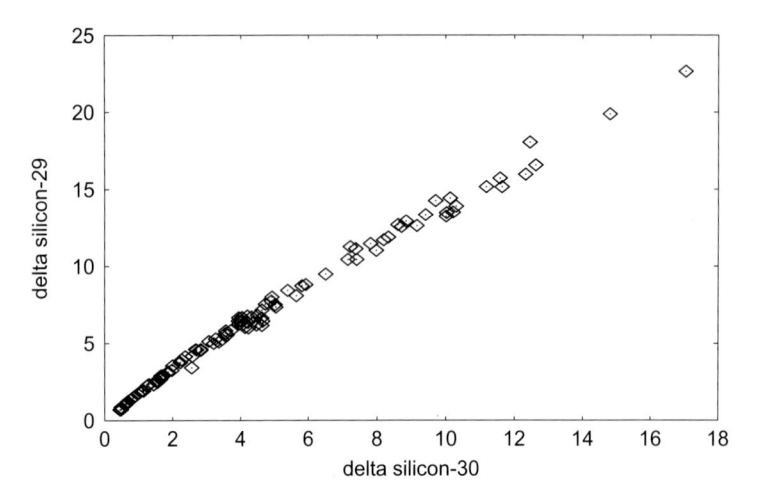

Fig. 17.9 A plot of $\delta^{29}_{14}Si$ against $\delta^{30}_{14}Si$ for various admixtures of processed material with original unprocessed material.

the values of $\delta^{29}_{14}Si$ and $\delta^{30}_{14}Si$ will always be positive, with the possible exception of $\delta^{29}_{14}Si$ at the very highest temperatures. A three-isotope plot of $\delta^{29}_{14}Si$ against $\delta^{30}_{14}Si$ is found by mixing up to 20% of material randomly from temperatures between 6×10^6 K and 1.0×10^8 K with unprocessed material. This is shown in Fig. 17.9 and it gives a slope of about 1.3, in agreement with observations.

We noted that a unit-slope oxygen three-isotope plot could give positive δ^{17}_8O and δ^{18}_8O values because SMOW, the Earth standard being used as a comparator, had itself been contaminated by the products of the nuclear reactions. It seems likely that this may also be true for the terrestrial standard of silicon isotope ratios, which would account for the negative $\delta^{29}_{14}Si$ and $\delta^{28}_{14}Si$ values found for some samples.

17.4.4.2. Carbon and nitrogen in silicon carbide

Now we consider two linked anomalies involving carbon and nitrogen. Carbon has two stable isotopes, $^{12}_6C$ and $^{13}_6C$. In terrestrial carbon, the ratio $^{12}_6C/^{13}_6C$ is 89.9 but the ratio in many silicon carbide samples is much lower, as low as 10, to give what is known as *heavy carbon*.

Trapped in the interstices of silicon carbide grains is nitrogen that can be released, together with other gases, by heating. Nitrogen has two stable isotopes, $^{14}_{7}N$ and $^{15}_{7}N$, and in the terrestrial atmosphere the ratio $^{14}_{7}N/^{15}_{7}N$ is 270. Most of the nitrogen samples from silicon carbide are *light nitrogen* with a ratio as high as 2,000 but, curiously, sometimes *heavy nitrogen* is found with a ratio as low as 50.

In considering the amounts of $^{13}_{6}C$, $^{14}_{7}N$ and $^{15}_{7}N$ in the products of the nuclear reactions, one has to take into account radioactive products with very short half-lives that give those isotopes as decay products. Thus $^{13}_{7}N$ decays to $^{13}_{6}C$ with a half-life of 9.97 minutes, $^{14}_{8}O$ decays to $^{14}_{7}N$ with a half-life of 70.6 seconds and $^{15}_{8}O$ decays to $^{15}_{7}N$ with a half-life of 70.6 seconds. By the time that vaporized material, away from the heart of the explosion and hence that has not undergone nuclear reactions, is contaminated by processed material, there is no distinction, for example, between original $^{13}_{6}C$ and that produced by the decay of $^{13}_{7}N$. However, $^{14}_{7}N$ is also a decay product of $^{14}_{6}C$ but in this case the half-life of the decay is 5,730 years so that in the period when minerals are forming, it behaves like carbon and later decays to produce $^{14}_{7}N$ trapped within the mineral — as long as the mineral is cool enough to retain it. Figure 17.10 shows the amounts of $^{12}_{6}C$, $^{13}_{6}C$, $^{14}_{7}N$, $^{15}_{7}N$ and $^{14}_{6}C$ produced by the nuclear reactions.

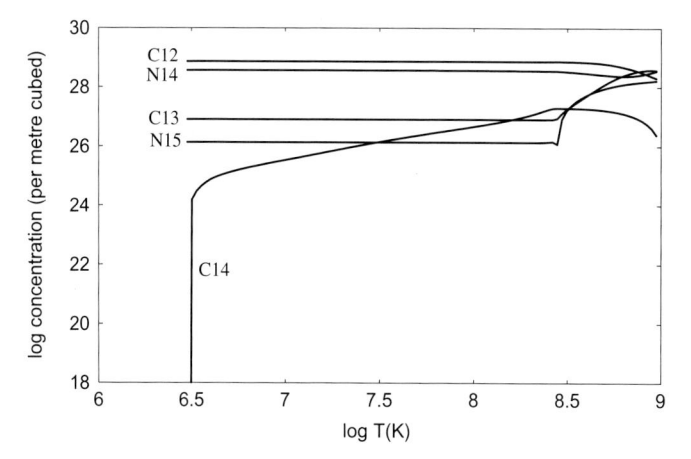

Fig. 17.10 The effective production of stable carbon and nitrogen isotopes (including decay products) and carbon-14.

We now consider the formation of a silicon carbide mineral with nitrogen gas trapped in interstices between grains. The silicon carbide will form from a mixture of unprocessed and some processed material but the amount by which the material had been processed would depend on which part of the collision region it had come from. The isotopic composition of the nitrogen released by heating the meteorite will depend not only on what was originally trapped in interstices between grains but also on the additional $^{14}_{7}$N released by the decay of $^{14}_{6}$C that will form part of the mineral lattice that comprises the grain. In Table 17.2 the result of taking various combinations of processed material and original material is shown with respect to the ratios $^{12}_{6}$C/$^{13}_{6}$C and $^{14}_{7}$N/$^{15}_{7}$N. The quantity p is the proportion of processed material in the mixture and r is the ratio of the mass of nitrogen in the meteorite to that of carbon. The results in the table illustrate the formation of heavy carbon (ratio less than 89.9) and of both light and heavy nitrogen (ratio greater than or less than 270). By varying the temperature from which the processed material is taken and the values of p and r within reasonable ranges, it is possible to find all the values for the isotopic ratios that have been observed and even ratios outside the observed limits.

17.4.4.3. Neon in silicon carbide

The measurements of neon isotopes in silicon carbide are of interest because they not only show an enhancement of $^{22}_{10}$Ne but also of $^{21}_{10}$Ne. A plot of $^{20}_{10}$Ne/$^{22}_{10}$Ne against $^{21}_{10}$Ne/$^{22}_{10}$Ne from different samples gives a nearly straight-line relationship. To explain this relationship we must consider the following scenario. When the silicon carbide

Table 17.2 Carbon and nitrogen isotopic ratios for various mixtures of materials.

Temperature of processed material (K)	p	r	$^{12}_{6}$C/$^{13}_{6}$C	$^{14}_{7}$N/$^{15}_{7}$N
2.79×10^8	3×10^{-10}	3×10^{-12}	89.7	1196
2.79×10^8	1×10^{-7}	3×10^{-9}	51.9	97.7
2.79×10^8	3×10^{-7}	1×10^{-10}	28.2	1794

grains form, a very tiny quantity of gases, including neon, becomes trapped in interstices within the grains. The neon gas that is trapped will be a mixture of neon produced by nuclear processing and non-processed neon from regions that were unaffected in their isotopic composition by the collision. Just as some neon becomes part of the silicon carbon grains, so these grains also contain traces of other elements, including sodium. Although the proportion of sodium in the grain will be extremely small — a trace — and $^{22}_{11}$Na will be only a small part of that, the decay of $^{22}_{11}$Na will add significantly to the $^{22}_{10}$Ne component. At a temperature of 2.86×10^8 K, the relative proportion of the stable neon isotopes is

$$^{20}_{10}\text{Ne} : {}^{21}_{10}\text{Ne} : {}^{22}_{10}\text{Ne} = 0.9281 : 0 : 0.0138 : 0.0581. \qquad (17.5)$$

At this temperature, about 1 part in 2,000 of the sodium present in the processed material is $^{22}_{11}$Na and after the passage of 20 years or so the product of its decay will have added to the $^{22}_{10}$Ne. We now assume that, in the material that contaminates the silicon carbide to variable extents, the $^{22}_{10}$Ne contribution by $^{22}_{11}$Na decay is 5.0 on the scale of proportions given by Eq. (17.5). This means that once the $^{22}_{11}$Na has decayed, the relative proportions of the three neon isotopes now becomes

$$^{20}_{10}\text{Ne} : {}^{21}_{10}\text{Ne} : {}^{22}_{10}\text{Ne} = 0.9281 : 0.0138 : 5.0581$$
$$= 0.1547 : 0.0023 : 0.8430, \qquad (17.6)$$

well on the way to becoming neon-E.

We now make the assumption that when the silicon carbide formed, it incorporated a mixture of normal neon, which we take as atmospheric composition, and products of the nuclear reaction that included neon and sodium in relative amounts characteristic of a temperature of 2.86×10^8 K. If the relative proportions of Eq. (17.3) and (17.6) are β and $1 - \beta$, then the relative amounts of the three isotopes, after $^{22}_{11}$Na decay, is

$$^{20}_{10}\text{Ne} : {}^{21}_{10}\text{Ne} : {}^{22}_{10}\text{Ne} = 0.9051\beta + 0.1547(1-\beta) : 0.0027\beta$$
$$+ 0.0023(1-\beta) : 0.0922\beta + 0.8430(1-\beta)$$

[Plot: Ne20/Ne22 vs Ne21/Ne22]

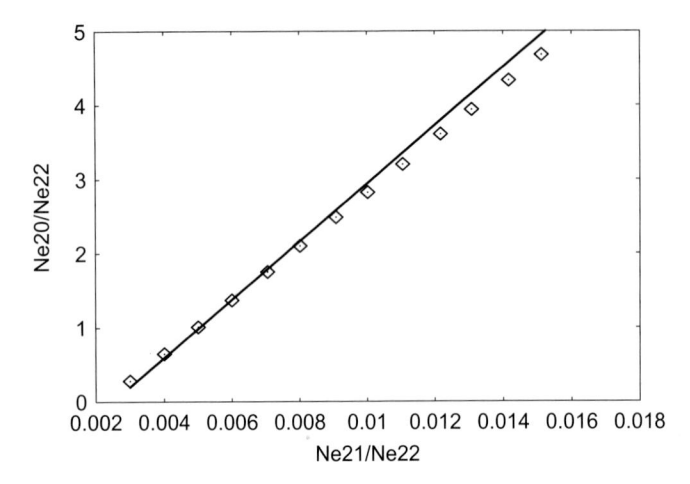

Fig. 17.11 The model points for the mixing on neon compositions based on Eq. (17.3) and (17.6). The proportion of composition (17.3) varies from 0.610 at the bottom point to 0.983 at the top point. The line is derived from observations of silicon carbide grains.

The value of β will vary from one silicon carbide sample to another. A plot of the ratios of $^{20}_{10}Ne/^{22}_{10}Ne$ and $^{21}_{10}Ne/^{22}_{10}Ne$ for various β is shown in Fig. 17.11. The American meteoriticists Zinner, Tang and Anders, who have made extensive studies of isotopic anomalies in silicon carbide grains in meteorites, found the linear relationship between these ratios.[4] According to which points they accepted, there were two possible, but very similar, straight lines. The equation of the average of the straight lines is

$$^{21}_{10}Ne/^{22}_{10}Ne = 0.00255(^{20}_{10}Ne/^{22}_{10}Ne) + 0.00249. \qquad (17.7)$$

The range of $^{20}_{10}Ne/^{22}_{10}Ne$ found by Zinner and his colleagues was between 0.23 and 5.0 that, in the present model, amounts to having β between 0.087 and 0.902. The upper points tend to fall below the line, a tendency also shown by the observations.

---

[4]Zinner, E., Tang, M. and Anders, E., 1989, *Geochim. Cosmochim. Acta*, **53**, 3273.

17.5. General Remarks Concerning Isotopic Anomalies

It will be noticed that the explanations for the various isotopic anomalies have involved different kinds of mixing process with processed material derived from different temperatures. It would, of course, be very convincing if there were just one composition of processed material that, mixed with some standard unprocessed material, explained all the anomalies — but that could not actually be so. The planetary collision gave rise to a wide variety of materials of different composition that mixed in a wide variety of ways. The patterns of development of the different isotopes are all very different and some kinds of mixtures of material give a discernable pattern, such as the three-isotope plot for silicon given in Fig. 17.9 that involves taking processed materials from lower temperatures. It is clear from Fig. 17.3 that mixtures of processed materials at the same temperatures that gave the silicon variations would show no anomalies for oxygen, since the isotopic composition of oxygen does not appreciably change until the temperature is much higher.

To test the validity of the model for isotopic anomalies I have proposed would involve finding the composition of single samples for a whole range of different isotopes for different elements and then seeing whether or not a particular mixture of materials at different temperatures could explain them all simultaneously. The data for doing such a test is not available, at least not in the public domain.

17.6. Summary

Isotopic anomalies are defined as departures from terrestrial isotopic composition that cannot be explained by mass-dependent fractionation. The anomalies found in meteorites fall into two main categories:

(1) Isotopic variations in individual samples that differ so markedly from terrestrial values that it seems almost inconceivable that they could come from the same source.

(2) Systematic variations of isotopic composition in grains of the same kind that are larger and of a different form from that expected by mass-fractionation.

The majority of isotopic anomalies in meteorites have been explained in terms of pre-solar grains, originating in distant sources, that have drifted through space and entered the Solar System, sometimes exchanging their anomalous isotopic content with normal solar-system material. The oxygen anomaly and those associated with silicon carbide grains have usually been explained in this way.

This chapter has shown that a planetary collision can trigger a series of nuclear reactions, starting with D-D reactions. If the material involved in the reactions is primarily volatile materials containing hydrogen, together with a small amount (about 1.6%) of normal cosmic gas consisting of hydrogen, helium and neon, and if the average D/H value is of order 0.02, which is towards the lower end of what recent observations allow, then temperatures can be reached at which nuclear reactions involving heavier elements can occur. These heavier elements are contained in the silicates that amount to about 24% of the total mass of the affected material.

At temperatures above 6×10^8 K, the processed material contains almost pure $^{16}_{8}$O that, added to SMOW in small concentrations, explains the oxygen anomaly observed in anhydrous material from carbonaceous chondrites and some chondrules. At temperatures above 2.5×10^8 K, quite a large amount of $^{26}_{13}$Al is produced that will then give rise to an excess of $^{26}_{12}$Mg proportional to the amount of aluminium in grains contained in CAIs. At similar temperatures there is a large increase in the amount of $^{25}_{12}$Mg present, which explains the large values of δ^{25}_{12}Mg in some CAIs.

Above about 2×10^8 K, a significant amount of $^{22}_{11}$Na is produced. In a meteorite grain containing sodium, even in trace amounts, if there is little or no neon initially present, then the neon expelled by heating will be neon-E, pure or nearly pure $^{22}_{10}$Ne.

Many isotopic anomalies have been observed for elements contained in silicon carbide grains. The three-isotope plot for silicon, with a slope of about 1.3, can be explained by mixing silicon processed to comparatively low temperatures, below 10^8 K, with terrestrial silicon. Mixing silicon from higher temperatures can explain other silicon anomalies that have been observed.

At temperatures above 3×10^8 K, there is a considerable increase in the production of both $^{13}_{6}$C and $^{15}_{7}$N that gives rise to heavy carbon and heavy nitrogen respectively. However, if there is very little nitrogen originally within the silicon carbide grains, then the decay of $^{14}_{6}$C, which is produced at quite low temperatures, to $^{14}_{7}$N can so enhance the $^{14}_{7}$N content that the result is light nitrogen.

Finally, the processed material at just below 3×10^8 K has an enhanced $^{21}_{10}$Ne content. When this is trapped in interstices of a silicon carbide grain, it also receives a boost in $^{22}_{10}$Ne content from the decay of trace amounts of sodium, containing $^{22}_{11}$Na. Admixtures of this neon with ordinary atmospheric neon can explain the observed linear relationship between $^{21}_{10}$Ne/$^{22}_{10}$Ne and $^{20}_{10}$Ne/$^{22}_{10}$Ne.

Chapter 18

Overview and Conclusions

18.1. What Constitutes a Good Theory?

It is natural for the originator of any theory to wish to present his creation as superior to any other previously given in the field and the only correct one. However, I must be cautious and objective — otherwise I would not be a credible scientist, which would be a poor basis from which to present a scientific theory. An American science historian, Stephen G. Brush, has made a study of theories of the origin of the Solar System and concluded that the 'lists of facts to be explained' presented by theorists do not provide a credible basis for choosing a best theory since the 'important features given by an author will be the ones with which his theory can best deal'.

There is an obvious danger in accepting the judgment of the originator of any new scientific concept or a work of art concerning the quality of what he has produced. In practice there is likely to be more honesty outside the scientific field than within it. There are gradations of literary quality and an author might willingly accept that his new book is quite good and very readable but perhaps not up to the standard of one that he, or someone else, had written previously. The same situation could occur in painting, sculpture or architecture but it is inconceivable that a scientist would do the same. If a theory is obviously not better, in the mind of its originator, than those that precede it, then he should not present it at all — but flawed theories do sometimes get presented and strongly defended.

Many individuals have pondered the problem of what defines a good scientific theory without having a theory of their own to promote. The earliest such individual of any fame and stature was an English Franciscan monk, William of Occam (1285–1349), who enunciated a principle usually referred to as *Occam's razor*. This was given in Latin, the *lingua franca* of educated people of his time, as '*Entia non sunt mutiplicanda praeter necessesitatem*'. The literal translation is 'Entities should not be multiplied beyond necessity' but the implication is that simplicity is a virtue in creating theories, that the number of assumptions should be kept to a minimum and that if more than one theory explains the facts, then the simplest is to be preferred. Of course, the important aspect of a theory is that it should 'explain the facts', which must have priority over simplicity. Others, usually scientists, throughout the ages, have enunciated the principle that was given in Occam's razor in different words. Thus Isaac Newton wrote in his *Principia*, 'We are to admit no more causes of natural things as are both true and sufficient to explain their appearances' and, more recently, Albert Einstein gave the advice, 'Make everything as simple as possible, but not more simple'.

The Austrian-born, later British, philosopher of science Karl Popper (1902–1994) made the following comments[1] about scientific theories:

1) It is easy to obtain confirmations, or verifications, for nearly every theory — if we look for confirmations.
2) Confirmations should count only if they are the result of *risky predictions*; that is to say, if unenlightened by the theory in question, we should have expected an event which was incompatible with the theory — an event which would have refuted the theory.
3) Every "good" scientific theory is a prohibition: it forbids certain things to happen. The more a theory forbids, the better it is.
4) A theory which is not refutable by any conceivable event is non-scientific. Irrefutability is not a virtue of a theory (as people often think) but a vice.

---

[1] Karl Popper, *Conjectures and Refutations*, London: Routledge and Keagan Paul, 1963.

5) Every genuine *test* of a theory is an attempt to falsify it, or to refute it. Testability is falsifiability; but there are degrees of testability; some theories are more testable, more exposed to refutation, than others; they take, as it were, greater risks.
6) Confirming evidence should not count *except when it is a genuine test of the theory*; and this means that it can be presented as a serious but unsuccessful attempt to falsify the theory. (I now speak in such cases as "corroborating evidence.")
7) Some genuinely testable theories, when found to be false, are still upheld by their admirers — for example by introducing *ad hoc* some auxiliary assumption, or by reinterpreting theory *ad hoc* in such a way that it escapes refutation. Such a procedure is always possible, but it rescues the theory from refutation only at the price of destroying, or at least lowering, its scientific status.

Not all philosophers of science agree with Popper and it is not clear that his criteria apply with equal force to all kinds of theory. In particular, it is difficult to assess to what extent these criteria and comments should apply to a theory of the origin of the Solar System. When Copernicus put forward his heliocentric theory, he did so on the basis of simplicity; it removed the need to explain the apparent to-and-fro motions of planets as seen from Earth, explained by Ptolemy as due to a complex set of motions involving epicycles, and substituted it with much simpler circular motion around the Sun. Essentially it was an improvement only in terms of Occam's razor, since it did not give much better agreement with the observed motions of the planets. It was simpler — and therefore to be preferred. Later Galileo showed that observations of Venus, sometimes as a small, almost-complete disk and sometimes as a much larger crescent, were incompatible with Ptolemy's theory and was explained by that of Copernicus. Although, for theological reasons, the 16th century Catholic Church did not accept the evidence, to an objective mind it would have been convincing proof of the superiority of Copernicus' theory over Ptolemy's. But I am not sure how well the theory as a whole would stand up to the criteria propounded by Popper.

Given the above considerations of what constitutes a plausible theory, I shall just summarise what has been written in the previous chapters and leave it to the reader to judge whether or not it is plausible and any better than its predecessors. Notice I have used the word 'plausible' and not 'correct' because it is impossible to assign 'correctness' as a label to any theory. Not only that, we can also say that any given scientific theory is much more likely to be wrong than to be correct — and I include the theory given in this book. Was Newton's theory of gravitation correct? No — because it was supplanted by Einstein's general theory of relativity and, who knows, that too may be supplanted in due course.

The arrogance of certainty is a corrosive influence in the search for truth — a search that is never-ending. I will conclude this section with another quote from Karl Popper: 'Whenever a theory appears to you as the only possible one, then take this as a sign that you have neither understood the theory nor the problem which it was intended to solve.'

18.2. Protostars and Stars

The basic premise of the Capture Theory is that, in a star-forming cool dense cloud, condensed stars, either on or close to the main sequence, coexist with both protostars and compressed regions of cloud that may or may not become a protostar and eventually a star. What evidence is there for that scenario?

The evidence for the existence of cool dense clouds is clear enough — they block out the light from the stars beyond them and form dark patches in the night sky, as seen in Fig. 2.1. Infrared observations reveal the temperatures of these clouds and also detect the presence of condensed, but cool, objects within the clouds that are identified as protostars. Another signature of protostar formation is the maser emission that comes from discrete sources within the clouds (Section 2.3). The exact mode of production of this maser radiation, characteristic of molecules such as water, carbon monoxide, carbon dioxide and many others, is not well understood but of its existence there is no doubt. An important deduction from the maser emission, detected by Doppler-effect frequency shifts, is that there is considerable turbulence within the clouds, with turbulent speeds of order $10\,\mathrm{km\,s^{-1}}$. Well-founded theory indicates that colliding turbulent streams produce regions of hot compressed gas. Other theory, equally well founded, indicates that such heated regions quickly cool, through dust radiation and by molecular, atomic and ionic cooling, on a timescale much shorter than that required for re-expansion of the compressed region. This leaves a cool compressed region of the cloud that, given enough time and if not disrupted by further collisions, would form a protostar and eventually a star.

There are about as many binary systems observed as there are individual stars and from theory by James Jeans, and also by numerical modelling, it can be shown that a likely outcome for a collapsing protostar is that at some stage it will bifurcate and form a binary pair (Section 2.5).

Stars that acquire more mass after they first form tend to spin more rapidly, as is observed for more-massive stars. However, stars that do not acquire extra mass will lose angular momentum due to the coupling of stellar-wind particles to the star's magnetic field lines, so explaining the low spin rates of low-mass late-type stars. This mechanism for losing angular momentum is particularly effective during the T-Tauri active period in the star's evolution.

The process of the collapse of a protostar, first to the form of a young stellar object (YSO) and then to become a main-sequence star, is well understood, as is the formation of the end-product, dependent on mass, which is one of a white dwarf, neutron star or black hole (Chapter 3). The process of forming an open cluster of Population I stars within a star-forming cloud takes tens of millions of years, thus ensuring the coexistence of compressed regions, protostars and condensed stars over a considerable period.

18.3. Creating the Conditions for the Capture-Theory Process

The formation of an open cluster of stars within a star-forming region, such as the Trapezium Cluster within the Orion Nebula (Fig. 4.1), takes place within a collapsing sub-cloud within the whole nebula. The gaseous sub-cloud, and the stars and protostars within it, are all taking part in this general inward motion during which new compressed regions, protostars, YSOs and main-sequence stars are continually being formed. As a consequence of this behaviour pattern, the number density of objects of various kinds within the sub-cloud steadily increases. These embryonic clusters are held together by the gravitational influence of the gas within which they are enclosed and they are referred to as embedded clusters (Section 4.1). In embedded clusters stellar number densities in the range of 100–10^5 pc^{-3} have been observed; in the Trapezium Cluster the stellar number density has been estimated to be in the range of 10^4–10^5 pc^{-3}. The embedded state lasts about 5 million years after which time the earliest formed massive stars become supernovae. Gas is driven out by the energy of these events and, freed of its gravitational influence, the cluster begins to expand, either to form an open cluster or, in 90% of cases, to expand indefinitely so that the contained stars become isolated field stars.

In a region of stellar number density of 10^4 pc^{-3}, the average distance between stars is about 9,600 au. If protostars are present, with an initial radii of 1,000 to 2,000 au, that stay in an extended state for 20,000 years or so, plus an unknown number, probably large, of compressed regions produced by turbulent collisions, then the condition is established for frequent interactions between condensed stars and low-density objects.

In the environment of a dense embedded cluster, frequent interactions between stars would be expected. Close interactions could lead to the break-up of binary systems and this idea is supported by the observed relationship between the frequency of binary systems and the local stellar density (Section 4.3). The direct formation of massive stars by the collapse of a single protostar mass is prevented by the radiation pressure of the forming core pushing material outwards and it has been suggested that accretion by stellar collisions in an embedded cluster could resolve this difficulty (Section 4.2).

18.4. The Capture-Theory Process

In developing his 1916 theory of planet formation from material drawn out of the Sun by a passing star, Jeans showed theoretically that a gaseous filament would be gravitationally unstable and break up to form a series of blobs, like beads on a string. If the blobs have a suitable combination of mass, density and temperature, then they can collapse to form planets. Within the embedded cluster the conditions exist for a rather distorted version of the Jeans mechanism. Instead of two condensed stars there is a condensed star and a protostar (or compressed region); instead of a condensed star being disrupted to form planets from a filament, the protostar is stretched into a filament and the planets are not retained but captured by the other body. The capture-theory process has been convincingly modelled and is extremely robust; planet formation and capture take place over a very wide range of parameters, including those parameters that would be relevant within embedded clusters (Section 6.1). Some planets are not captured but escape into the galaxy to form the free-floating planets that have been observed.

It was observed in 2008 that a very young planet associated with the star HL Tau may have been produced by the close interaction of HL Tau with a nearby binary system, XZ Tau, one member of which is a protostar. This suggests that the planet may be the outcome of a capture-theory interaction — although an alternative proposed origin was suggested.[2]

The initial orbits produced by the capture-theory process are very extended, with apastron distances of order 1,000–2,000 au, and with high eccentricities — typically 0.9. Some of the protostar material forms planetary condensations, some is captured to form a disk

---

[2]Greaves, J.S., Richards, A.M.S., Rice, W.K.M. and Muxlow, T.W.B., 2008, *Mon. Not. R. Astr. Soc.*, **391**, L74.

around the star and some impinges directly on the star and is absorbed. The disk material acts as a resisting medium and the numerical modelling of orbital evolution shows that orbits decay and round off on a timescale of about 1 million years, within the expected lifetime of the disks (Section 7.4). The final semi-major axes of the orbits depend on the mass of the disk and its duration, and accounts for the wide range now known to exist for exoplanets since direct visual images of exoplanets have revealed some of them at more than 100 au from planets. If the star is very active while the orbits are decaying, then the resultant slowing-down of the medium motion may be such as to produce final planetary orbits with high eccentricity, consistent with observations for some exoplanets (Section 7.5).

18.5. The Frequency of Planetary Systems

From the number density of stars, the characteristics of protostars and knowledge of average relative speeds of stars in embedded clusters, it is possible to calculate how often capture-theory interactions should occur and what proportion of solar-type stars should thereby acquire a family of planets. The conclusion from such an analysis is that for dense embedded clusters, about one half of solar-type stars should have accompanying planets but the proportion would be lower for less dense clusters (Section 8.2).

However, a dense stellar environment is not just favourable to the formation of planetary systems but also to their disruption. Immediately after formation, with planets on very extended orbits, the probability per unit time of disruption is high. Once the orbits have completely evolved, the probability is negligible. A Monte Carlo-style simulation of interactions between stars and planetary systems during the period of orbital evolution has shown that between one-third and two-thirds of the planets would be lost but that most stars would retain all or some of their original complement of planets (Section 8.3). The estimated proportion of solar-type stars with planetary companions from theory comfortably accommodates the estimate from observations.

18.6. Satellite Formation

The simulations of planet formation show that protoplanetary blobs produced in the filament condense to form dense central bodies surrounded by disks containing about one-half of the total mass (Fig. 6.2). Application of the theory developed by solar-nebula theorists for producing planets, when applied to the material of a planetary disk indicates that the processes of forming a dust carpet, producing condensations, known as satellitesimals, in the carpet and the accumulation of satellitesimals to form satellites takes place on a timescale of order hundreds of thousands of years (Section 9.5). This is also the timescale for round-off and decay of the satellite orbits due to motion within the disk. There are no timescale problems with this mechanism on the small scale of satellite formation, such as occurs when planets are to be formed in the same way from a circumstellar disk. There are also no angular momentum difficulties since the relationship of relative angular momentum of primary and secondary bodies for planet-satellite systems is completely different from that for the Sun-planet system (Table 9.1).

18.7. The Tilts of Spin Axes of the Planets and Stars

Due to the gravitational field of the circumstellar disk, the net force on a planet in its evolving orbit is not centrally directed towards the star. This creates a gyroscopic effect so that the orbit undergoes precession, the instantaneous rate of which depends on the characteristics of the orbit at any time. Although the orbits, with various inclinations, do not intersect in general, because their precessions are at different rates they do so occasionally. When they intersect — or nearly intersect — there is a possibility that planets could closely approach and tidally interact with each other. The net effect of tidal interactions is to impart angular momentum to planets, which tilts their spin axes away from their original directions. This mechanism is particularly effective when the planets are at the protoplanet stage and still fairly extended. The various planetary tilts of the major planets can be explained in this way; in particular, the numerical simulation of a close passage between an early Jupiter and Uranus, originally with spin axes perpendicular to their orbital planes, has reproduced the extreme tilt of the Uranus axis (Section 11.2). Such close interactions could also be responsible for the loss of much of the atmospheres of Uranus and Neptune, converting them from gas giants into ice giants.

In the nature of the capture-theory process, there is no systematic relationship between the original direction of the spin axis of the star and the plane of the star-protostar motion. Material from the protostar absorbed by the star, and the additional absorption of small solid particles in the disk due to the Poynting–Robertson effect, adds angular momentum to the star that pulls its spin axis closer to the normal to the orbital plane of the planets. Depending on the relative magnitudes of the original and acquired angular momentum of the star, its final spin axis can point in almost any direction but with a strong tendency to make an angle less than 90° with the angular momentum vector of the planetary orbits (Section 7.8). In the case of the Sun this angle is 7°. Many exoplanets have been recorded with orbits inclined at considerable angles to the plane perpendicular to the stellar spin axis, but still within 90°. An observation in August 2009 showed a planet with an inclination, relative to the spin axis of the star, of 150°. Such a result is to be expected for a small minority of planetary systems from the Capture Theory.

18.8. A Planetary Collision — Earth and Venus

The initial planets formed by the Capture Theory would all have had the characteristics of gas-giant major planets with iron and silicate cores and with an atmosphere accounting for most of the mass. Just as the precession of evolving orbits gave the possibility of close interactions between planets, it also gave a smaller possibility of an actual planetary collision. A model of an initial Solar System with six major planets has been postulated for which the probability that some pair of them may have collided has been estimated at about 0.1. Support for the possibility of such an event also comes from a NASA report in August 2009 of the detection of the residue of a planetary collision a few thousand years ago around a young star, HD 172555.

A collision between two planets, Bellona, with twice the mass of Jupiter, and Enyo, with 22% more mass than Saturn, has been simulated using smoothed-particle hydrodynamics (Section 11.5). An important factor in the event was the presence of a layer of material close to the planetary cores, derived from grains of solidified hydrogen-containing materials, with an enhanced D/H ratio due to the action of grain-surface chemistry (Section 11.4). The temperature generated by the collision is sufficient to initiate D-D nuclear reactions, which set off an extensive series of nuclear reactions involving heavier elements. The energy thus generated enhances the effect of the collision in dispersing planetary material (Appendix AD).

A major outcome of the collision is the residues of the two planetary cores, both of masses similar to those of the larger terrestrial planets, moving apart from a location in the present vicinity of the Earth to take up independent heliocentric orbits. In one simulation these were found to take up orbits that would round off close to those of the Earth and Venus. The residues of Bellona and Enyo are identified as the Earth and Venus respectively.

18.9. The Moon, Mars and Mercury

The explanation of the relationship of the Moon to the Earth has long been a subject of speculation. In recent times the favoured explanation has been that Moon formation was due to a collision of Earth with a body with about the mass of Mars. The Moon was then formed from material coming from the two bodies that was left in orbit around Earth. The Capture Theory offers an alternative scenario.

Bellona and Enyo, like the existing major planets, would have had satellite families. One possible outcome of the planetary collision for a satellite is that its parent planet retains it. It is proposed that the Bellona fragment, that formed Earth, retained a satellite of Bellona — the Moon. This explanation has the advantage that the Moon is not regarded as special in any way and was formed like all other large satellites.

The hemispherical asymmetry of surface features and the internal structure of the Moon support this explanation (Section 12.4). The side facing the collision would have been heavily bombarded, so abrading away a considerable thickness of crust. Seismometers on the Moon have confirmed a thinner crust on the near side. Subsequent impacts by large bodies excavated large basins all over the Moon's surface but only on the bombarded hemisphere was magma able to break through to create mare basins. The arrangement of mass within the Moon that caused one hemisphere to be tidally locked facing Bellona would, through tidal forces, have locked the same hemisphere towards Earth.

Mars and Mercury are best understood as very large one-time satellites of Bellona, a planet with twice the mass of Jupiter. The density of Mars links it more comfortably with some of the larger satellites than with the larger terrestrial planets and the low mass of Mercury is more like that of a satellite — although its high density requires an explanation (Section 13.1).

Like the Moon, Mars has hemispherical asymmetry, being divided into northern volcanic plains and southern highlands. Since it went into an independent heliocentric orbit, and was not tidally linked to a planet, there was no external body to influence the arrangement of its surface features relative to its spin axis. It underwent a process similar to continental drift that, to satisfy a minimum energy condition while conserving angular momentum, required material to move as far as possible from the spin axis. An analysis shows that the present arrangement of surface features closely satisfies this condition (Appendix AI).

The iron core of Mercury is similar in size to that of Mars and, in common with suggestions made in the past, it is proposed that Mercury was once very similar to Mars but had much of its mantle removed. If Mercury was a satellite very close to Bellona, then the planetary collision, giving heavy bombardment by debris, offers a plausible mechanism for removing large quantities of mantle material (Section 13.3).

18.10. Neptune, Triton and Pluto

Triton, the large retrograde close satellite of Neptune, and Pluto, a dwarf planet with an orbit apparently linked in some way to that of Neptune, are both small bodies with curious features. The Capture Theory offers a scenario that explains the relationship of these three bodies and their main features. The starting point is Pluto as a regular satellite of Neptune and Triton as an escaped satellite from the collision on an orbit that ranged out well beyond the orbit of Neptune. On an inward journey towards the Sun, Triton collided with Pluto, ejecting it on a path similar to its present one. The sideswipe collision set Pluto into a retrograde spin and also sheared off the material that produced the satellites Charon, Nix and Hydra. For its part Triton was captured into a highly eccentric retrograde orbit around Neptune. Tidal effects between a planet and satellite in a retrograde orbit round off the orbit very quickly (Section 14.3).

Another satellite of Neptune, Nereid, has an orbit of eccentricity 0.76, the largest of any satellite in the Solar System. This could have been caused by perturbation due to Triton as it approached its meeting with Pluto but Nereid is more probably a captured body.

18.11. Small Bodies of the Solar System

The dwarf planet Ceres, once regarded as the largest asteroid, is most easily interpreted as an escaped satellite that happened to end up between Mars and Jupiter. The three dwarf planets within the Kuiper Belt also have the characteristics of escaped satellites that were thrown into very extended orbits. The major planets were also initially on very extended orbits and over the course of time, as the planetary orbits evolved, a close interaction between a dwarf planet and a major planet could deflect the dwarf planet into an orbit that kept it beyond the present orbit of Neptune (Section 15.1.2).

Cometary material is identified as the debris from the colliding planets, containing a considerable amount of volatile material, that came together to form solid bodies. Coming from the outer regions of the planetary cores, it would have been thrown out furthest and it too would have interacted with the major planets in their evolving orbits. Cometary material deflected into the Kuiper Belt region would be relatively safe from disturbance by any other solar-system body, especially after the major planets had settled into their final orbits. However, perturbation mainly by Neptune on bodies in the inner region of the Kuiper Belt provides a steady supply of short-period comets. Approaches of stars or Giant Molecular Clouds close to the Solar System perturb the comet system outwards; at first this augmented the Oort cloud and then maintained it once it had become established (Section 15.2).

Asteroids were derived from material closer to the centres of the colliding planets that contained little or no volatile material. Some would be condensed, or even solid, material that left a planet as a coherent mass; others would be accumulations of grains and chondrules that assembled under gravitational forces and became compressed into a rocky form. Those asteroids that kept clear of planets during the early stages of orbital evolution of the planets and then ended up in a safe zone, such as between Mars and Jupiter, have survived to form the present population (Section 15.2.1).

The asteroids of various types collide from time to time and the fragments so produced that go into Earth-crossing orbits constitute the meteorites that fall to Earth.

18.12. The Characteristics of Meteorites

The general classes of meteorites — stones, irons and stony-irons — reflect the types of asteroids from which they come and reveal something about the events which led to the formation of those asteroids (Chapter 16). The presence of chondrules show that high-temperature silicate vapours existed from which the glassy spheres condensed. The non-equilibrated minerals within the chondrules indicate that, although the temperature at one stage was very high, cooling from that temperature was extremely fast. This scenario is entirely consistent with a planetary collision (Appendix AM).

Achondrites appear to be similar to terrestrial surface rocks and would have originated in regions away from the heart of the collision. Chondritic meteorites derive from amalgamated material consisting of chondrules in a matrix of compressed grains. Carbonaceous chondrites were formed from later condensing material that contained a high proportion of volatiles — although they also contain refractory CAIs.

Iron meteorites, which contain a great deal of nickel, are noted for showing Widmanstätten figures, nickel-poor kamacite plates surrounded by nickel-rich taenite. From the size of the plates, cooling rates of 1–10 K per million years are estimated, indicating that the source bodies were of asteroid size.

The appearance of stony-irons depends on their origin. Pallasites, where stony globules are enclosed in a metal matrix, are the result of the solidification of material within which molten silicate material was rising in a gravitational field within molten metal. The other type of stony-irons, mesosiderites, appear to be the product of silicate and iron material being splattered together — something likely to happen in the violent environment of a planetary collision.

Perhaps the most intriguing property of meteorites is that many of them contain isotopic anomalies that, in the past, have often been explained by the presence of pre-solar grains that travelled from distant stars and became part of the forming Solar System. The presence of a sufficient deuterium content in the hydrogen contained within the collision region enables D-D nuclear reactions to raise the temperature to à level at which a reaction chain involving many other light elements is initiated. A calculation, including the effect of 274 nuclear reactions, plus reverse reactions, on a mixture of materials likely to be present in the collision region, gives an outcome that explains isotopic anomalies associated with oxygen, magnesium, aluminium, carbon, nitrogen, silicon and neon (Chapter 17).

18.13. Conclusions

I started this chapter with a discussion of what constitutes a good theory. The reader must judge whether or not the ideas concerning planet formation and the evolution of the Solar System presented here constitute a good theory, by which I mean one that is plausible.

Of course, there may be flaws in the science or flaws in the reasoning but if I thought so I would have not written this book.

Something I *would* claim for the ideas I have presented is that they have coherence. There are a series of causally related events and processes that go from the turbulence within star-forming regions to the various events that shape the details of the Solar System. It is an historical narrative, not a series of *ad hoc* unconnected short stories. It also connects with observations quite well. The Capture Theory has comfortably accommodated recent observations that have caused grief to the standard model — for example:

1. There is no problem with a large range of orbital radii for exoplanets since a planet's initial orbit takes it to a large distance from its star and the extent to which its orbit decays depends on the characteristics of the resisting medium. These characteristics are very variable.
2. The occasional incompatible senses of stellar spin and planetary orbit are also easily accommodated since there is no reason for them to be initially compatible and the final spin of the star depends on a combination of initial spin angular momentum and that subsequently acquired.

Another feature of the capture-theory and planetary-collision models I have presented is that postulates that were made from time to time seem to be supported to a greater or lesser extent by subsequent theory or observations. Examples of this are:

3. Dormand and I introduced the idea of a planetary collision in 1977. Its plausibility depended on the precession of the evolving elliptical orbits and the accompanying probability of about 0.1 deduced from mathematical analysis. The detection of the outcome of a planetary collision around a young star with newly formed planets, announced by NASA in August 2009, gives support to the plausibility of such a hypothesis.
4. The announcement in 2008 of the detection of a recently formed planet in the vicinity of HL Tau, with the suggestion that it may have been caused by the influence of another star, XZ Tau, passing

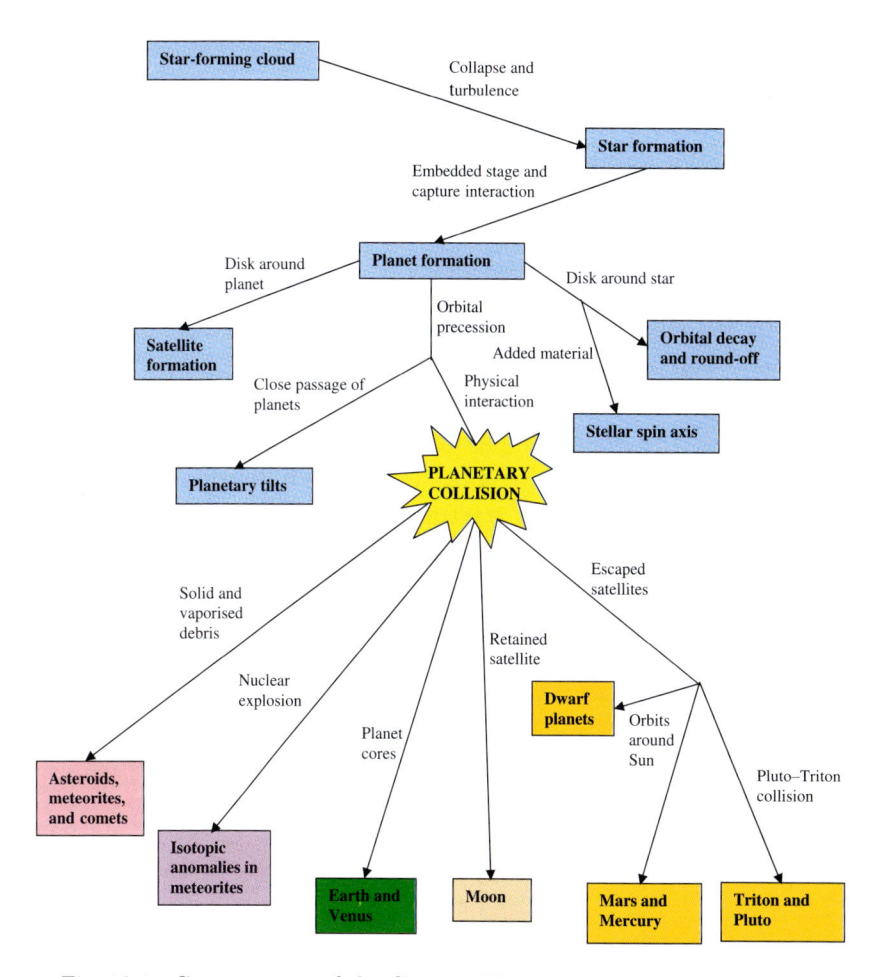

Fig. 18.1 Consequences of the Capture Theory and a planetary collision.

close to it 1,600 years ago, may indicate that a capture-theory process took place.

5. When Paul Holden and I presented our ideas on the formation of isotopic anomalies in 1995, we could only provide a handwaving justification that a collision would give a temperature high enough to trigger D-D reactions. The later numerical analysis, described in Appendix AD, has supported the assumption we made.
6. The D/H ratio Paul Holden and I took was based on the assumption that the planetary atmosphere of at least one of the colliding

planets would be similar to that of Venus. Observations, starting in 2002, have shown the presence of highly enhanced D/H ratios in the icy grains contained in molecular clouds and protostars due to the action of grain surface chemistry. Protoplanets produced from the material of protostars or condensed regions would also have these high D/H ratios in the icy grains within them.

Of course, these observations and theory, and how they seem to relate to the capture-theory and planetary-collision models, prove nothing — but at the very least they are not in conflict with the models and give no reason to modify or abandon them.

Following the adage 'a picture is worth a thousand words', I conclude my narrative with Fig. 18.1, which illustrates the flow of ideas I have described, starting from the formation of stars. It illustrates how the fall-out from a single event — a planetary collision — explains so many disparate features of the Solar System. The blue items are those that might be expected for any planetary system. The remainder, the consequences of the planetary collision, relate only to the Solar System.

Appendix A

Angular Momentum

Angular momentum is a quantity that describes a property of a body orbiting or spinning around some axis — like a child's top, for example.

Consider a small particle of mass m moving in a circular path around an axis AB, as shown in Fig. A.1. The angular momentum associated with the rotation of the particle around the axis is

$$J = mr^2\omega \tag{A1}$$

where r is the distance of the particle from the axis and ω is the angular velocity of the rotation, usually expressed in radians per second. A radian is a natural unit of angle and is the angle subtended at the centre of a circle by an arc of the circle equal in length to its radius. One radian is equivalent to $57.29°$.

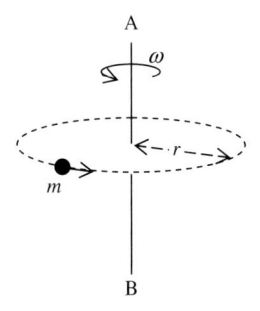

Fig. A.1 A particle in circular motion around an axis.

If we now consider a large body in rotational motion around an axis, then its total angular momentum can be found by dividing it up into a large number of smaller bodies (particles) and adding together the quantities given by (A1) for each of them. If we are dealing with a rigid body, then the value of ω will be the same for each of the small bodies into which it is divided. However, if it is not rigid, as would be the case if it were liquid or gaseous, then ω might be different from one particle to another.

Angular momentum is a *vector quantity*, meaning that it has both a magnitude and a direction. The angular momentum of the particle shown in Fig. A.1 is along the direction BA, *i.e.* it can be represented as an arrow of length J pointing upwards. If the direction of the rotation were reversed, then it would be an arrow of length J pointing downwards.

When adding vector quantities, their directions must be taken into account. A two-dimensional illustration is given in Fig. A.2 where the vector sum, **J**, is given of five individual vectors $\mathbf{dJ}_1$, $\mathbf{dJ}_2$, $\mathbf{dJ}_3$, $\mathbf{dJ}_4$ and $\mathbf{dJ}_5$. Note that it is customary to represent vectors in bold type.

Angular momentum has the important property that in any isolated system, not acted on by forces from outside, it remains constant. Within the system itself there may be a great number of physical interactions — collisions or those due to gravity — but these will not affect the angular momentum of the system as a whole, although the angular momenta of various parts of the system might change.

An illustration of the principle of conservation of angular momentum may be seen within the area of exhibition ice-skating. An

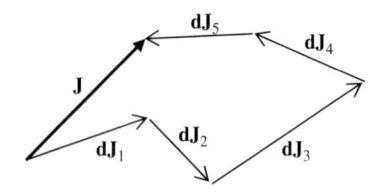

Fig. A.2   **J** is the vector sum of the five vectors $\mathbf{dJ}_1$ to $\mathbf{dJ}_5$.

ice skater may sometimes be seen pirouetting gracefully with arms outstretched. Suddenly she draws her arms in close to her body and she spins much more rapidly. The reason for this is seen in (A1). Since her contact with the outside world is mainly through her skates, and frictional forces between the skates and the ice are small, she is an approximation to an isolated system. For this reason her angular momentum will be very nearly conserved when she draws in her arms. For each part of the mass contained in her arms, the value of r, the distance from the spin axis, is reduced, so to keep angular momentum constant the value of ω must increase; that is, she spins more rapidly.

Appendix B

Equipotential Surfaces of a Tidally Distorted Star

Figure B.1 shows the tidally distorted star with its mass, M, concentrated at point S and the external star represented by its mass, M', at point T. The gravitational potential at point P due to the mass M at S and the mass M' at T is represented by Ω_P, which is given by

$$\Omega_\mathrm{P} = -\frac{GM}{r} + V_\mathrm{T} \tag{B1}$$

where G is the gravitational constant and V_T is the tide-raising potential due to M'.

While the potential at P due to M' is GM'/r', to find the part giving a tidal force we have to subtract that part of the potential that is giving the acceleration (force per unit mass) of S, where all the mass is concentrated, towards T. This part would come from a

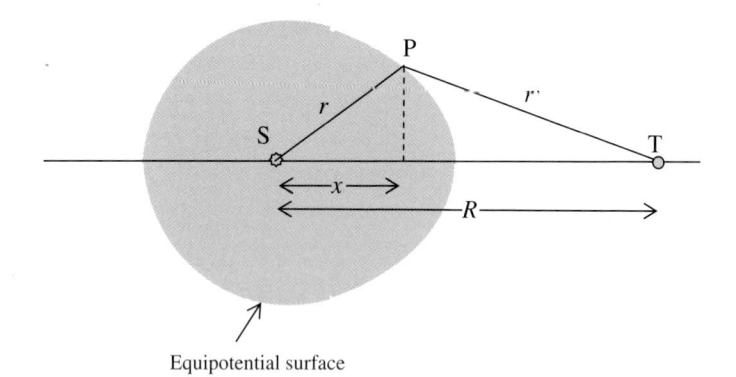

Fig. B.1 An equipotential surface of a tidally distorted star.

potential $GM'x/R^2$ that, when differentiated with respect to x, gives the required acceleration.

Thus the total tide-raising potential at point P is

$$\Omega_{\rm P} = -\frac{GM}{r} - \frac{GM'}{r'} + \frac{GM'x}{R^2} \tag{B2}$$

The equipotential surface is one for which the potential given by (B2) has the same value for all points P on the surface.

Appendix C

The Instability of a Gaseous Filament

Jeans' analysis of the problem of the break-up of a uniform gaseous filament was based on a detailed consideration of the physical processes involved. In Jeans' description of the instability, the stream takes on a wavelike variation of density along its length and the distance, l, between the condensations is the wavelength of the variation. Assuming that the factors controlling the wavelength are: c, the speed of sound in the gas which controls the speed with which a disturbance would move through the filament, G, the gravitational constant, and ρ, the density of the gas, then it is possible to derive the form of the relationship giving l by dimensional analysis. This involves deriving a relationship that gives the same dimensions on the two sides of the equation without considering the detailed physics.

The equation given by Jeans was

$$l = \left(\frac{\pi}{\gamma G \rho}\right)^{1/2} c \tag{C1}$$

where γ is the ratio of the principal specific heats of the gas. Dimensional analysis gives this result but without the numerical constant $(\pi/\gamma)^{1/2}$.

Since

$$c = \left(\frac{\gamma k T}{\mu}\right)^{1/2} \tag{C2}$$

where k is Boltzmann's constant, T the absolute temperature of the gas and μ its mean molecular mass, then, combining (C1) and (C2),

$$l = \left(\frac{\pi kT}{G\rho\mu}\right)^{1/2}. \tag{C3}$$

Appendix D

The Jeans Critical Mass

In deriving the Jeans critical mass one can use an important relationship known as the *Virial Theorem* (Appendix P), which is a special case of a more general theorem given by the French mathematician and theoretical physicist Henri Poincaré. In the present application it says that for a body in equilibrium, neither expanding nor contracting,

$$2E + V = 0 \tag{D1}$$

where E is the total translational kinetic energy of the material in the system and V its potential energy.

We now consider a uniform sphere of gas of radius R, density ρ and absolute temperature T. For a mass of gas without any motion of the body as a whole, the translational kinetic energy is just the thermal motion of the individual molecules of the gas. This is $\frac{3}{2}kT$ per molecule of gas, where k is Boltzmann's constant. For a mass M of gas, the number of molecules is M/μ where μ is the mean molecular mass. From this we find

$$E = \frac{3}{2}kT \times \frac{M}{\mu}. \tag{D2}$$

The potential energy of a uniform spherical sphere of mass M and radius R is given by

$$V = -\frac{3}{5}\frac{GM^2}{R}. \tag{D3}$$

From (D1), (D2) and (D3) we find the Jeans critical mass as

$$M_J = \frac{5kTR}{G\mu}.\tag{D4}$$

Expressing R in (D4) in terms of M_J and ρ and then rearranging gives

$$M_J = \left(\frac{375k^3T^3}{4\pi G^3\mu^3\rho}\right)^{1/2}.\tag{D5}$$

Figure IX shows the Jeans critical mass for various combinations of density and temperature for $\mu = 4 \times 10^{-27}$ kg, corresponding to a mixture of hydrogen atoms, hydrogen molecules and helium such as occurs in many astronomical bodies.

Appendix E

The Lynden-Bell and Pringle Mechanism

The basis of the Lynden-Bell and Pringle mechanism is that, for a rotating system in which energy is being lost but angular momentum is remaining constant, inner material will move further inward while outer material will move further outward. This amounts to a transfer of angular momentum from inner material to outer. That this is so can be shown very simply. Consider two bodies in circular orbits, with radii r_1 and r_2 around a central body of much greater mass. The energy, E, and angular momentum, J, of the system are

$$E = -C\left(\frac{1}{r_1} + \frac{1}{r_2}\right) \tag{E1}$$

and

$$J = K(\sqrt{r_1} + \sqrt{r_2}) \tag{E2}$$

where C and K are two positive constants. For small changes in r_1 and r_2 the changes in E and J are

$$\delta E = C\left(\frac{1}{r_1^2}\delta r_1 + \frac{1}{r_2^2}\delta r_2\right) \tag{E3}$$

and

$$\delta J = \frac{1}{2}K\left(\frac{1}{\sqrt{r_1}}\delta r_1 + \frac{1}{\sqrt{r_2}}\delta r_2\right). \tag{E4}$$

If angular momentum remains constant then, from (E4),

$$\delta r_1 = -\sqrt{\frac{r_1}{r_2}}\delta r_2 \tag{E5}$$

and substituting this in (E3) gives

$$\delta E = \frac{C}{\sqrt{r_2}}\left(\frac{1}{r_2^{3/2}} - \frac{1}{r_1^{3/2}}\right)\delta r_2. \tag{E6}$$

Given that δE is negative, it is clear that if $r_2 < r_1$ then δr_2 must be negative, that is to say, the inner body moves inwards and hence, from (E5), the outer body moves outwards.

Appendix F

Grains in Molecular Clouds

The vast majority of the dust grains in cold molecular clouds, and in the objects produced in those clouds, are less than 1 micron in diameter and often much smaller. They reveal themselves in two ways. The first is by blocking out light because they are opaque and the second is by preferentially scattering blue light so that objects viewed through a dust cloud appear to be reddened. Scattering is much more effective with sub-micron particles than with larger particles and, for a given mass of material, small particles offer a much larger absorption cross-section.

From absorption and scattering studies it has been concluded that the distribution function for the numbers of particles with different diameters, D, is of the form

$$n(D) = KD^{-3.5} \tag{F1}$$

where K is a constant. The number of particles with diameter between D and $D + dD$ is $n(D)dD$. This distribution tails off very sharply for higher values of D. On the assumption that dust particle diameters vary between 0.01 and 5.0 microns, suggested by some observations, Fig. F.1 shows the proportion of all particles with diameters from 5 microns down to some specified diameter. The proportion has to be shown on a log scale because it is so sharply peaked but it will be seen that the proportion of particles with diameters greater than 1 micron is about 10^{-5}.

A dust particle has an absorption cross-section $\pi D^2/4$ and the total absorption cross-section for particles of diameter between D

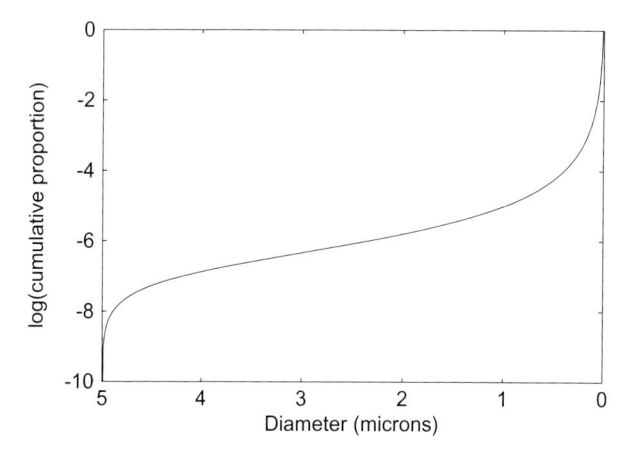

Fig. F.1 Proportion of dust particles with diameters from 5 microns down to some specified diameter.

and $D + dD$ is

$$dA = n(D)dD \times \pi D^2/4 = \pi K D^{-1.5} dD/4. \quad (F2)$$

The total absorption cross-section for particles with diameters between D_1 and D_2 is

$$A(D_1, D_2) = \frac{\pi K}{4} \int_{D_1}^{D_2} D^{-1.5} dD = \frac{\pi K}{2} \left(\frac{1}{D_1^{1/2}} - \frac{1}{D_2^{1/2}} \right). \quad (F3)$$

From this the total absorption cross-section for particles between the top limit of diameter, D_T (5 microns), and diameter D as a proportion of the total cross-section for all particles is

$$P_A(D) = \frac{A(D, D_T)}{A(D_L, D_T)}, \quad (F4)$$

where D_L is the lower limit for a diameter, 0.01 micron. This function of D is shown in Fig. F.2. This is less extreme than Fig. F.1 but still shows that absorption is dominated by the smallest particles. Particles greater than 1 micron in diameter account for only about 6% of all absorption.

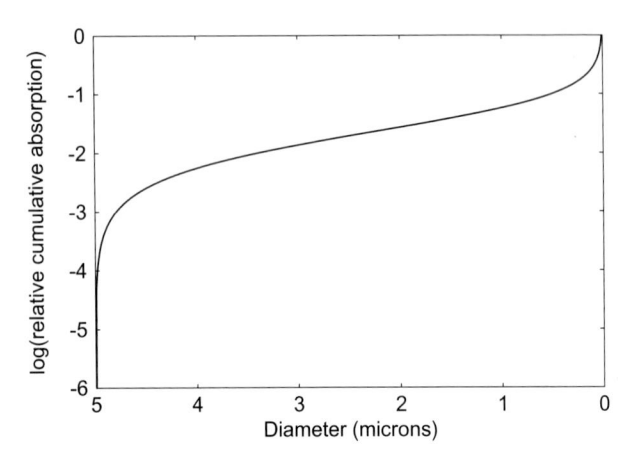

Fig. F.2 The proportional cumulative cross-section from the largest diameter down to D.

Finally we come to contributions of total mass. The mass of the particles with diameter between D and $D + dD$ is

$$dM = CD^{-1/2}dD \qquad (F5)$$

where C is a constant and, by steps similar to those that gave the absorption cross-section results, the proportion of mass of particles

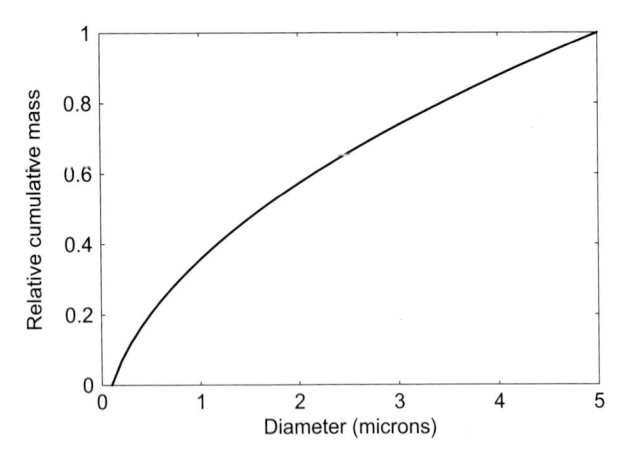

Fig. F.3 The proportional cumulative mass from the smallest dust particles to some diameter D.

with diameters between the lower limit D_L and some value D is

$$P_M(D) = \frac{\sqrt{D} - \sqrt{D_L}}{\sqrt{D_T} - \sqrt{D_L}}. \tag{F6}$$

This distribution is shown in Fig. F.3, which shows quite clearly that most of the mass of the dust is in particles with diameter greater than 1 micron, those most difficult to detect visually.

There are many uncertainties in this analysis, in particular the upper limit for the size of a dust particle and also the lower limit, to which the results shown in Figs. F.1 and F.2 are quite sensitive. There is also some doubt as to whether the distribution (F1) is valid over the complete diameter range. Nevertheless, the results found here do suggest that there could be a substantial contribution of mass by larger particles that would settle to form a dust carpet on a timescale short enough to satisfy the requirement of the lifetime of circumstellar disks.

Appendix G

The Structure of a Spiral Galaxy

Galaxies are of two main types, *elliptical*, in which the stars are located within an ellipsoidal volume without distinctive internal features, and *spiral*, in which the main structure is within a disk-like region and consists of concentrations of stars within spiral arms emanating from a central region. The Milky Way is a spiral galaxy and we can see what it is like by looking at similar galaxies in different orientations. Figure G.1 shows the Pinwheel Galaxy, a fine example of a spiral galaxy seen along the normal to the *galactic plane*. The spiral arms are very clear. At the centre of the galaxy is a bright region, the *nucleus*, within which the number density of stars is very high. The nucleus can also be seen in Fig. G.2 that shows the Sombrero Galaxy, a spiral galaxy viewed along a line making a small angle with

Fig. G.1   The Pinwheel Galaxy (HST/NASA).

Fig. G.2 The Sombrero Galaxy (HST/ESA/NASA).

the galactic plane. This view shows another feature of spiral galaxies, the *halo*, a diffuse flattened sphere of stars and stellar clusters that encompasses most of the disk.

The Solar System is situated in a spiral arm about two-thirds of the way out from the galactic centre.

Appendix H

The Centre of Mass and the Orbits of Binary Stars

Consider a system of N particles for which the ith particle has mass m_i and coordinates (x_i, y_i, z_i). The centre of mass of the system is at (X, Y, Z) where

$$X = \frac{\sum_{i=1}^{N} m_i x_i}{\sum_{i=1}^{N} m_i}, \quad Y = \frac{\sum_{i=1}^{N} m_i y_i}{\sum_{i=1}^{N} m_i}, \quad Z = \frac{\sum_{i=1}^{N} m_i z_i}{\sum_{i=1}^{N} m_i}. \quad \text{(H1)}$$

If, for example, the system of particles were in a uniform gravitational field, then it would be in perfect balance about a fulcrum placed at the centre of mass.

In the case of a binary system, the centre of mass is situated on the line joining the two stars (Fig. H.1). If the stars at S_1 and S_2 have masses M_1 and M_2 respectively, then, with the centre of mass at C,

$$\frac{S_1 C}{S_2 C} = \frac{M_2}{M_1}. \quad \text{(H2)}$$

From (H2) it will be seen that the centre of mass is closer to the more massive of the two stars.

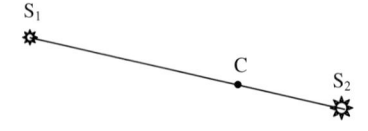

Fig. H.1 Two stars and their centre of mass.

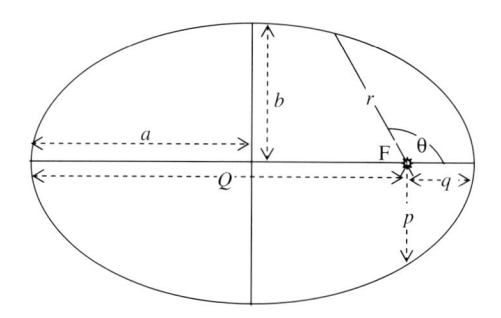

Fig. H.2 A general ellipse.

For close binary stars, because of their mutual tidal effects, their orbits around the centre of mass tend to be close to circular. However, in general, orbits are elliptical; an elliptical orbit is shown in Fig. H.2.

I shall describe the orbit in terms of a planet, of negligible mass, in orbit around the Sun. The Sun is at point F, a *focus* of the ellipse, a is the *semi-major axis* and b the *semi-minor axis*. The quantities a and b are related through e, the *eccentricity* of the ellipse, by

$$b^2 = a^2(1 - e^2). \tag{H3}$$

The ellipse may be described in terms of polar coordinates, with the origin at F by

$$r = \frac{a(1 - e^2)}{1 + e\cos\theta}, \tag{H4}$$

where θ is the angle shown in Fig. H.2. The *semi-latus rectum* of the ellipse, p, is the value of r when $\theta = \pm\pi/2$ or

$$p = a(1 - e^2). \tag{H5}$$

The closest distance of the planet from the Sun, the *perihelion distance*, q in Fig. H.2, is given by putting $\theta = 0$ in (H4) and is

$$q = a(1 - e) \tag{H6}$$

while the furthest distance, the *aphelion distance*, Q, is given by putting $\theta = \pi$ in (H4) and is

$$Q = a(1 + e). \tag{H7}$$

332 *On the Origin of Planets*

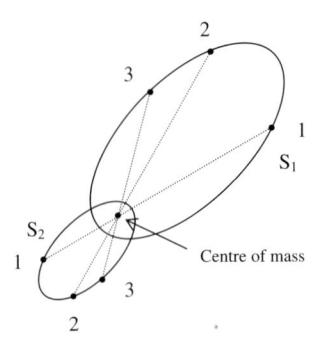

Fig. H.3 The elliptical orbits of the two stars with corresponding positions.

Returning now to binary systems, in Fig. H.3 the orbits of the two stars are shown with their positions at three times. The stars move such that the centre of mass is on their line of centres with a position satisfying (H2). The centre of mass is also the focus of each of the orbits, corresponding to point F in Fig. H.2.

Just as a final comment on centres of mass, it is often stated that the Earth moves in an elliptical, almost circular, orbit around the Sun, but this is not strictly true. What does move in this orbit is the centre of mass of the Earth–Moon system and the Earth and the Moon are both in orbit around this point. Thus the motion of the Earth around the Sun is a sinuous fluctuation around an elliptical path with roughly 13 fluctuations per year.

Appendix I

The Doppler Effect

Anyone attending a Formula I Grand Prix race, or even watching a race on television, will have experienced the Doppler effect. As the cars approach, so the pitch, or frequency, of the engine noise is higher that it would be if the engine were running at the same speed at rest. Then, as the car goes through its nearest point and recedes, the pitch falls noticeably.

This phenomenon, known as the Doppler effect, was first described by the Austrian physicist, Christian Doppler (1803–1853). According to Doppler's equation for describing the frequency shift, if the vehicle approaches at 20% of the speed of sound (*i.e.* at about 240 kilometres per hour), the frequency heard will be increased by 20% and, similarly, if the vehicle recedes at 20% of the speed of sound, the frequency heard will be 20% less. For any wave motion, the speed of the wave, v, is related to its frequency, f, and its wavelength, λ, by the relationship

$$v = f\lambda \tag{I1}$$

and since the speed of sound is independent of frequency, this means that an increase in frequency gives a decrease in wavelength and *vice versa*.

The Doppler effect applies to the observations of wave motions of any kind. The quantity that is most easily measured in the case of light is its wavelength for which very precise instruments, interferometers, are available. If a source of light is moving along the line of sight at speed v with respect to an observer, then the radiation of wavelength λ that it emits will be seen by the observer with wavelength $\lambda + d\lambda$ where

$$\frac{d\lambda}{\lambda} = \frac{v}{c}, \qquad (I2)$$

in which c is the speed of light, $299{,}792\,\text{km}\,\text{s}^{-1}$. If the source is moving away from the observer, then v is positive and the observed wavelength is longer than that emitted. The change is towards the red end of the spectrum and the light is said to be *red-shifted.* Conversely, if the motion is towards the observer, then $d\lambda$ is negative, the observed wavelength is less than that emitted and the light is said to be *blue-shifted.*

Fortunately, nature gives some very precise wavelengths on which to make measurements, the spectral lines shown in Fig. 1.3. These lines provide well defined wavelength markers that can be used to detect small changes of wavelength due to Doppler shifts. With the most precise interferometers, speeds of approach or recession can be measured to within a accuracy of about $1\,\text{m}\,\text{s}^{-1}$.

To be absolutely correct, Eq. (I2) applies to light only for speeds v that are much less than the speed of light — which is valid for all the cases in which we are interested. However, some very distant galaxies are moving away from Earth at a considerable fraction of the speed of light and to estimate the wavelength shift in such cases one must use the *relativistic Doppler effect*, which is different, and somewhat more complicated, than (I2).

Appendix J

Atomic Energy Levels and Stellar Spectra

Atoms consist of a tiny nucleus, containing protons with a positive charge and uncharged neutrons, surrounded by electrons equal in number to the protons and each having a negative charge equal in magnitude to a proton charge. In 1913 the Danish physicist Niels Bohr (1885–1962) explained the discrete spectral lines of the hydrogen spectrum by showing that its single atomic electron could only have a set of discrete energies that were proportional to $-1/n^2$ where n is an integer. This condition on the energies comes from the assumption that the electrons were in circular orbits around the nucleus, just a proton for hydrogen, and that the angular momentum of the electron in its orbit was of the form $nh/(2\pi)$ where h is Planck's constant. If an electron jumps from a higher energy state with energy E_2 to a lower energy state with energy E_1, a *photon* (a packet of light energy) is emitted with a frequency ν such that

$$h\nu = E_2 - E_1. \tag{J1}$$

This model agreed with observations for hydrogen but not for more complex atoms. In 1926 the German physicist Erwin Schrödinger (1887–1961) developed the theory of wave mechanics in which material particles were considered to have wave-like properties. The result of this development, the Schrödinger wave equation, can be applied to atomic electrons and gives the Bohr energies for hydrogen and valid results for many-electron atoms in general.

Figure J.1 shows the energy levels for hydrogen. Since the energy levels are all negative, the lowest energy is for $n = 1$.

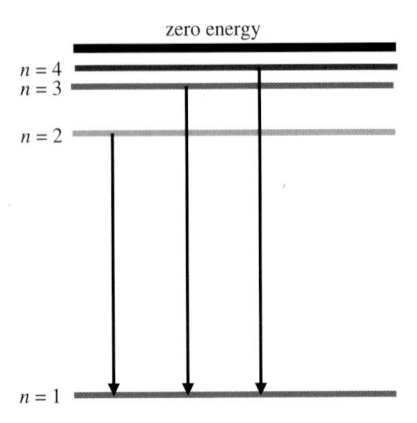

Fig. J.1 The energy levels for hydrogen.

The process of the absorption of light by the material above the photosphere depends on the following mechanism. If an electron in an atom, or ionized atom, in the material is at an energy level E_1 and is encountered by a light photon with energy *precisely* that required to raise it to energy E_2, i.e. of energy given by (J1), then it has a high probability of being absorbed and raising the energy of the electron to the higher level. This means that the light traversing the material selectively loses photons corresponding to the energies of allowed electronic transitions and the corresponding wavelengths are weakened, or even missing, in the transmitted light. This gives rise to the Fraunhofer lines.

If hydrogen is raised to a high temperature, collisions between the atoms have the effect of pushing electrons into higher energy states or may even eject electrons from the atom. We can illustrate the general principle of how the intensities of spectral lines change with stellar temperatures by considering the Hγ line shown in Fig. 1.9. The absorption of light for this line is due to the transfer of an electron from level $n = 2$ to level $n = 5$. Starting with the coolest M-type stars, the temperature is too low to excite the electrons of many hydrogen atoms to the level $n = 2$, hence there are very few transitions from that state and the absorption line is absent. At a temperature corresponding to a G0 star, there are a sufficient number of hydrogen atoms with electrons in the $n = 2$ state for the Hγ line to appear faintly. As the temperature further increases, so more and more atoms with electrons with $n = 2$ are present in the absorbing material and the absorption line gets stronger and stronger up to class A1, after which it gradually becomes weaker. This is because at the highest temperatures most electrons have been excited to a *higher* level than $n = 2$ and the higher the temperature, the fewer electrons at the $n = 2$ level able to give the Hγ transition and so the line becomes weaker.

A similar situation occurs for other absorption lines but since the energies will be different, so will be the temperature conditions corresponding to the peak intensity and the temperatures at which they first appear and then fade away. By comparing line strengths for many lines, quite accurate stellar temperature estimates can be made.

Appendix K

Stellar Masses from Observations of Binary Systems

In the following treatment of spectroscopic binary systems, two simplifying assumptions will be made. The first assumption is that the orbits of the stars are circular, which will certainly be true for close binaries where tidal dissipation effects tend to produce near-circular orbits. The second assumption is that the line of sight to the binary is in the plane of its orbit, which means that twice per orbital period the motions of the stars will be such that one of them is moving directly towards, and the other directly away from, the observer.

In Fig. K.1 the two stars, moving around their centre of mass, are moving directly towards and directly away from the observer. Their speeds relative to the centre of mass (Appendix H) are v_1 and v_2 and the component of the speed of the centre of mass along the line of sight is V. The star radial velocity components relative to the observer are $v_{1,1} = V - v_1$ and $v_{1,2} = V + v_2$. One-half orbital period

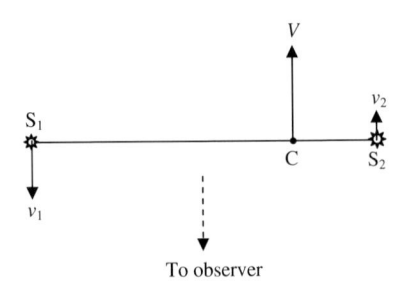

Fig. K.1 The velocity, V, of the centre of mass and the velocities of the stars, v_1 and v_2, with respect to the centre of mass.

later, the radial velocity components are $v_{2,1} = V + v_1$ and $v_{2,2} = V - v_2$. Using Doppler-shift (Appendix I) measurements, $v_{1,1}$, $v_{1,2}$, $v_{2,1}$ and $v_{2,2}$ can be found directly from the individual stars for a visual binary and from the splitting of spectral lines for a spectroscopic binary. From just three of the four speed measurements, V, v_1 and v_2 can be found — the fourth measurement can be used as a check on accuracy.

Having measured the period of the stellar orbits, it is possible to find the distance between the two stars and their individual masses. First the period P gives the angular speed in the orbits and then one has

$$\frac{2\pi}{P} = \omega = \frac{v_1}{r_1} = \frac{v_2}{r_2} \qquad (K1)$$

from which r_1 and r_2 can be found. The ratio of the distances gives the ratio of the stellar masses by

$$\frac{M_1}{M_2} = \frac{r_2}{r_1}. \qquad (K2)$$

The laws of orbital motion give

$$\omega^2 = \frac{G(M_1 + M_2)}{(r_1 + r_2)^3}, \qquad (K3)$$

which enables the sum of the masses to be found. A combination of (K2) and (K3) gives the individual masses.

A basic problem in determining masses is that the line of sight will, in general, be inclined to the plane of the orbits by some angle α and this means that the speeds of the stars will be underestimated by a factor of $\cos(\alpha)$. This will also be the factor by which the distances will be underestimated (Eq. K1), leading to an underestimate of masses by a factor $\cos^3(\alpha)$ (Eq. K3). However there is a circumstance when true masses can be found, which is for an eclipsing binary when one star passes in front of the other; the occurrence is shown by the reduction in the brightness of the combined stars when one eclipses the other. When this happens it is known that the line of sight is in the orbital plane and the analysis we have given here will yield the correct masses.

For visual binaries the passage of the stars in their orbits can be directly observed. Since the stars are seen individually, then, if they are main-sequence stars, their spectral class can be determined, so giving an estimate of mass. The position of the centre of mass can be found as the intersection of lines joining the two stars at several points in their orbits (Fig. H.3). This enables the ratio of the masses to be found from a relationship such as (K2) where r_1 and r_2 are the distances of the stars at some time. The ratio can give a confirmatory check on the estimates of mass found from the spectral classes.

Appendix L

Smoothed-Particle Hydrodynamics

Smoothed-particle hydrodynamics is a computational technique in which a fluid body is represented by a set of particles.[1] Associated with the ith particle is a mass m_i, a velocity $\mathbf{v}_i$, a position $\mathbf{r}_i$ and specific internal energy (internal energy per unit mass) u_i. At each point of the fluid being modelled, the properties are a weighted average of the properties of the neighbouring particles, with the weight function, the *kernel*, monotonically decreasing with distance. The kernel falls to zero at some distance and is a function of h, the *smoothing length*. Ideally there should be at least 20 particles within the averaging region and 40–50 particles are commonly used. The smoothing length is adjusted at each time step for each particle to ensure that the desired number of particles is obtained for the averaging.

The kernel, $W(\mathbf{r}, h)$, is a normalised function so that

$$\int W(\mathbf{r}, h) dV_{\mathbf{r}} = 1, \tag{L1}$$

where the integral is taken over the kernel volume. As an example of its use, the density at point j in the fluid is given by

$$\rho_j = \sum_i m_i W(r_{ij}, h) \tag{L2}$$

---

[1] Lucy, L.B., 1977, *Astron. J.*, **82**, 1013; Gingold, R.A. and Monaghan, J.J., 1977, *Mon. Not. R. Astr. Soc.*, **181**, 375.

where the points i are those in the kernel region and r_{ij} is the distance from point j to point i. For the value of a general quantity, q, at point j we write

$$q_j = \sum_i \frac{m_i}{\rho_i} q_i W(r_{ij}, h). \tag{L3}$$

Thus the velocity of the material at point j is estimated as

$$\mathbf{v}_j = \sum_i \frac{m_i}{\rho_i} \mathbf{v}_i W(r_{ij}, h). \tag{L4}$$

The property of differentiability of the kernel is important when it is necessary to calculate the divergence or gradient of some property. To find ∇q then, we evaluate

$$\nabla q_j = \sum_i \frac{m_i}{\rho_i} q_i \nabla W(r_{ij}, h) \tag{L5}$$

and to find $\nabla \cdot \mathbf{s}$, where $\mathbf{s}$ is a vector quantity we evaluate

$$\nabla \cdot \mathbf{s}_j = \sum_i \frac{m_i}{\rho_i} \mathbf{s}_i \cdot \nabla W(r_{ij}, h). \tag{L6}$$

A problem with methods of this kind is that when there are shocks the results can be physically unrealistic with large oscillations in the region of the shock. The behaviour of the system can be improved by introducing *artificial viscosity* that broadens the shock and spreads its effect over a small volume. The equations that have to be solved are then

$$\frac{d\mathbf{r}_j}{dt} = \mathbf{v}_j \tag{L7}$$

and

$$\frac{d\mathbf{v}_j}{dt} = -\nabla \phi_j - \frac{1}{\rho_j} \nabla P_j + \nu_j^a \tag{L8}$$

where ϕ_j is the gravitational potential, P_j the pressure, ν_j^a the artificial viscosity and j now refers to one of the particles. If the contribution of particle i on particle j is not the same as the contribution of particle j on particle i, then momentum will not be conserved. A symmetrical form of the second term on the right hand side of (L8) that solves this problem and also introduces artificial viscosity is

$$\sum_i \left(\frac{P_i}{\rho_i^2} + \frac{P_j}{\rho_j^2} + \Pi_{ij} \right) \nabla W \left(r_{ij}, \frac{h_i + h_j}{2} \right). \quad (L9)$$

In (L9)

$$\Pi_{ij} = \begin{cases} \dfrac{-\alpha \overline{c_{ij}} \mu_{ij} + \beta \mu_{ij}^2}{\overline{\rho_{ij}}} & \mathbf{v}_{ij} \cdot \mathbf{r}_{ij} \leq 0 \\ 0 & \mathbf{v}_{ij} \cdot \mathbf{r}_{ij} > 0 \end{cases} \quad (L10a)$$

where

$$\mu_{ij} = \frac{\overline{h_{ij}} \mathbf{v}_{ij} \cdot \mathbf{r}_{ij}}{|\mathbf{r}_{ij}|^2 + \eta^2}, \quad (L10b)$$

$\overline{c_{ij}}, \overline{\rho_{ij}}$ and $\overline{h_{ij}}$ are the means of the sound speeds, densities and smoothing lengths for particles i and j, the numerical factors α and β are usually taken as 1 and 2 respectively and $\eta^2 \sim 0.01 \overline{h_{ij}}^2$ is included to prevent numerical divergences.

The gravitational potential at point j due to the surrounding points is usually of the form

$$\phi_j = -\sum_i \frac{G m_i}{(r_{ij}^2 + \xi^2)^{1/2}} \quad (L11)$$

that allows for the fact that each mass represents a distribution of mass rather than a point mass.

Updating the specific internal energy is done via

$$\frac{du_j}{dt} = \sum_i \frac{1}{2} m_i \left(\frac{P_i}{\rho_i^2} + \frac{P_j}{\rho_j^2} + \Pi_{ij} \right) \mathbf{v}_{ij} \cdot \nabla W \left(r_{ij}, \frac{h_i + h_j}{2} \right). \quad (L12)$$

Smoothed-particle hydrodynamics can be applied to matter in any state if the appropriate state function is used. For a gas with the normal ratio of specific heats γ, the equations

$$P = (\gamma - 1)\rho u \qquad \text{(L13)}$$

and

$$P = \frac{\rho k T}{\mu}, \qquad \text{(L14)}$$

where k is Boltzmann's constant, μ the mean molecular mass and T the absolute temperature, complete the requirements for a smoothed-particle hydrodynamics simulation.

Appendix M

Free-Fall Collapse

Our model for free-fall collapse is that for a uniform sphere of mass M, initial radius r_0 and initial uniform density ρ collapsing from rest under the force of gravity alone. The effect of pressure forces is not included; it is as though the model consisted of a uniform distribution of small masses throughout the spherical volume.

Figure M.1 shows a small particle at point P, at distance x from the centre of the sphere, experiencing an acceleration inwards due to the gravitational attraction of the shaded region. Its acceleration is given by

$$\frac{d^2x}{dt^2} = -\frac{4}{3}\pi\rho Gx. \tag{M1}$$

Since the equation is linear in x, the acceleration and the velocity at different points will always be proportional to the distance of the point from the centre. The implication of this is that at all times

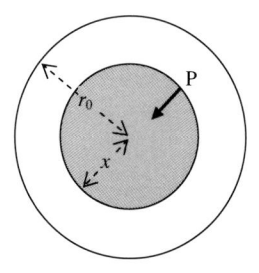

Fig. M.1 The acceleration of a particle at P is due to only the gravitational effect of the shaded region.

during the collapse, the density remains uniform throughout the sphere.

The total energy of any particle must remain uniform throughout the collapse. Equating the potential energy for a boundary particle at the beginning of the collapse to the total energy at some subsequent time, when the radius of the sphere is r, we have

$$-\frac{GM}{r_0} = -\frac{GM}{r} + \frac{1}{2}\left(\frac{dr}{dt}\right)^2, \qquad (M2)$$

or

$$\frac{dr}{dt} = -\left\{2GM\left(\frac{1}{r} - \frac{1}{r_0}\right)\right\}^{\frac{1}{2}}. \qquad (M3)$$

The negative sign is taken on the right-hand side because the sphere is collapsing and r is reducing with time. Substituting $r = r_0 \sin^2 \theta$ gives

$$\frac{d\theta}{dt} = -\left(\frac{GM}{2r_0^3}\right)^{\frac{1}{2}} \frac{1}{\sin^2 \theta}. \qquad (M4)$$

If in time t the radius of the sphere is r_t, then, from (M4)

$$t = -\left(\frac{2r_0^3}{GM}\right)^{\frac{1}{2}} \int_{\pi/2}^{\theta_t} \sin^2 \theta \, d\theta, \qquad (M5)$$

where $\theta_t = \sin^{-1}\{(r_t/r_0)^{1/2}\}$.

Integrating the right-hand side and expressing in terms of r_t gives

$$t = \left(\frac{r_0^3}{2GM}\right)^{\frac{1}{2}} \left\{\frac{\pi}{2} - \sin^{-1}\left[\left(\frac{r_t}{r_0}\right)^{\frac{1}{2}}\right] + \left(\frac{r_t}{r_0}\right)^{\frac{1}{2}}\left(1 - \frac{r_t}{r_0}\right)^{\frac{1}{2}}\right\}. \qquad (M6)$$

The time for complete collapse, known as the *free-fall time*, t_{ff}, is found by making $r_t = 0$ in (M6), giving

$$t_{ff} = \frac{\pi}{2}\left(\frac{r_0^3}{2GM}\right)^{\frac{1}{2}} = \left(\frac{3\pi}{32\rho G}\right)^{\frac{1}{2}}. \qquad (M7)$$

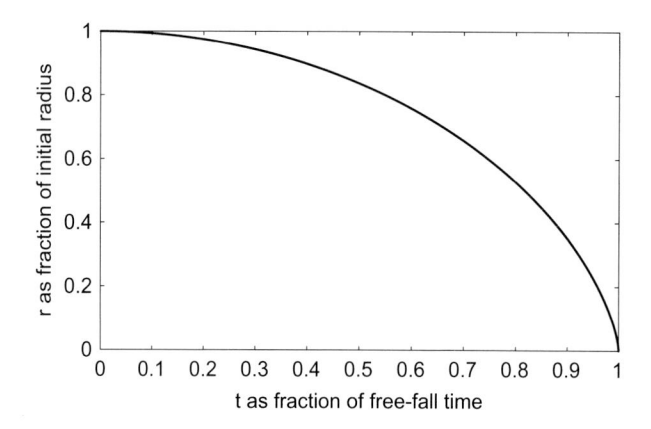

Fig. M.2 The relationship between radius and time in free fall of a uniform sphere.

From (M6) and (M7),

$$\frac{t}{t_{ff}} = 1 - \frac{2}{\pi}\left\{\sin^{-1}\left[\left(\frac{r_t}{r_0}\right)^{\frac{1}{2}}\right] - \left(\frac{r_t}{r_0}\right)^{\frac{1}{2}}\left(1 - \frac{r_t}{r_0}\right)^{\frac{1}{2}}\right\}. \quad \text{(M8)}$$

Figure M.2 shows the variation of r_t/r_0 with time expressed in units of the free-fall time. It will be seen that the collapse is very slow at first but gradually speeds up and is very rapid in the final stages. For an initially diffuse gas sphere of considerably greater mass than the Jeans critical mass, the first stage of the collapse will be similar to free fall. The free-fall time will be a reasonable estimate of the time taken to collapse to a high density.

Appendix N

Fragmentation and Binary Characteristics

As an example to illustrate the formation and characteristics of a binary system produced by fragmentation, we shall take the case considered in Chapter 2 of a protostar with mass $M = 3 \times 10^{30}$ kg, radius $R = 3 \times 10^{14}$ m (2,000 au) and spin speed $\omega_s = 5 \times 10^{-14}$ radians per second. The total angular momentum of a sphere, with density a function of distance from the centre, spinning on its axis is of the form

$$J_{sph} = \alpha M R^2 \omega_s, \tag{N1}$$

where α is called the *moment-of-inertia factor*. For a uniform sphere, to which the initial protostar approximates, $\alpha = 0.4$; for bodies with their mass concentrated towards the centre, which is true for most astronomical bodies, α is less than 0.4. Inserting numerical values, we find $J_{sph} = 5.4 \times 10^{45}$ kg m^2 s^{-1}.

When the final binary system is formed, its angular momentum must equal J_{sph} and will be contained in two main types of motion. The first is the orbits of the two stars, taken as point masses, about their centre of mass and the second is the spin of the stars about their spin axes. The second contribution is normally negligible compared with the first, since, despite having spin rates much higher than the orbital angular velocity of the binary, the final stars have dimensions very much smaller than their distance apart. We shall assume that the binary, considered as two point masses, takes up all the original angular momentum.

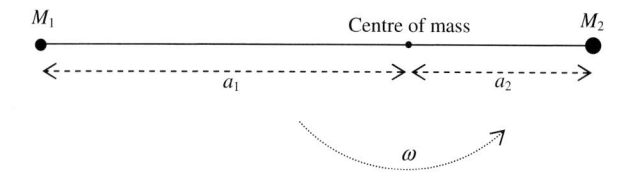

Fig. N.1 A binary system.

Figure N.1 represents the binary system with stars, of mass M_1 and M_2 at distances a_1 and a_2 from the centre of mass, in circular orbits with angular speed ω.

The mass M_1 is in a circular orbit of radius a_1 around the centre of mass. Equating the centripetal acceleration with the centripetal force per unit mass,

$$a_1\omega^2 = \frac{GM_2}{(a_1+a_2)^2}. \tag{N2}$$

The total angular momentum associated with the orbits of both stars is

$$J = \left(M_1 a_1^2 + M_2 a_2^2\right)\omega = M_2 a_2 (a_1+a_2)\omega. \tag{N3}$$

In (N3) we have used

$$M_1 a_1 = M_2 a_2 \tag{N4}$$

which comes from the definition of the centre of mass.

Squaring both sides of (N3) and substituting for ω^2 from (N2), we have

$$J^2 = GM_1 M_2^2 a_2. \tag{N5}$$

We now use the relationships

$$\frac{M_2}{a_1} = \frac{M_1}{a_2} = \frac{M_1+M_2}{a_1+a_2}$$

to give

$$a_1 = \frac{M_2}{M_1+M_2}(a_1+a_2) \quad \text{and} \quad a_2 = \frac{M_1}{M_1+M_2}(a_1+a_2). \tag{N6}$$

Substituting in (N5) for a_1 and a_2 from (N6)

$$J^2 = \frac{GM_1^2 M_2^2 a}{M_1 + M_2} \qquad (N7)$$

in which $a = a_1 + a_2$ is the separation of the stars in the binary system.

Equating J^2 to J_{sph}^2 we find

$$a = \frac{J_{sph}^2 (M_1 + M_2)}{GM_1^2 M_2^2}. \qquad (N8)$$

With $M_1 = 10^{30}$ kg and $M_2 = 2 \times 10^{30}$ kg, this gives $a = 3.28 \times 10^{11}$ m ≈ 2.19 au. From (N6) we find $a_1 = 2.19 \times 10^{11}$ m and then from (N2) $\omega = 7.52 \times 10^{-8}$ radians s^{-1}. The period of the binary orbit is

$$P = \frac{2\pi}{\omega} = 8.37 \times 10^7 \text{ s} = 2.64 \text{ years}. \qquad (N9)$$

Appendix O

Spin Slowing Due to a Stellar wind

The case we consider is that of a star with mass M, radius R, moment-of-inertia factor α and initial angular speed ω_0, losing mass at a rate m per unit time with the lost mass coupled to the magnetic field out to a distance βR. The assumption we make is that over the period of loss we are considering, the mass and the radius of the star change so little that we can take them as constant.

The angular momentum of the star is

$$J = \alpha M R^2 \omega \tag{O1}$$

so that

$$\frac{dJ}{dt} = \alpha M R^2 \frac{d\omega}{dt}. \tag{O2}$$

The rate of loss of angular momentum of the star equals the rate of gain of angular momentum of the lost material. The stellar wind leaving the surface per unit time has angular momentum $mR^2\omega$ and is decoupled from the magnetic field line with angular momentum $\beta^2 m R^2 \omega$. Hence

$$\frac{dJ}{dt} = -(\beta^2 - 1) m R^2 \omega = \alpha M R^2 \frac{d\omega}{dt}. \tag{O3}$$

This gives

$$\frac{d\omega}{dt} = -\frac{m(\beta^2 - 1)}{\alpha M} \omega \tag{O4}$$

the solution of which is

$$\omega = \omega_0 \exp\left(-\frac{m(\beta^2-1)}{\alpha M}t\right). \tag{O5}$$

Figure 2.7 shows the value of ω/ω_0 for $\beta = 3$, $\alpha = 0.055$ and $m/M = 10^{-8}$ year^{-1}.

Appendix P

The Virial Theorem and Kelvin–Helmholtz Contraction

The Virial Theorem applies to any system of particles with pair interactions for which the overall distribution of particles does not vary with time, although the particles are moving around. The theorem states that

$$2K + \Omega = 0 \tag{P1}$$

where K is the total translational kinetic energy, *i.e.* the energy associated with motion from place to place, not including energy with motions such as spinning of individual particles, and Ω is the potential energy. Here we show the validity of the theorem for a system of gravitationally interacting bodies.

We take a system of N bodies for which the ith body has mass m_i, coordinates (x_i, y_i, z_i) and velocity components $(\dot{x}_i, \dot{y}_i, \dot{z}_i)$. We define the *geometrical moment of inertia* as

$$I = \sum_{i=1}^{N} m_i (x_i^2 + y_i^2 + z_i^2). \tag{P2}$$

Differentiating I twice with respect to time and dividing by two, we get

$$\frac{1}{2}\ddot{I} = \sum_{i=1}^{N} m_i (\dot{x}_1^2 + \dot{y}_i^2 + \dot{z}_i^2) + \sum_{i=1}^{N} m_i (x_i \ddot{x}_i + y_i \ddot{y}_i + z_i \ddot{z}_i). \tag{P3}$$

The first term on the right-hand side is $2K$; the second can be transformed by noting that $m_i \ddot{x}_i$ is the x component of the total force on the body i due to all the other particles or

$$m_i x_i \ddot{x}_i = \sum_{j=1(j\neq i)}^{N} G m_i m_j \frac{x_i(x_j - x_i)}{r_{ij}^3}, \tag{P4}$$

where r_{ij} is the distance between particle i and particle j.

Combining the force on i due to j with the force on j due to i, the second term on the right-hand side of (P3) becomes

$$\sum_{i=1}^{N} m_i(x_i \ddot{x}_i + y_i \ddot{y}_i + z_i \ddot{z}_i)$$

$$= -\sum_{pairs} G m_i m_j \frac{(x_i - x_j)^2 + (y_i - y_j)^2 + (z_i - z_j)^2}{r_{ij}^3}$$

$$= -\sum_{pairs} \frac{G m_i m_j}{r_{ij}} = \Omega. \tag{P5}$$

Equation (P3) now appears as

$$\frac{1}{2}\ddot{I} = 2K + \Omega, \tag{P6}$$

an equation originally due to Henri Poincaré (1854–1912).

If the system stays within the same volume with the same general distribution of matter, at least in a time-averaged sense, then $d^2 I/dt^2 = 0$ and the Virial Theorem is verified.

The Virial Theorem has a wide range of applicability and can be applied to the motions of stars within a cluster of stars or, as we shall do here, to an individual star where the translational kinetic energy is the thermal motion of the particles constituting the material. The total energy of a star, E, may be expressed as the sum of the thermal energy and the potential energy, the latter being negative in sign. This gives

$$E = K + \Omega, \tag{P7}$$

in which E must be negative if the star is to be stable and not fly apart.

Combining (P1) and (P7),

$$K = -E. \tag{P8}$$

If the star is radiating energy, then its total energy must be decreasing, *i.e.* E, which is negative, is becoming more negative. The implication of this is that K, a measure of the thermal energy, is increasing, which is to say that the temperature is increasing.

Appendix Q

The Lifetime of Stars on the Main Sequence

The estimation of lifetimes on the main sequence depends on a number of rather rough approximations. The first of these is the amount of hydrogen that will be consumed in the core region up to the time that the star leaves the main sequence, m_H, which is proportional to the mass of the star, M_*. This can be expressed as

$$m_\mathrm{H} = CM_*, \qquad (Q1)$$

where C is a constant.

Empirically it is found that the luminosity of a star is related to its mass by

$$L_* = KM_*^{7/2} \qquad (Q2)$$

where K is a constant. Actually the power relationship varies for different stellar mass ranges but (Q2) gives an approximate fit over the whole range.

The luminosity of a star is proportional to the rate that energy is being generated that, in its turn, is proportional to the rate of consumption of hydrogen in the core. Thus we can write

$$\frac{dm_H}{dt} = AL_* = BM_*^{7/2} \qquad (Q3)$$

where A and B are constants.

The duration of the main sequence, t_{ms}, is now taken as the mass of hydrogen available to be consumed in the core divided by the rate

Table Q.1   Test of equation (Q4).

Star mass (solar units)	$t_{ms}M_*^{5/2}$ (10^6y $M_\odot^{5/2}$)
15	8714
9	6075
5	5590
3	5455
2.25	6834
1.5	5511
1.25	6998
1	10000
0.5	8839

of consumption so that

$$t_{ms} = m_H \div \frac{dm_H}{dt} = CM_* \div BM_*^{7/2} = FM_*^{-5/2} \qquad (Q4)$$

where F is another constant. For this relationship to be true the product $t_{ms}M_*^{5/2}$ should be a constant. The actual value is shown in Table Q.1 for the stellar masses and lifetimes given in Table 3.1.

The magnitude of the second column is reasonably constant, taking into account the crudeness of the assumptions made in the analysis.

Appendix R

The Eddington Accretion Mechanism

The theory described here was first given by the British astrophysicist, Arthur Eddington (1882–1944), to describe the accretion of gas flowing past a spherical body. If a uniform stream of gas is moving at relative speed V that is not too great, then, on striking the spherical body, it will lose its energy of motion and be accreted. Alternatively, if V is very high then it may not lose enough energy to be accreted and will bounce off the body and escape. The same consideration will apply for a projectile moving relative to the spherical body; here we consider the conditions for the accretion of a projectile.

Figure R.1 shows a spherical body of mass M and radius R with various possible paths of a projectile moving at speed V. These paths are deflected as shown by the gravitational attraction of the body. It is clear that the limiting path is that marked OP where the projectile arrives at the surface tangentially at P with speed V_P. The distance D is the *accretion radius* and the accretion cross-section, $A = \pi D^2$. From conservation of angular momentum of the projectile,

$$VD = V_P R$$

or

$$V_P = \frac{D}{R} V. \tag{R1}$$

From the conservation of energy,

$$\frac{1}{2}V_P^2 = \frac{1}{2}V^2 + \frac{GM}{R}. \tag{R2}$$

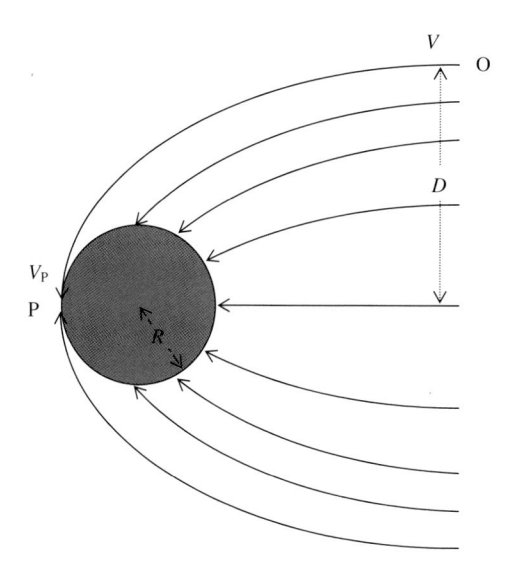

Fig. R.1 Possible paths of a projectile. The accretion radius is D.

Substituting for V_P from (R1) in (R2) and rearranging,

$$D^2 = R\left(R + \frac{2GM}{V^2}\right). \tag{R3}$$

If V is small compared with the escape speed, then the second term in the brackets on the right-hand side will dominate.

As an example of the application of this equation, we consider a forming stellar core of mass 10^{31} kg (5 solar masses) and radius 10^{10} m (about 14 times the solar radius). The relative speed between the stars in an embedded cluster is about $1\,\mathrm{km\,s^{-1}}$ and inserting these values into (R1) we find $D = 3.65 \times 10^{12}$ m. The accretion cross-section is $(D/R)^2 = 133{,}225$ times greater than the geometrical cross-section.

Appendix S

The Mass and Orbit of an Exoplanet

Here we deal with the case of Doppler-shift measurements of the motion of a star with a single planet moving in a circular orbit. Measurements give the following three quantities — the speed of the star in its orbit, v_*, the period of the orbit, P, and the mass of the star, M_* — deduced from its spectral class. Assuming that the mass of the planet, M_p, is very small compared with the mass of the star, the angular speed of the star, ω, is given by

$$\omega^2 = \frac{4\pi^2}{P^2} = \frac{GM_*}{R^3} \tag{S1}$$

where R the radius of the planetary orbit. Since M_* and P are known, then so is the radius of the orbit. From this we find the speed of the planet in its orbit, v_p, from

$$v_p = R\omega \tag{S2}$$

and hence the mass of the planet from

$$\frac{M_p}{M_*} = \frac{v_*}{v_p}. \tag{S3}$$

Appendix T

Radiation Pressure and the Poynting–Robertson Effect

The Force Due to Radiation Pressure

Photons possess both energy and momentum. The energy of a photon of electromagnetic energy of frequency ν is $h\nu$, where h is Planck's constant, and its momentum is $h\nu/c$, where c is the speed of light. This gives the relationship linking energy, E, and momentum, p, for electromagnetic radiation as

$$p = \frac{E}{c}. \tag{T1}$$

For the Sun with luminosity $L_\odot$, the energy flux (energy per unit area per unit time) at distance r from the Sun is

$$Q_E = \frac{L_\odot}{4\pi r^2}, \tag{T2}$$

and, from (T1), the momentum flux is, correspondingly,

$$Q_m = \frac{L_\odot}{4\pi r^2 c}. \tag{T3}$$

If a grain absorbs radiation then it gains the momentum of the radiation. The force on the grain will equal the rate of change of momentum. Hence a spherical grain of radius a will experience a radial force due to radiation

$$F_{gr} = \pi a^2 Q_m = \frac{L_\odot a^2}{4 r^2 c}. \tag{T4}$$

This force on the grain will oppose that due to gravity, which is

$$F_{gg} = \frac{GM_\odot \times \frac{4}{3}\pi a^3 \rho}{r^2}, \tag{T5}$$

where ρ is the density of the grain material. The ratio of F_{gr} to F_{gg}, which is independent of distance from the Sun, is given by

$$\frac{F_{gr}}{F_{gg}} = \frac{3L_\odot}{16\pi GM_\odot c\rho a}. \tag{T6}$$

Taking approximate values $L_\odot = 4 \times 10^{26}$ W, $M_\odot = 2 \times 10^{30}$ kg and $c = 3 \times 10^8$ m s^{-1}, then for a grain of radius $1\,\mu$m,[2] and density 3×10^3 kg m^{-3}, the ratio is 0.20 and for a grain of radius $0.2\,\mu$m, the ratio would be almost unity. Thus very tiny grains will experience a net outward force pushing them away from the Sun; somewhat larger grains would have the attractive force of the Sun partially counterbalanced by that due to radiation pressure.

The Poynting–Robertson Effect

A second effect due to the absorption of radiation, the Poynting–Robertson effect, is linked to the motion of the absorbing body around the Sun. The momentum vector corresponding to the absorbed radiation is in a radial direction. If the body is in thermal equilibrium, then it must be re-radiating energy at the same rate as it is being absorbed. However, this energy radiation is isotropic *with respect to the frame of reference of the body* and hence, in the Sun's reference frame, the radiated energy has a net momentum component in a tangential direction, *i.e.* along the direction of motion of the body. From conservation principles the gain of momentum by the radiation in this direction must be balanced by an equal loss of momentum by the radiating body. We now find the magnitude of this effect for a body in a circular orbit.

The tangential rate of change of momentum (corresponding to a force) can be found by multiplying the rate of radiated energy (equal to the rate of absorbed energy), expressed in equivalent mass

---

[2] $1\,\mu$m (1 micrometre) is 10^{-6} m.

terms through Einstein's relationship, by the speed of the body in its orbit, v. This gives

$$F_t = L_\odot \frac{a^2 v}{4r^2 c^2}, \tag{T7}$$

This force exerts a torque that changes the angular momentum of the body at a rate

$$\frac{dh}{dt} = -F_t r = -L_\odot \frac{a^2 v}{4rc^2}. \tag{T8}$$

If the mass of the body is m, then its angular momentum is

$$h = mr^2 \omega = mr^2 \sqrt{\frac{GM_\odot}{r^3}} = m\sqrt{GM_\odot r} \tag{T9}$$

so that, by differentiating (T9) with respect to t,

$$\frac{dh}{dt} = \frac{1}{2} m \sqrt{\frac{GM_\odot}{r}} \frac{dr}{dt} = \frac{1}{2} mv \frac{dr}{dt}. \tag{T10}$$

From (T8) and (T10) and by expressing m in terms of the radius of the grain, a, and its density, ρ, we find

$$\frac{dr}{dt} = -\frac{3L_\odot}{8\pi rac^2 \rho}. \tag{T11}$$

This can be integrated to give the total time for the particle to go from distance r_0 to r_f as

$$t = -\frac{8\pi ac^2 \rho}{3L_\odot} \int_{r_0}^{r_f} r\, dr = \frac{4\pi ac^2 \rho}{3L_\odot}(r_0^2 - r_f^2). \tag{T12}$$

For a spherical body of radius 1 cm and density $3 \times 10^3 \,\mathrm{kg\, m^{-3}}$ in the vicinity of the Earth, the time to be absorbed by the Sun, *i.e.* going from distance 1 au to zero, is 6.4×10^{14} s or 2.0×10^7 years. For a grain of radius 0.5 mm, the time is 1 million years. Since the time is proportional to a it seems that even bodies as large as 1 metre in radius or greater would have been absorbed by the Sun from the vicinity of the Earth during the lifetime of the Solar System.

Appendix U

Active Stars and Their Effect on a Stellar Disk

The interstellar medium contains about 1–2% by mass of solid material in the form of grains, mostly of diameter between 0.1 and 1.0 μm. The absorption and scattering of radiation is mostly mediated by dust at temperatures below 1,500 K, above which temperature silicate grains will evaporate. The dust is heavily coupled to the gas in the medium so that finding the forces on the dust grains is tantamount to finding the force on the complete medium. In what follows I shall assume that only absorption and scattering by dust is responsible for the forces on the medium due to radiation and the stellar wind. If there is appreciable absorption or scattering by the gas component of the medium, then this just means that the calculated forces will be underestimated.

The Cross-Section of Dust per Unit Volume of the Medium

We consider a medium of density ρ containing 1% by mass of spherical dust particles of radius a consisting of material of density σ. The number of dust particles per unit volume is

$$n_d = \frac{0.03\rho}{4\pi a^3 \sigma}. \tag{U1}$$

This gives the cross-section of dust particles per unit volume

$$A_d = \pi a^2 n_d = \frac{0.03\rho}{4a\sigma}. \tag{U2}$$

The Effect of Stellar Radiation

The absorption and scattering of radiation is strongest at ultraviolet wavelengths that are a considerable part, but not all, of the energy radiated by stars of solar mass or thereabouts. If a photon is absorbed by a dust particle, then the particle's gain of momentum is that of the photon, $h\nu/c$, where h is Planck's constant, ν the frequency of the photon and c the speed of light. In the case of scattering, if the photon is scattered in its original direction of motion, then the momentum of the dust particle will not change. If it is backscattered along its original path, then, to conserve momentum, the dust particle will gain momentum $2h\nu/c$. We shall take the average change of momentum from a single photon interaction as $h\nu/c = E/c$ where E is the energy of the photon.

The intensity of the radiation (energy per unit time per unit area) from a star of luminosity L_* at a distance r is

$$I = \frac{L_*}{4\pi r^2}. \tag{U3}$$

Hence, the force (rate of change of momentum) on a unit volume of the medium as a result of the interaction of the radiation with dust grains is

$$F_v = \alpha \frac{L_*}{4\pi r^2 c} \frac{0.03\rho}{4a\sigma} \tag{U4}$$

where α is a factor, less than unity, that allows for the different efficiency, with frequency, of interaction of radiation with dust particles. This gives a force per unit mass

$$F_m = \alpha \frac{L_*}{4\pi r^2 c} \frac{0.03}{4a\sigma}. \tag{U5}$$

This force per unit mass acts outwards and counters the force per unit mass due to the star at the same distance which is

$$F_* = \frac{GM_*}{r^2}. \tag{U6}$$

The fraction of the gravitational field neutralised by the effect of the luminosity of the star is

$$\eta_L = \frac{F_m}{F_*} = \alpha \frac{0.03 L_*}{16\pi c a \sigma G M_*}. \tag{U7}$$

For a solar-mass star we take $\alpha = 0.5$, σ for silicate grains is about 3×10^3 kg m^{-3} and for an average dust particle we use $a = 2.5 \times 10^{-7}$ m. With the mass of the Sun 2×10^{30} kg and its luminosity 4×10^{26} W, we find $\eta_L = 0.0040$.

During the Kelvin–Helmholtz stage of the Sun's evolution, when it was moving towards the main sequence, its luminosity would have been considerably higher than at present. The Italian astrophysicists D'Antona and Mazzitelli (Section 7.5) estimated that 3.6×10^6 years after embarking on the Kelvin–Helmholtz path, the luminosity of the Sun would have been 60 times the present value, falling to 10 times the present value after 5.2×10^6 years. This would give a reduction in the effective gravitational field of the Sun of between 24% and 4% in this period.

The Effect of a Stellar Wind

The strength of a stellar wind is often given in terms of the rate of loss of mass of the star expressed in the unit of solar masses per year. The present rate of mass loss due to the solar wind is about $3 \times 10^{-14} M_\odot$ per year so that during its lifetime at that rate it will have lost 0.015% of its mass. However, many stars have a much higher rate of loss.

If the rate of loss is $\varepsilon M_\odot$ per year, then the momentum flux (momentum of particles passing through a unit area per unit time) at a distance r from a star is

$$P_r = \frac{\varepsilon M_\odot V_w}{4\pi r^2 t_y}, \tag{U8}$$

where V_w is the speed of the stellar wind and t_y is one year, normally expressed in seconds. The force per unit mass exerted on the medium by the stellar wind is

$$F_{sw} = \frac{P_r A_d}{\rho} = \frac{\varepsilon M_\odot V_w}{4\pi r^2 t_y} \frac{0.03}{4a\sigma}. \tag{U9}$$

This gives the proportion of the gravitational field neutralised by the stellar wind as

$$\eta_{sw} = \frac{0.03\varepsilon M_\odot V_w}{16\pi a\sigma t_y GM_*}. \tag{U10}$$

For the present Sun, V_w is about $4 \times 10^5 \,\mathrm{m\,s^{-1}}$ that, combined with values previously given, gives $\eta_{sw} = 4.5 \times 10^{-6}$. Active stars may have much stronger stellar winds — some astrophysicists have suggested $\varepsilon = 10^{-7}$ — but the range 10^{-10} to 10^{-8} is probably more typical for very active young stars. This would give η_{sw} in the range 0.015 to 1.5, the latter figure indicating that the medium is being strongly repelled away from the star.

Variation of β with Time

The effects of radiation pressure and the stellar wind add together to give a net proportion of the gravitational field neutralised as

$$\eta = \eta_L + \eta_{sw}. \tag{U11}$$

Since the times of capture-theory interactions are not correlated in any way with stellar activity, any pattern of rising and falling activity is possible during the period of the evolution of a planetary orbit. The example used in Chapter 7 is of a rising activity that saturates at a high level. It is of the form

$$\eta(t) = \eta_{\max}\{1 - \exp(-\delta t)\} \tag{U12}$$

with $\delta = 10^{-6}$ year^{-1}.

With part of the gravitational field neutralised, the medium moves more slowly. The Keplerian speed at distance r from a star of mass M_* is

$$v_r = \sqrt{\frac{GM_*}{r}}. \tag{U13}$$

With a fraction η of the effect of the stellar mass removed, the orbital speed is

$$v_r' = \sqrt{\frac{GM_*(1-\eta)}{r}}. \tag{U14}$$

Appendix V

The Structure and Decay of a Stellar Disk

There are two kinds of areal density distributions used in modelling a medium in the form of a disk.

The Gaussian Distribution

The general form for a Gaussian distribution of areal density as a function of distance from the star, with a peak at r_{peak} and a standard deviation σ, is

$$\rho(r) = C \exp\left\{-\frac{(r-r_{peak})^2}{2\sigma^2}\right\}. \tag{V1}$$

The total mass of the material in the disk is given by

$$M_{disk} = 2\pi C \int_0^\infty r \exp\left\{-\frac{(r-r_{peak})^2}{2\sigma^2}\right\} dr. \tag{V2}$$

Given the mass, this equation serves to determine the value of the constant C. There are standard tables available that enable the integral in (V2) to be evaluated but it is quite straightforward to evaluate it numerically.

The Exponential Distribution

In this distribution it is assumed that the density is a maximum at $r = 0$, which clearly it is not since the star itself has a finite radius.

However, ignoring that discrepancy, the distribution is of the form

$$\rho(r) = \rho(0)\exp(-\alpha r) \qquad (V3)$$

where $\rho(0)$ is the density when $r = 0$. The constant α determines the rate at which the density falls off. In a distance $1/\alpha$ the density is reduced by a factor $e^{-1} = 0.3679$.

The total mass of the medium in the disk is given by

$$M_{disk} = 2\pi\rho(0)\int_0^\infty r\exp(-\alpha r)dr = \frac{2\pi\rho(0)}{\alpha^2}. \qquad (V4)$$

If the total mass of the disk and the fall-off factor, α, are fixed, then (V4) enables the maximum density, $\rho(0)$, to be found.

The Fall-Off Perpendicular to the Disk

The form of the density along a line perpendicular to the disk is a Gaussian distribution with a peak in the mean plane of the disk and standard deviation

$$\sigma_z = \frac{\pi c}{2\Omega} \qquad (V5)$$

where c is the speed of sound in that region of the disk and Ω is the angular speed of a body in a circular orbit at the point in the mean plane. The distribution of density perpendicular to the disk then has the form

$$\rho(z) = \frac{\rho(0)}{\sqrt{2\pi\sigma_z^2}}\exp\left\{-\frac{z^2}{2\sigma_z^2}\right\}, \qquad (V6)$$

where $\rho(0)$ is the areal density at the location and ρ_z is the density at distance z from the mean plane. In the mean plane ($z = 0$), the density is $\rho(0)/\sqrt{2\pi\sigma_z^2}$.

The Decay of the Disk

The density at all points of the disk decays with time in an exponential fashion. The density at any point at time t, ρ_t, is given

by
$$\rho_t = \rho_0 \exp(-\gamma t) \tag{V7}$$
where ρ_0 is the original density and γ is the decay constant. If $\gamma = 10^{-6}$ years^{-1}, then the density decays to $e^{-1} = 0.3679$ of its original value in 10^6 years.

Appendix W

The Formation of Exoplanets

The computer program that was used to estimate the probability that a particular protostar would be disrupted to produce planets required several parameters to be set. These were chosen as follows:

Selecting Protostars and Stars

For stars in the range of $0.1\,M_\odot$ to $10\,M_\odot$, the mass frequency distribution (number of stars per unit mass range) has the form, derived from observations,

$$f(M) = KM^{-2.4} \qquad (W1)$$

where we take M in solar-mass units. For disruptable protostars we are only interested in the mass range of 0.25 to $0.75\,M_\odot$. To select a mass at random from the distribution (W1) within the chosen mass range, we make use of the *transformation method*.[3]

The total number of stars in the mass range is given by

$$N_T = K \int_{0.25}^{0.75} M^{-2.4} dM = \frac{K}{1.4}(0.25^{-1.4} - 0.75^{-1.4}) = \frac{K}{1.4} \times 5.46846.$$

The total number of stars in the mass range from 0.25 to M is, similarly,

$$N_M = \frac{K}{1.4}(0.25^{-1.4} - M^{-1.4}).$$

---
[3] Woolfson, M.M. and Pert, G.J., 1999, *An Introduction to Computer Simulation* (Oxford: Oxford University Press), pp. 132–133.

Now a random number r is selected from a uniform distribution in the range of 0 to 1; there are standard programs for doing this. Setting up an equation

$$r = \frac{N_M}{N_T}$$

and solving for M gives a value of mass for the protostar, M_p, selected at random from the distribution (W1). This value of M_p is given by

$$M_p = \left(\frac{1}{6.96440 - 5.46846r}\right)^{1/1.4}. \tag{W2}$$

By an exactly similar process, the 1,000 individual stars in the simulated region of the embedded cluster can be allocated masses in the range of 0.8 to 2.0 solar masses. In this case the values of mass are similarly chosen by

$$M_* = \left(\frac{1}{1.36670 - 0.98777r}\right)^{1/1.4}. \tag{W3}$$

The radius of the protostar is chosen on the basis that it has a temperature of 20 K, consists of material of molecular mass 4×10^{-27} kg and has the Jeans critical mass. From Eq. (D4) the radius of the protostar is given as

$$R = \frac{GM_P\mu}{5kT}. \tag{W4}$$

During the period when the interaction of the protostar with the stars is being followed, it is assumed that the protostar radius remains constant at $R'_p = 0.9R_p$.

Positioning Stars

Theory developed by Jeans, basically that given in Appendix B, shows that the critical centre-to-centre distance for disruption of the protostar is given by

$$R_c = C\left(\frac{M_*}{M_p}\right)^{1/3} R_p \tag{W5}$$

where the constant C varies slightly with the ratio of masses but will be somewhat above 2.0 for the mass ranges of interest here. Taking the limiting distance for disruption as 1.5, R_p is a conservative choice since computational work shows that a larger distance could be justified.

The stellar number density is chosen in the range of 10^4–10^5 pc^{-3}, a density range that is believed to exist for about 5 million years within an embedded cluster. The free-fall time for the protostars we are considering, which may be found from Eq. (M7), is comparatively short, of the order 20,000 years, so the assumption that the stellar number stays constant during the collapse of a protostar is valid. The selection of the stellar number density was again made using a random number, r, in the range of 0 to 1. The number density was taken as

$$n = 10^{4+r}. \tag{W6}$$

This selection tends to favour the smaller densities in the range; one-half of the densities selected will be less than $10^{4.5} = 3.16 \times 10^4$ pc^{-3}. Having selected n, the radius of the sphere within which the 1,000 stars are contained, is given as

$$R_s = \left(\frac{750}{\pi n}\right)^{1/3} \text{pc} = 3.086 \times 10^{16} \left(\frac{750}{\pi n}\right)^{1/3} \text{m}. \tag{W7}$$

For the maximum value of n, this would be 4.12×10^{15} m which would take a protostar travelling at 2,000 m s^{-1} over 60,000 years to traverse if travelling in a straight line. This is several times the usual period we are considering, during which a protostar could be disrupted, so it is clear that the part of the embedded cluster being simulated is more than sufficient for the purpose in hand. The stars are placed in random positions within the spherical volume, once again by a process using numbers chosen from a random distribution in the range of 0 to 1. The protostar starts at the centre of the spherical volume.

The root-mean-square speeds of stars within embedded clusters have been observed to vary from cluster to cluster but in the range of 500 to 2,000 m s^{-1}. For each of the trial runs a random speed was chosen within that range and applied to all stars and the protostar. The method of choosing a random direction for a star's motion was equivalent to choosing a random position in a sphere — the process used for placing the stars.

With the initial configuration selected, the motions of all the bodies, the 1,000 stars and the protostar, were followed using a standard Runge–Kutta integration procedure.[4] After each time step, the distance of the protostar from all the stars was checked. If any distance was less than $1.5R'_p$, then a capture-theory event is assumed to have taken place unless subsequent motion takes it within a distance $0.5\ R_p$, which may produce a collision of the star and protostar. If no capture event occurs after one-half of the free-fall time for the protostar has elapsed, then it is assumed that the protostar will continue to collapse to form a star.

---

[4]Woolfson, M.M. and Pert, G.J., 1999, *An Introduction to Computer Simulation* (Oxford: Oxford University Press), pp. 88–89.

Appendix X

Disrupting a Planetary System

We consider a planet in an orbit with semi-major axis a and eccentricity e around a parent star of mass M_P. A passing star, of mass M_S, the speed of which was V_∞ relative to the parent star when they were far apart, moves past the parent star with closest approach distance D. Given that the planet can be anywhere in its orbit at the time of closest approach and that the plane of the planet's orbit can have any orientation with respect to the stellar orbit, we wish to find the probability that the planet is removed from its parent star.[5]

The speed V_D of the passing star relative to the parent star when they are closest together can be found from conservation of energy by

$$\frac{1}{2}V_D^2 = \frac{G(M_P + M_S)}{D} + \frac{1}{2}V_\infty^2. \qquad (X1)$$

The starting time of the passing star, prior to the time of closest approach, was taken as the earlier of 100,000 years before closest approach or when the distance between the stars was $4a(1+e)$, which is four times the furthest distance of the planet from the star. At this starting distance the effect of the passing star on the planet will be negligible. The starting point can be found analytically, or numerically by placing the passing star at distance D from the parent star, giving it a speed V_D normal to the line joining the stars, and integrating backwards in time either for 100,000 years or until the distance between the stars is $4a(1+e)$, whichever comes later.

---

[5] Woolfson, M.M., 2004, *Mon. Not. R. Astr. Soc.*, **348**, 1150.

The approach used to allow for the planet being anywhere in its orbit at the time of closest approach was to take six positions in the orbit equally spaced in time. This can be done analytically by means of *Kepler's equation*.[6] Finally, to allow for the random orientation, the line connecting the star to the planet and, correspondingly, the planet's velocity are randomly reoriented in space by means of the *Euler rotation matrix* that enables this to be carried out by the selection of three random angles. With a rectangular Cartesian system established, the rotation can be decomposed into a rotation α about the x-axis followed by a rotation β around the y-axis and finally a rotation γ around the z-axis. Positive angles α, β and γ rotate y towards z, z towards x and x towards y, respectively. This transforms a point (x, y, z) to the point (x', y', z') where

$$\begin{pmatrix} x' \\ y' \\ z' \end{pmatrix} = \begin{pmatrix} \cos\gamma & -\sin\gamma & 0 \\ \sin\gamma & \cos\gamma & 0 \\ 0 & 0 & 1 \end{pmatrix} \begin{pmatrix} \cos\beta & 0 & -\sin\beta \\ 0 & 1 & 0 \\ \sin\beta & 0 & \cos\beta \end{pmatrix}$$

$$\times \begin{pmatrix} 1 & 0 & 0 \\ 0 & \cos\alpha & -\sin\alpha \\ 0 & \sin\alpha & \cos\alpha \end{pmatrix} \begin{pmatrix} x \\ y \\ z \end{pmatrix}. \qquad (X2)$$

Velocity components are similarly transformed. A combination of six positions in the orbit and 150 random orientations gives 900 different configurations for a passage with a particular combination (a, e, D). For each of the 900 cases the motions of the passing star and planet, relative to the parent star, were followed by a four-step Runge–Kutta integration routine with variable time control that gave a maximum error of 1 metre per time step — more accuracy than was required but not expensive computationally. The integration was followed until the passing star reached the same distance from the parent star as that from which it started. One of three possible outcomes was recorded for each calculation — one that the planet was removed from the parent star, the next that the orbit was hardened and, finally, that

---

[6] Woolfson, M.M., 2000, *The Origin and Evolution of the Solar System* (Bristol: Institute of Physics), p. 4.

the orbit was softened. For each (a, e, D) combination, the proportion of the 900 trials giving each outcome was noted (sum of probabilities equals unity). We indicate the probability of disruption as $p(a, e, D)$.

A selection of 30 values of a was taken to cover the range from 0.5 to 1,500 au by

$$a_n = 0.5 \times 3000^{(n-1)/29} \text{ au} \quad n = 1 \text{ to } 30 \tag{X3}$$

so that each value of a differed from the previous one by a factor of $3000^{1/29} = 1.318$.

For the purpose of this calculation five representative orbital decay and round-off curves were used as shown in Fig. X.1.

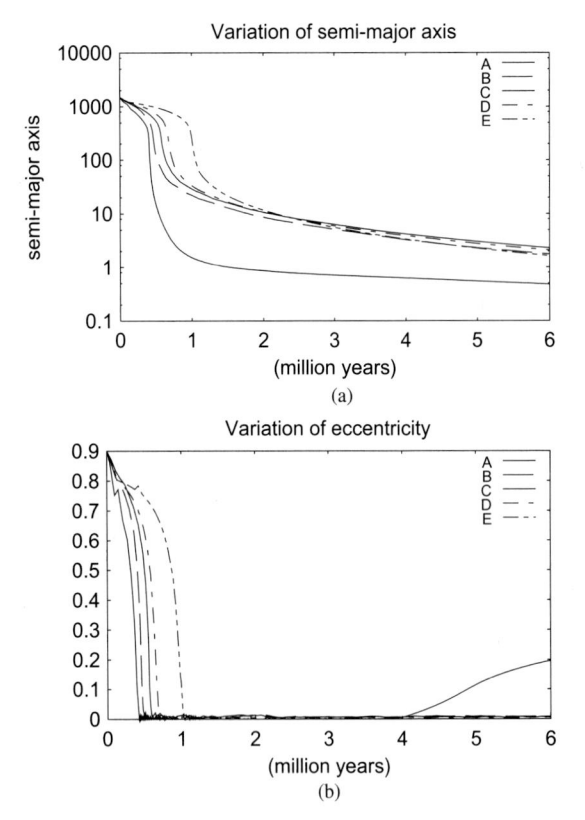

Fig. X.1 The five representative orbital evolutions (a) semi-major axis (b) eccentricity.

The values of e were selected at intervals of 0.1 that covered the range for each value of a indicated by the five curves. Finally the values of D were chosen to cover a wide range from being so close, that removal of the planet was almost inevitable, to being so far that removal of the planet was extremely unlikely. The range covered was from 0.5 to 10,361 au with

$$D = 0.5 \times 3000^{(n-1)/29} \text{ au}, \quad n = 1 \text{ to } 37. \tag{X4}$$

The probabilities of removal of the planet and of the hardening of its orbit were recorded in three-dimensional arrays for the different values of (a, e, D). By interpolation the proportions could be found for any intermediate value. It turned out, rather surprisingly, that for a given a and D the probability of removal of the planet was virtually independent of the value of e; this was unexpected since the larger the value of e, the further the planet moves from its parent star and the more vulnerable it would seem to be.

It is now possible to determine the probability of removal of the planet per stellar passage, taking into account the relative probabilities of having different values of D. Figure X.2 shows the path of a passing star starting at a large distance from the parent star.

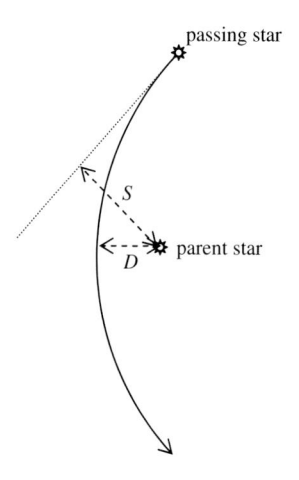

Fig. X.2 The orbit of the passing star relative to the parent star.

Because the path curves in towards the parent star, the *gravitational cross-section* for a closest approach D is given by

$$A_D = \pi S^2 = \pi D \left(D + \frac{2G(M_P + M_S)}{V_\infty^2} \right). \quad (X5)$$

From (X5) the cross-section for a closest distance between D and $D + dD$ is

$$dA_D = 2\pi \left(D + \frac{G(M_P + M_S)}{V_\infty^2} \right) dD. \quad (X6)$$

This quantity is the area through which passing stars must travel in order to give a nearest passage between D and $D + dD$. If at time t the number density of stars in the embedded cluster is $n(t)$, then in time δt the expected number of stars that will pass within those distance limits — or, equivalently, if the expected number is less than unity, the probability that a star will pass between those limits — is

$$P_D dD = 2\pi n(t) V_\infty \left(D + \frac{GM}{V_\infty^2} \right) \delta t dD. \quad (X7)$$

We now consider a planet produced at a particular time t_0 within a period of 10^7 years that covers the time of greatest stellar number density in the embedded cluster, as illustrated in Fig. 8.2. We divide the interval between t_0 and the end of the dense period into a large number of small time intervals δt. If in the jth time interval, at time t_j, the values of a and e are found from one of the orbital evolutions illustrated in Fig. X.1, then the probability that in the interval δt the planetary system is disrupted is

$$2\pi n(t_j) V_\infty \left(D + \frac{GM}{V_\infty^2} \right) p(a_j, e_j, D) \delta t dD. \quad (X8)$$

Hence the probability of disruption for all possible values of D in the jth time interval is

$$P(t_j) = 2\pi n(t_j) V_\infty \delta t \int_0^\infty \left(D + \frac{GM}{V_\infty^2} \right) p(a_j, e_j, D) dD. \quad (X9)$$

This expression can be evaluated numerically for any given variation with time of stellar number density and of planetary orbital parameters. The intervals δt must be chosen small enough so that all $P(t_j)$ are less than unity, and preferably much less. If this condition is satisfied, then in the jth time interval the probability that the planet is *retained* is $\{1 - P(t_j)\}$ and the probability that the planet is retained for all m time intervals of interest is

$$Q = \prod_{j=1}^{m} \{1 - P(t_j)\}. \tag{X10}$$

The results of this calculation are presented in Table X.1, which show the retention probabilities for various times, taken from the peak density shown in Fig. 8.2 for each of the orbital parameter evolutions shown in Fig. X.1. For the purpose of this calculation, I took $M_P = M_S = M_\odot$ and $V_\infty = 1000\,\mathrm{ms}^{-1}$.

To convert these results into the expected proportion of planets that are retained overall requires knowledge of the rate of protostar, and hence of planet, formation as the cluster develops. A theoretical study of star formation in a galactic cluster that I carried out in 1979,[7] and an observational study of young clusters carried out by I.P. Williams and A.W. Cremin in 1969,[8] both indicate that the rate of star formation increases with time before finally tailing off.

Table X.1 Retention probabilities for the orbital variations shown in Fig. X.1 for different formation times of the planets.

t_0 (10^6 years)	$n(t_0)$ (pc^{-3})	A	B	C	D	E
0	1.00×10^5	0.196	0.090	0.051	0.041	0.013
1	8.33×10^4	0.271	0.151	0.097	0.086	0.039
2	4.93×10^4	0.475	0.343	0.271	0.257	0.173
3	2.19×10^4	0.724	0.631	0.572	0.560	0.479
4	8.03×10^3	0.890	0.848	0.819	0.813	0.771
5	2.67×10^3	0.962	0.947	0.936	0.934	0.918

---

[7] Woolfson, M.M., 1979, *Phil. Trans. R. Soc. Lond.*, **291**, 219.
[8] Williams, I.P. and Cremin, A.W., 1969, *Mon. Not. R. Astr. Soc.*, **144**, 359.

The retention probabilities given in Table V.1 probably cover the circumstances for the majority of planets being produced by the capture-theory mechanism. The average of all the entries in the table, 0.48, gives support to the conclusion that between one-third and two-thirds of planets are retained.

Appendix Y

From Dust to Satellitesimals

The Rate of Growth of a Dust Particle by Brownian Motion

Very tiny dust particles of mass m, radius s and density ρ_s in an environment at temperature T will have a mean speed of motion

$$\bar{v} = \left(\frac{3kT}{m}\right)^{1/2} = 3\left(\frac{kT}{4\pi s^3 \rho_s}\right)^{1/2}, \tag{Y1}$$

where k is Boltzmann's constant $(1.38 \times 10^{-23}\,\mathrm{J\,K^{-1}})$.

For a medium with density ρ, a fraction f of which is dust, the mass of dust per unit volume is $f\rho$. Assuming that the dust particle accretes all the mass of dust in the volume its cross-section sweeps out as it moves, then its rate of increase of mass is

$$\frac{dm}{dt} = \pi s^2 \bar{v} f \rho = 3\pi s^{1/2} f \rho \left(\frac{kT}{4\pi \rho_s}\right)^{1/2}. \tag{Y2}$$

Since $m = \frac{4}{3}\pi s^3 \rho_s$, we have

$$\frac{dm}{dt} = 4\pi s^2 \rho_s \frac{ds}{dt}$$

so that

$$\frac{ds}{dt} = \frac{3f\rho}{4}\left(\frac{kT}{4\pi \rho_s^3}\right)^{1/2} s^{-3/2}. \tag{Y3}$$

The radius of the particle at any time comes from the solution of this differential equation:

$$s = \left\{ s_0^{5/2} + \frac{15}{8} f\rho \left(\frac{kT}{4\pi \rho_s^3} \right)^{1/2} t \right\}^{2/5}. \tag{Y4}$$

This expression is similar to, but different from that given by Weidenschilling, Donn and Meakin.[9]

The Speed of a Small Spherical Particle Falling Towards the Mean Plane

To determine the way that the various factors affect the speed of the particle, we make a simplifying assumption that all gas molecules are moving either parallel or anti-parallel to the motion of the particle. Figure Y.1 shows a dust particle and gas molecules striking it both on the side facing its direction of motion and on the rear side. The speed of the gas molecules, c, is much greater than the speed of the dust particles, v.

In one second the molecules striking the front surface of the dust particle, of radius s, will have come from within a distance $c+v$ and they will number

$$n_f = \frac{1}{2}\pi s^2 (c+v) \frac{\rho}{\mu} \tag{Y5}$$

where μ is the mass of a gas molecule. The factor of $\frac{1}{2}$ is due to the fact that only one-half of the gas molecules in the volume will

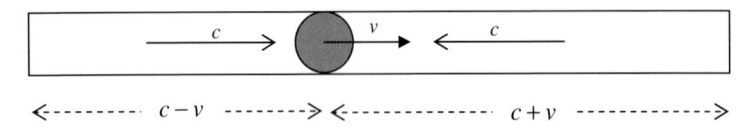

Fig. Y.1 Gas molecules striking the dust particle from the forward and backward directions.

---

[9]Weidenschilling, S.J., Donn, B. and Meakin, P., 1989, *The Formation and Evolution of Planetary Systems*, eds. H.A Weaver and L. Danley (Cambridge: Cambridge University Press).

be travelling towards the particle. They will strike the particle with speed $c+v$ and rebound with the same speed. Hence the momentum they impart to the particle is

$$p_f = 2\mu n_f(c+v) = \pi s^2 \rho (c+v)^2. \tag{Y6}$$

The momentum imparted by the rear-striking molecules, p_b, striking with speed $c-v$, is found similarly so the force on the dust particle, the rate of change of momentum, is given by

$$F_{gas} = p_f - p_b = \pi s^2 \rho\{(c+v)^2 - (c-v)^2\} = 4\pi s^2 \rho c v. \tag{Y7}$$

This force opposed the motion of the particle. Now we find the force due to the planet; Figure Y.2 shows the way this force originates.

The total force on the dust particle is

$$F_R = \frac{GM_P}{R^2}\frac{4}{3}\pi s^3 \rho_s \tag{Y8}$$

where M_P is the mass of the planet and its component towards the mean plane is

$$F_z = F_R \frac{z}{R} = \frac{4}{3}\frac{GM_P}{R^3}\pi s^3 \rho_s z. \tag{Y9}$$

For the particle to fall at a constant speed, $F_{gas} = F_z$, which gives

$$v = \frac{GM_p \rho_s z s}{3R^3 \rho c}. \tag{Y10}$$

Because of the simplifying assumption that all gas molecules are moving along the line of motion of the particle, (Y10) is not correct but does give the dependence on the various parameters correctly. A more detailed analysis gives (Y10) without the factor 3 in the divisor. Because z is normally much less than R, it is valid to take R as the

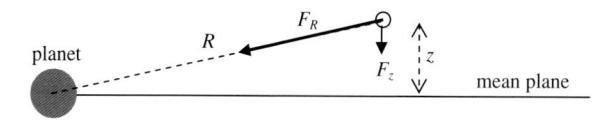

Fig. Y.2 The force on a dust particle due to the planet. The component towards the mean plane is F_z.

distance to the point in the mean plane to which the dust is falling. We then have

$$\frac{GM_P}{R^3} = \Omega^2 \tag{Y11}$$

where Ω is the angular speed of the disk at distance R. Removing the factor 3 in the divisor of (Y10) and using (Y11), we find

$$v = \frac{\Omega^2 \rho_s z s}{\rho c}, \tag{Y12}$$

the expression given by Weidenschilling, Donn and Meakin.

Density to Withstand Tidal Disruption

In Fig. 9.5 we take the radius of the spherical blob to be a. Then, with the blob mass M_b,

$$F_{gA} = F_{gB} = \frac{GM_b}{a^2} = \frac{4}{3}\pi G \rho_s a. \tag{Y13}$$

We also have

$$F_{PA} = \frac{GM_P}{(R-a)^2} \quad \text{and} \quad F_{PB} = \frac{GM_P}{(R+a)^2}. \tag{Y14}$$

Thus Eq. (9.5) gives

$$\frac{8}{3}\pi G \rho_s a = GM_P \left\{ \frac{1}{(R-a)^2} - \frac{1}{(R+a)^2} \right\}. \tag{Y15}$$

Using $R \gg a$ and the binomial approximation, this gives for the critical value of ρ_s

$$\rho_s = \frac{3M_P}{2\pi R^3}. \tag{Y16}$$

Appendix Z

From Satellitesimals to Satellites

Safronov's theory[10] was originally developed for describing planet formation but here we apply it to satellite formation. When the random motions become constant the mean speed of satellitesimals, v, is of the same order as the escape speed from the largest satellitesimal, v_e. This gives

$$v^2 = \frac{2Gm_L}{\beta r_L} = \frac{8G\pi r_L^2 \rho_s}{3\beta} \tag{Z1}$$

where m_L and r_L are the mass and radius of the largest satellitesimal and β is a factor somewhere in the range of 4 to 10. In the analysis that follows, it is assumed that the original satellitesimals and the satellites formed from them have the same density ρ_s.

Assuming that all bodies adhere when they collide, we must take into account the focusing effect of the accreting body (Appendix R). In Fig. Z.1 a small body is shown approaching the accreting body from an infinite distance with speed v such that the direction of motion, without deflection, gives D as the closest approach to the centre of the accreting body. The body is on a critical path so that it strikes the accreting body at P.

If the speed of the small body is v' at P, then, from conservation of angular momentum,

$$rv' = Dv. \tag{Z2}$$

---

[10] Safronov, V.S., 1972, *Evolution of the Protoplanetary Cloud and Formation of the Earth and Planets* (Jerusalem: Israel Program for Scientific Translations).

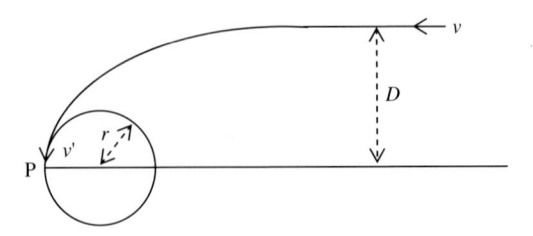

Fig. Z.1 A small body joining an accreting body on a critical path.

From conservation of energy considerations and (Z1),

$$v'^2 = v^2 + \frac{8G\pi r^2 \rho_s}{3} = v^2\left(1 + \frac{r^2}{r_L^2}\beta\right). \tag{Z3}$$

Combining (Z2) and (Z3) gives the accretion cross-section as

$$\pi D^2 = \pi r^2 \left(1 + \frac{r^2}{r_L^2}\beta\right). \tag{Z4}$$

The rate of mass increase of a satellitesimal is

$$\frac{dm}{dt} = \pi D^2 v \rho_{sat} = \pi v \rho_{sat} r^2 \left(1 + \frac{r^2}{r_L^2}\beta\right) \tag{Z5}$$

where ρ_{sat} is the space-average density of solid material, including satellitesimals, in the region through which the accreting satellitesimal is moving.

For the largest satellitesimal in the same region, $r = r_L$, so for its rate of growth we have

$$\frac{dm_L}{dt} = \pi v \rho_{sat} r_L^2 (1 + \beta) \tag{Z6}$$

and the relative rate of growth of the largest satellitesimal compared to that of radius r is

$$\frac{dm_L}{dm} = \frac{r_L^2}{r^2} \frac{(1+\beta)}{(1+\beta r^2/r_L^2)}. \tag{Z7}$$

The right-hand side can be expressed in terms of ratios of the two masses since

$$\frac{r_L^2}{r^2} = \left(\frac{m_L}{m}\right)^{2/3},$$

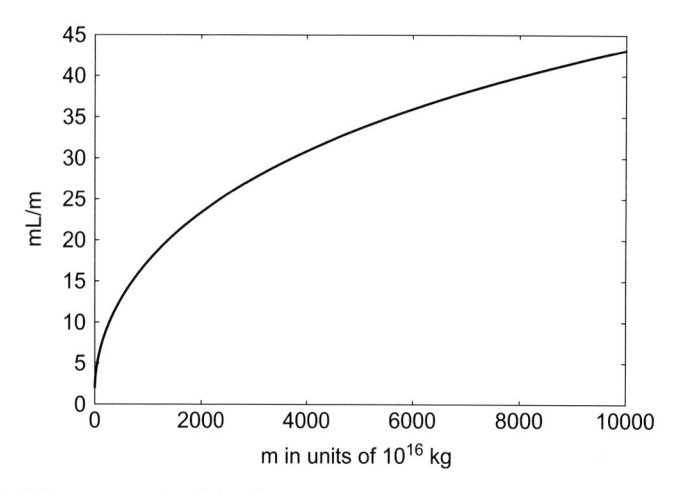

Fig. Z.2 The mass ratio of the dominant satellitesimal to a general satellitesimal as they both grow.

giving

$$\frac{dm_L}{dm} = \left(\frac{m_L}{m}\right)^{2/3} \frac{(1+\beta)}{\left\{1+\beta\left(m/m_L\right)^{2/3}\right\}}. \quad (Z8)$$

This differential equation can be solved numerically and a solution is given in Fig. Z.2 where the initial values of m and m_L are 10^{17} and 2×10^{17} kg respectively and $\beta = 8$. The final value of m is 10^{21} kg, by which time m_L has become 4.31×10^{22} kg. The figure shows the ratio m_L/m plotted against m and it will be seen that the ratio continually increases so the more massive satellitesimal proportionately increases its mass more rapidly. If ever the ratio of dm_L/dm became equal to m_L/m, then thereafter the ratio of the masses stays constant. It can be shown from (Z8) that this will occur when the ratio is 512 — although this would happen when both masses have become unrealistically high.

It is possible to estimate the time it takes for the largest satellite to form in any particular region. Although dm_L/dt, as given by (Z6), is not a constant by taking r_L at its final value, we have an upper bound for the rate of increase of mass. The time of formation can be expressed as

$$\tau_S = \frac{m_L}{dm_L/dt} = \frac{4r_L\rho_s}{3v\rho_{sat}(1+\beta)}. \tag{Z9}$$

To find an estimate for ρ_{sat} we consider a particle in a circular orbit of radius r for which the speed in the orbit is $2\pi r/P$, where P is the orbital period. If particles have a random speed up to v, which will be the mean relative speed of satellitesimals, then material at a distance r from the planet will have an orbital inclination up to an angle

$$\phi = \frac{vP}{2\pi r}. \tag{Z10}$$

The material at a distance r will be spread out over a distance

$$h = 2r\phi = \frac{vP}{\pi}. \tag{Z11}$$

Hence the average density of solid material at that distance will be

$$\rho_{sat} = \frac{\rho_{ad}}{h} = \frac{\pi\rho_{ad}}{vP}, \tag{Z12}$$

where ρ_{ad} is the areal density of solid material (satellitesimals in this case) at distance r from the planet. From (Z9) and (Z12),

$$\tau_s = \frac{4r_L\rho_s P}{3\pi\rho_{ad}(1+\beta)}, \tag{Z13}$$

which is the equation given by Safronov.

The results given here for the growth of satellitesimals are not precise and must be regarded only as approximate and indicative. Factors that have been neglected are the variation of radius of the satellitesimal and the exhaustion of available material to accrete as the satellitesimal grows.

Appendix AA
The Tidal Heating of Io

Elastic Hysteresis

A perfectly elastic body, say, in the form of a uniform rod, has the property that if it is stretched, then the amount by which it is stretched, the extension, is proportional to the stretching force. The relationship between the extension and stretching force is shown in Fig. AA.1(a) as the line OP. If the stretching force is F and the corresponding extension is ε, then the total work done in the stretching process is $\frac{1}{2}F\varepsilon$, which is the shaded area under the line OP in the figure. If the force F is gradually reduced, then points on the line OP again represent the state of the rod on the way back to the original state O. The work done *by* the rod as it contracts just equals that required to stretch it so that in the process of stretching and contracting mechanical energy is conserved.

However, real materials are not perfectly elastic and when stretched and then allowed to contract, their paths on the (F, ε) diagram will be somewhat as shown in Fig. AA.1(b). The stretching is along the upper path, A, and work done in stretching is the area under OP along the path A. Similarly the work done *by* the material in contracting is the area under OP along the path B. The difference — the loss of mechanical energy — represented by the shaded area appears as heat within the material. Physically what has happened is that internal friction between different grains within the material has generated heat and, hence, from the principle of conservation of energy, it cannot do as much work in contraction as was done on it in expansion. This phenomenon is known as hysteresis

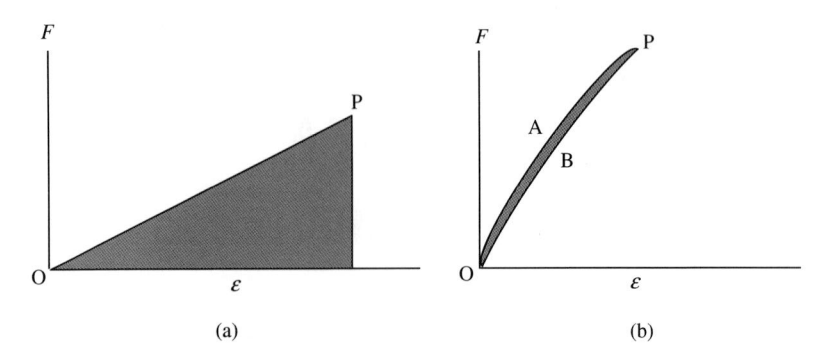

Fig. AA.1 (a) The path stretching and contraction for a perfectly elastic body. (b) The path of stretching (A) and of contraction (B) for an imperfectly elastic body.

and a similar phenomenon occurs with magnetic materials subjected to an alternating magnetising field.

The heating efficiency of the hysteresis cycle is described by a *quality factor*, Q, defined as

$$Q = 2\pi \frac{\text{energy stored in the extension}}{\text{energy lost in one cycle}}. \quad \text{(AA1)}$$

Highly compressed material in the interior of a planet is highly elastic and the Q value could be of order 1,000, so less than 1% of the mechanical energy is transformed into heat energy per cycle. On the other hand, for material in a less compressed state, closer to the surface, Q could be about 100 so that 6% of the mechanical energy expended in stretching is transformed into heat.

Tidal Stress in Io

We can make an order-of-magnitude estimate of the input of tidal energy into Io by a rough calculation. Figure AA.2 shows a representation of Io, with radius a, divided by a diametrical plane perpendicular to the orbital radius vector. Each hemisphere experiences forces away from the centre of the satellite due to Jupiter's tidal force and so the satellite is stretched. To estimate the total stretching force we first calculate the force per unit mass at P relative to O, where P is distant $\frac{1}{2}a$ from O. The force per unit

The Tidal Heating of Io

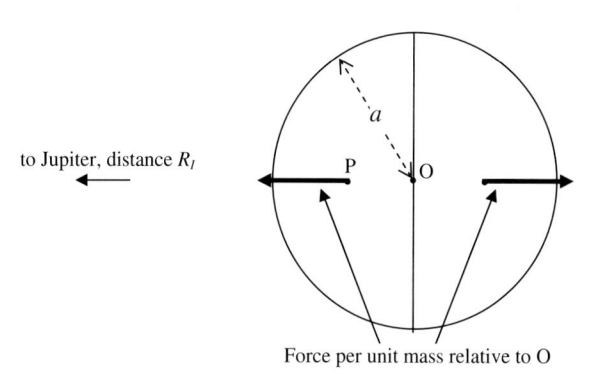

Fig. AA.2 Approximation to tidal force acting on Io divided into two hemispheres.

mass at P relative to O, due to both gravitational and rotational forces, is

$$A = \frac{GM_J}{\left(R - \frac{1}{2}a\right)^2} - \frac{GM_J}{R^2} - \left(R - \frac{1}{2}a\right)\omega^2 + R\omega^2 \quad (AA2)$$

where M_J is the mass of Jupiter, R the Jupiter–Io distance and ω is Io's angular speed in its orbit. We can eliminate ω^2 by

$$\omega^2 = \frac{GM_J}{R^3}$$

and, since $a \ll R$, we can use the binomial theorem approximation to give

$$A = \frac{3GM_J a}{2R^3}. \quad (AA3)$$

Assuming that this is the average force per unit mass for the nearer hemisphere, the total force on it towards Jupiter, F, is obtained by multiplying A by half the mass of Io, giving

$$F = \frac{3GM_I M_J a}{4R^3} \quad (AA4)$$

where M_I is the mass of Io. The force on the far hemisphere is equal and opposite. The next stage of the approximation is to take the satellite as a cube of side $2a$ being stretched along one principal direction. If Young's modulus for the satellite material is Y, then the extension is

$$\varepsilon = \frac{F \times \text{length}}{Y \times \text{area}} = \frac{F}{2aY}$$

and the energy to stretch it is

$$E = \frac{1}{2}F\varepsilon = \frac{F^2}{4aY} = \left(\frac{3GM_IM_J}{4}\right)^2 \frac{a}{4Y} \times \frac{1}{R^6}. \quad \text{(AA5)}$$

Since Io is in an eccentric orbit, the distance R varies between $R_I(1+e)$ and $R_I(1-e)$ where R_I is the mean orbital radius of Io and e the eccentricity of its orbit. Hence the total change in energy between the extreme positions is

$$\Delta E = \left(\frac{3GM_IM_J}{4}\right)^2 \frac{a}{4YR_I^6} \left\{\frac{1}{(1-e)^6} - \frac{1}{(1+e)^6}\right\}. \quad \text{(AA6)}$$

For $e \ll 1$,

$$\frac{1}{(1-e)^6} - \frac{1}{(1+e)^6} \approx 12e$$

so that

$$\Delta E = \frac{27}{16}(GM_IM_J)^2 \frac{ae}{R_I^6 Y}. \quad \text{(AA7)}$$

From (AA1) the heat energy generated per orbital period of Io, P, is $2\pi\Delta E/Q$ and hence the mean rate of energy generation in Io is

$$W = \frac{2\pi\Delta E}{QP} = (GM_IM_J)^2 \frac{27\pi ae}{8R_I^6 QPY}. \quad \text{(AA8)}$$

We now consider values for the quantities Q and Y in (AA8). Taking $Q = 1,000$ for a body the size of Io and $Y = 10^{11} \, \text{N m}^{-2}$, a typical figure for meteorite material, and other values as known, we find $W = 1.26 \times 10^{14}$ W. This is within the quoted estimated value, based on the volcanic activity, of between 6×10^{13} and 1.6×10^{14} W.

Appendix AB

The Trojan Asteroids

The orbits of the two sets of three Saturn satellites, (Tethys, Telesto, Calypso) and (Dione, Helene, Polydeuces), and of the set consisting of Jupiter and the two groups of Trojan asteroids are an example of a stable dynamical system. We shall describe it here in terms of the Trojan asteroids. Figure AB.1(a) shows the arrangement of the Sun, Jupiter and a Trojan asteroid at the 60° trailing position. The asteroid is taken as of negligible mass so the centre of mass of the system, C, is on the line SJ.

The Sun and Jupiter, of masses $M_\odot$ and M_J respectively, both orbit around the centre of mass C and, assuming circular orbits, the angular velocity will be given by

$$\omega^2 = \frac{G(M_\odot + M_J)}{R^3} \tag{AB1}$$

where R is the Sun–Jupiter distance. The Trojan asteroid, T, is so situated that SJT forms an equilateral triangle. For the asteroid to orbit in equilibrium, it too must move in a circular orbit about C with the angular velocity indicated by (AB1). For this to be so, a necessary condition is that the resultant of the gravitational fields of the Sun and Jupiter at T must be directed along TC.

In the triangle STC, $ST = R$ and $SC = M_J R/(M_J + M_\odot)$. Hence we can find a relationship involving the angle α (in degrees) as

$$\frac{SC}{\sin \alpha} = \frac{ST}{\sin(120 - \alpha)}$$

396   On the Origin of Planets

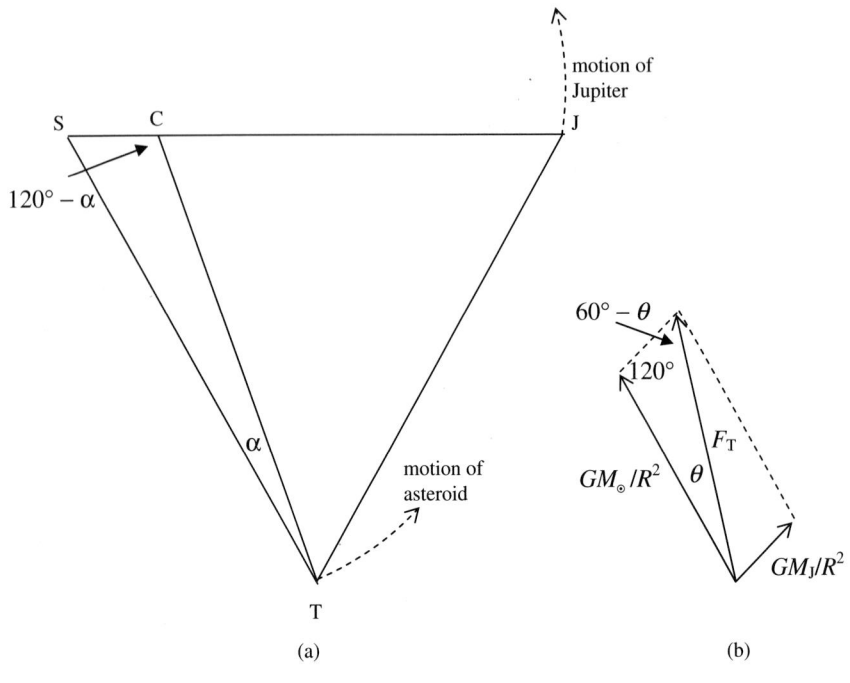

Fig. AB.1 (a) Positions of the Sun (S), Jupiter (J), a Trojan asteroid (T) and the centre of mass of the system (C). (b) Forces on the asteroid due to the Sun and Jupiter and the resultant F_T.

or

$$\frac{M_J}{\sin\alpha} = \frac{M_\odot + M_J}{\sin(120-\alpha)}. \qquad \text{(AB2)}$$

This gives

$$M_J\{\sin(120-\alpha) - \sin\alpha\} = M_J \sin(60-\alpha) = M_\odot \sin\alpha. \qquad \text{(AB3)}$$

In Fig. AB.1(b) the fields at T due to the Sun and Jupiter are shown and, since the asteroid is equidistant from the other two bodies, these fields are proportional to $M_\odot$ and M_J respectively. The resultant field makes an angle θ with the direction TS given by

$$\frac{M_J}{\sin\theta} = \frac{M_\odot}{\sin(60-\theta)}. \qquad \text{(AB4)}$$

Comparison of (AB3) and (AB4) gives that $\alpha = \theta$ so that the resultant field at T points towards C, the centre of mass of the system. The magnitude of the field, found from Fig. AB.1(b), is

$$F_T = \frac{G}{R^2}(M_\odot^2 + M_J^2 + M_\odot M_J)^{1/2}. \tag{AB5}$$

The distance TC ($= a$) is found from the triangle SCT as

$$a^2 = R^2 + \frac{M_J^2}{(M_\odot + M_J)^2}R^2 - \frac{M_J}{M_\odot + M_J}R^2$$

$$= R^2 \frac{M_\odot^2 + M_J^2 + M_\odot M_J}{(M_\odot + M_J)^2}. \tag{AB6}$$

For a circular orbit at distance a with the angular velocity given by (AB1), the required field is

$$a\omega^2 = R\frac{(M_\odot^2 + M_J^2 + M_\odot M_J)^{1/2}}{M_\odot + M_J}\frac{G(M_\odot + M_J)}{R^3}$$

$$= \frac{G}{R^2}(M_\odot^2 + M_J^2 + M_\odot M_J)^{1/2} \tag{AB7}$$

that is the field given in (AB5).

It has been shown that the three bodies all rotate about the centre of mass at the same angular velocity so that what we have is a state of equilibrium. What has not been shown is that it is *stable* equilibrium, so that if the Trojan asteroid wandered from the position given in Fig. AB.1(a), then the forces acting on it would return it to where it came from. A straightforward but lengthy analysis can show that the system is stable but we can also show the stability numerically. A computer program was run which simulated the Trojan asteroid system.[11] Jupiter was placed in a circular orbit around the Sun with a orbital radius of 5.2 au and a Trojan asteroid of zero mass was placed in either a trailing or leading position. The asteroid was placed about 0.2 au from the ideal Trojan position and its velocity was also slightly different from the correct one for an

---
[11] Woolfson, M.M. and Pert, G.J., 1999, *An Introduction to Computer Simulation* (Oxford: Oxford University Press), p. 15.

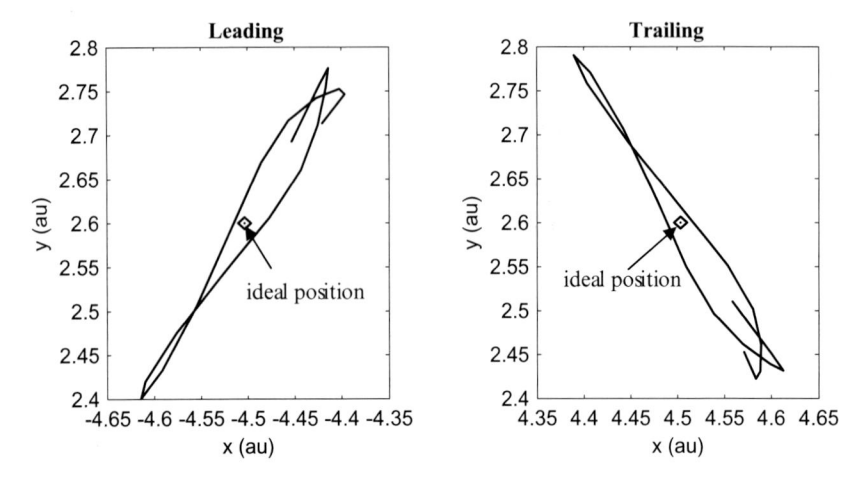

Fig. AB.2 Migration of a Trojan asteroid around the ideal position over a period of 200 years.

ideally placed asteroid. Figure AB.2 shows the motion of the asteroids relative to the true position. To display this in a comprehensible form the rotation of the system has been removed by always placing Jupiter on the y-axis. The program was run for a simulated 200 years and it will be seen that both the leading and trailing asteroids oscillate about the ideal position. This is what happens for the large numbers of asteroids occupying extensive regions around the ideal leading and trailing points for a Trojan asteroid.

Appendix AC

Orbital Precession

The complications of orbital precession are best investigated by numerical analysis, which John Dormand and I carried out in 1977.[12] Figure AC.1 shows the geometry of an orbit related to the mean plane of the disk and a fixed direction in space.

Normally, in defining a planetary orbit, it is customary to take the *ecliptic*, the plane of Earth's orbit, as a reference plane and a fixed direction in space contained in the ecliptic, called the First Point of Aries. This is the direction of the line joining the Earth to the Sun at the time of the vernal (spring) equinox, when the Sun is crossing the ecliptic, moving from south to north. Here we define our

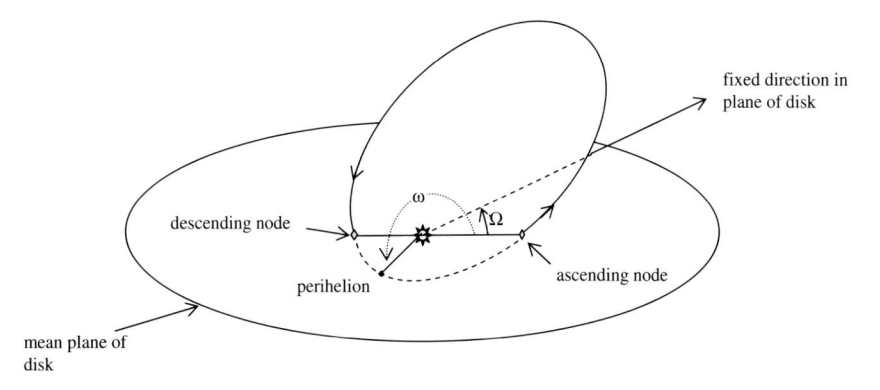

Fig. AC.1 The longitude of the ascending node, Ω, and the argument of the perihelion, ω.

---

[12]Dormand, J.R. and Woolfson, M.M., 1977, *Mon. Not. R. Astr. Soc.*, **180**, 243.

reference plane as the mean plane of the disk and take an arbitrary fixed direction within the plane.

The orbit crosses the reference plane at two points, at the *ascending node* where it crosses from south to north and the *descending node* where it crosses from north to south. Since a disk may have any orientation, we have to define one side of it as north and the other side as south. The line joining the ascending and descending nodes is called the *line of nodes*. The first component of the precession is the rotation of the line of nodes, defined by changes in the angle Ω, described as the *longitude of the ascending node*. It is this rotation that gives the motion of the aphelion indicated in Fig. 11.2.

The angle Ω is between two lines contained in the disk. Now we define the *argument of the perihelion*, ω, as the angle between two lines in the plane of the orbit — the line joining the Sun to the ascending node and the line joining the Sun to the perihelion of the orbit. This is another angle that varies with time and forms the second component of the overall precession. This motion means that the perihelion moves from the south to the north side of the plane and back again, as does the aphelion in the opposite sense. To complicate matters, the rotation of the argument of the perihelion sometimes changes direction.

It is impossible to give an analytical expression for the motion of a planet, the orbit of which is undergoing precession, decay and round-off, but it is a fairly simple matter to follow the motion by numerical analysis.

Appendix AD

The Temperature Generated by Colliding Planets

The structures of major planets consist of an iron and silicate core and an atmosphere, mainly of hydrogen and helium but with other gases present. When the planet is at the protoplanet stage, grains settle towards the centre, segregating by density with the densest silicate and iron grains at the centre, followed by grains of frozen materials including water, ammonia and methane. This structure persists when the protoplanet collapses to form a young planet so that above the dense core is a deuterium-rich layer of hydrogen-containing molecules.

The atmosphere of a planet increases greatly in both density and temperature from the surface to the interior. Since density gradually tails off as one moves outwards from the centre, it is customary to define the surface as the level at which the pressure is 1 bar, 10^5 newtons per square metre. For a planet such as Jupiter the density there is of order 0.2 kg m^{-3} and the temperature 125 K, although that temperature is due to stellar radiation and would be much lower if the planet were isolated. There are no precise estimates of the density and temperature at the base of the atmosphere but $3 \times 10^4 \text{ kg m}^{-3}$ and 20,000 K are in the right ballpark.

To begin the consideration of the result of the conditions generated by a planetary collision, we shall first take a much simpler case — the collision of two similar streams of gas of uniform density and temperature. The advantage of this simple case is that an analytical solution is available via a set of equations known as the Rankine–Hugoniot equations which gives a starting point for dealing with the more complex case of planetary collisions.

Uniform Streams and the Rankine-Hugoniot Equations

Figure AD.1 depicts the head-on collision of two similar uniform streams of gas, of pressure p and density ρ, each moving relative to the collision interface with a speed v. A region of compressed and stationary gas is formed, separated from the uncompressed gas by two shock fronts, each of which has speed u relative to the collision interface. The pressure and density of the compressed material are p_c and ρ_c respectively.

The jump in parameters on crossing the shock front is described by the following three Rankine–Hugoniot (RH) equations which express, respectively, conservation of mass, momentum and energy:

$$\rho(v+u) = \rho_c u \qquad \text{(AD1a)}$$

$$p + \rho(v+u)^2 = p_c + \rho_c u^2 \qquad \text{(AD1b)}$$

and

$$\frac{\gamma}{\gamma-1}\frac{p}{\rho} + \frac{(v+u)^2}{2} = \frac{\gamma}{\gamma-1}\frac{p_c}{\rho_c} + \frac{u^2}{2}, \qquad \text{(AD1c)}$$

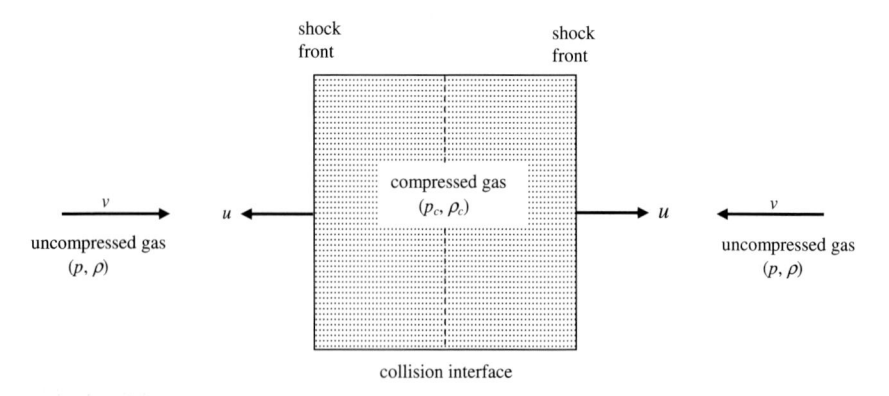

Fig. AD.1 A schematic representation of the head-on collision of two similar uniform gas streams.

where γ is the usual ratio of specific heats. From these equations and the normal perfect-gas equations, u is found as

$$u = \left\{\left(\frac{\gamma+1}{4}\right)^2 v^2 + c^2\right\}^{1/2} + \left(\frac{\gamma-3}{4}\right)v. \quad (AD2)$$

The speed of sound, c, in the uncompressed gas is given by $\sqrt{\gamma kT/\mu}$, where T is the temperature of the uncompressed gas and μ is the mean mass of the gas molecules. The compression of the gas is given by

$$\frac{\rho_c}{\rho} = \frac{v+u}{u}. \quad (AD3)$$

Pressure may be expressed as

$$p = \frac{\rho kT}{\mu} = (\gamma-1)\rho\varepsilon \quad (AD4)$$

where k is Boltzmann's constant and ε is the specific internal energy (SIE), the internal energy per unit mass. From this and previous equations we find the SIE of the compressed material as

$$\varepsilon_c = \frac{u}{v+u}\varepsilon + \frac{uv}{\gamma-1}. \quad (AD5)$$

These equations form a complete analytical solution of the RH equations and, in particular, give the change of SIE (or temperature) produced across the shock.

Figure AD.2 shows the outcome of the head-on collision of two gas streams, each with a speed of $50\,\mathrm{km\,s^{-1}}$, 400 s after first contact. The gas has a density of $10\,\mathrm{kg\,m^{-3}}$, mean molecular mass of 4×10^{-27} kg and temperature of 1,000 K with $\gamma = \frac{5}{3}$. The speed of the gas streams corresponds to Mach 20.85. The SIE for the shocked gas is $1.258\times 10^9\,\mathrm{J\,kg^{-1}}$, corresponding to a temperature of 2.430×10^5 K, well below the level that would initiate D-D nuclear reactions.

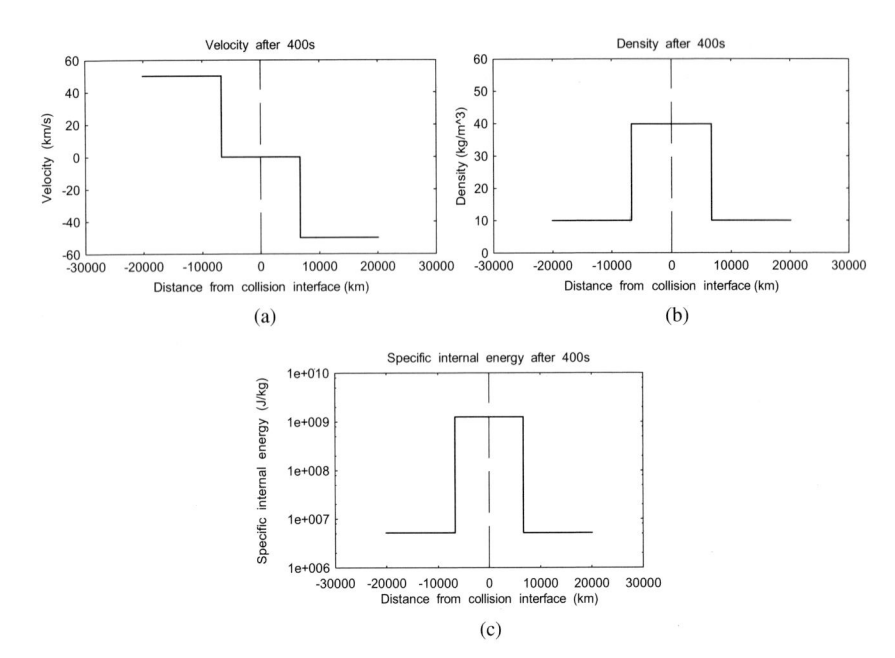

Fig. AD.2 Results from the Rankine–Hugoniot equations after 400 s for (a) velocity (b) density and (c) specific internal energy. For the SIE, 1e + 007 represents 1×10^7.

Uniform Streams and the Von Neumann–Richtmyer Computational Method

The system illustrated in Fig. AD.1 can also be analysed by computational approaches and one such is the Von Neumann–Richtmyer (VNR) method.[13] This is a Lagrangian leapfrog finite-difference method for solving the set of equations

$$\frac{dx}{dt} = u \qquad \text{(AD6a)}$$

$$\frac{du}{dt} = -V\frac{dp}{dx} \qquad \text{(AD6b)}$$

$$\frac{d\varepsilon}{dt} = -p\frac{dV}{dt} \qquad \text{(AD6c)}$$

---

[13] Woolfson, M.M. and Pert, G.J., 1999, *An Introduction to Computer Simulation* (Oxford: Oxford University Press), pp. 243–247.

where p, u, ε and V are now the pressure, velocity, SIE and specific volume $(1/\rho)$ of the gas at position x. Numerical solutions of these equations, especially when shocks are involved, give large spurious oscillations. To deal with this problem Von Neumann and Richtmyer introduced an artificial viscosity, which has the effect of spreading the effect of the shock over several cells. It appears as an extra pressure and is of the form

$$q = \begin{cases} -\zeta \rho (\Delta x)^2 \left(\dfrac{du}{dx}\right)^2 & \dfrac{du}{dx} < 0 \\ 0 & \dfrac{du}{dx} \geq 0 \end{cases} \qquad \text{(AD7)}$$

where Δx is the length of the local Lagrangian cell and ζ is a constant, of order unity, the best value of which can be found by trial and error. The artificial viscosity is included by replacing p by $p + q$ in equations (AD6b) and (AD6c).

A VNR computer program was run with two gas streams having the characteristics that gave the results shown in Fig. AD.2. There were 2,000 Lagrangian cells, each of length 50 km. The results, illustrated in Fig. AD.3, show important differences from those given by RH. There is a sharp decrease in the indicated density at the collision interface and a sharp increase in the SIE, less evident because of the logarithmic scale being used but quite large. However, the pressure, proportional to the product $\varepsilon\rho$, is well behaved and constant in the shocked region. This form of behaviour of the VNR equations with artificial viscosity is well known and has been thoroughly investigated. Errors of the order of 100% in SIE are frequently encountered and a number of correction techniques have been suggested, many quite complicated to apply.

Here I describe a simple and computationally cheap procedure that has been found to work well in a range of applications in reducing the gross errors at the interface, as shown in Fig. AD.3, and in transferring the error into other forms that may be less significant in interpreting and applying the computational results. After every step of the calculation and at each lattice point i, the quantity g_i is

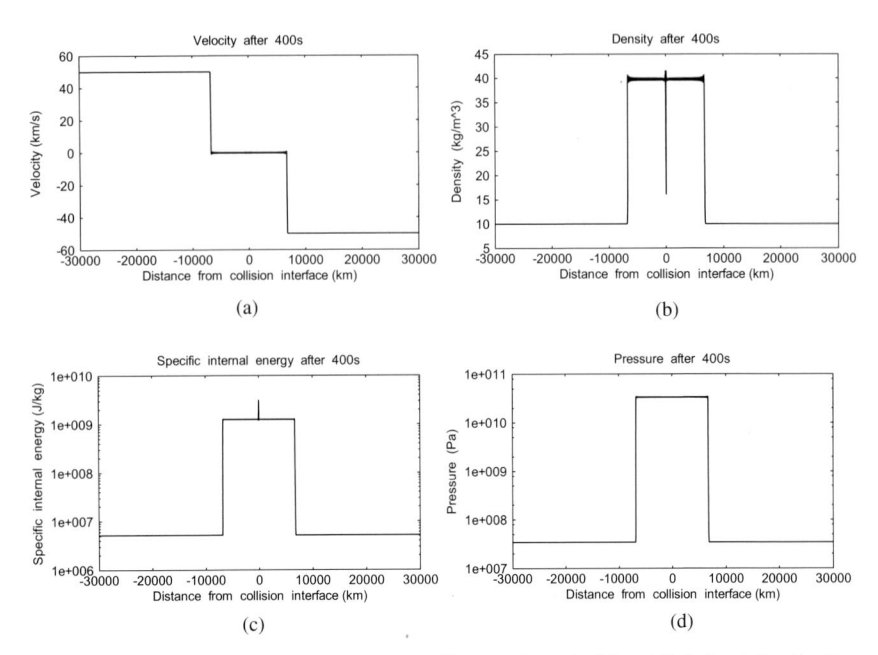

Fig. AD.3 The VNR results corresponding to those in Fig. AD.2 for (a) velocity, (b) density, (c) specific internal energy and (d) pressure.

found where

$$g_i \text{ is the minimum of } \varepsilon_i/\varepsilon_{i-1} \text{ and } \varepsilon_i/\varepsilon_{i+1}. \quad \text{(AD8a)}$$

From the value of g_i, the quantity η_i is found from

$$\eta_i = \alpha + (1-\alpha)\exp(-g^2). \quad \text{(AD8b)}$$

The correction of the aberration at the interface involves the transfer of SIE between neighbouring elements. The amount of internal energy transferred depends on the timestep through the quantity

$$\chi_i = 1 - (1-\eta_i)\Delta t/0.003 \quad \text{(AD8c)}$$

where the timestep Δt is in seconds and restricted to being not greater than 0.003 s in this example.

The form of χ, the spread factor, as a function of g is shown in Fig. AD.4 for $\alpha = 0.99$ and $\Delta t = 0.003$ s. The specific internal energy

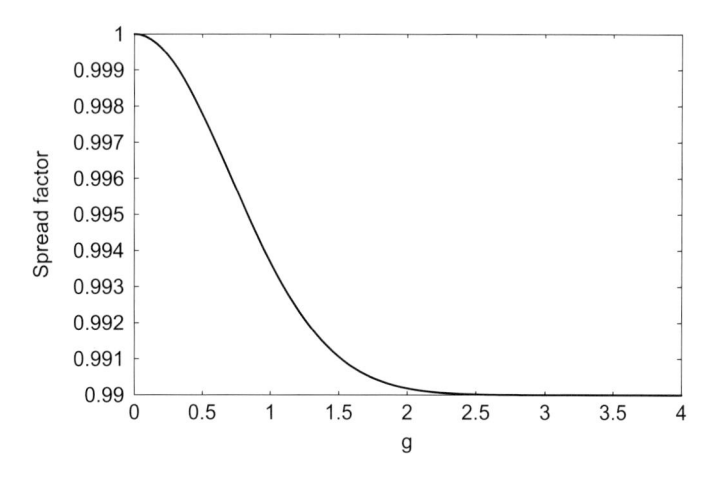

Fig. AD.4 The relationship between g and χ for $\alpha = 0.99$.

is then transformed by

$$\varepsilon_i' = \chi_i \varepsilon_i + 0.5 \left\{ (1 - \chi_{i-1}) \frac{m_i}{m_{i-1}} \varepsilon_{i-1} + (1 - \chi_{i+1}) \frac{m_i}{m_{i+1}} \varepsilon_{i+1} \right\} \tag{AD8d}$$

where m_i is the mass associated with point i. At each step of the calculation this procedure spreads the specific internal energy over three points while maintaining conservation of internal energy. The proportion of ε_i that is spread to neighbouring points is greatest when ε_i is large compared with both of its neighbours. If it is only large compared to one neighbour, then this is recognised as a probable shock front and the spread is low or zero. For $\chi = 1$ there is no spread.

The problem illustrated in Fig. AD.3 was run with the modified VNR program, again using 2,000 cells each of length 50 km and taking $\alpha = 0.995$. The results after 100, 200 and 400 s, as distributions of SIE, are shown in Fig. AD.5(a). They are very similar to the analytical results found from the RH equations except for a very small blip at the interface. The SIE has been plotted at its actual value, and not on a logarithmic scale, so that the small size of the blip can be better appreciated. The density distribution shows that

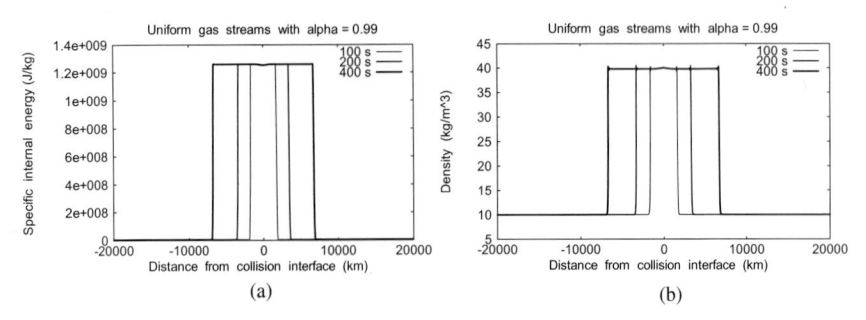

Fig. AD.5 (a) Specific internal energy and (b) density after 100, 200 and 400 s with modified VNR.

the anomaly at the collision interface has been largely removed and it is also clear that the correction operates over the whole period from the start of the collision.

Limits of Specific Internal Energy

To a first approximation the increase in the specific internal energy of the compressed material over that of the uncompressed material of the system displayed in Fig. AD.1 equals the specific kinetic energy of the oncoming streams of material. This gives

$$\varepsilon_c = \varepsilon + \frac{1}{2}v^2 = \frac{kT}{(\gamma-1)\mu} + \frac{1}{2}v^2. \qquad \text{(AD9)}$$

where ε_c and ε are the specific internal energies in the compressed and uncompressed regions respectively. For the parameters that gave Fig. AD.2, $\varepsilon_c = 1.255 \times 10^9 \, \text{J kg}^{-1}$, which is very close to the value previously given.

It must not be thought that Eq. (AD9) gives a limit to the SIE for any situation in which gas streams collide, that is, it is simply a matter of converting kinetic energy into internal energy. To understand why this is not so we must consider the detailed process that goes on when the gas streams collide. The gas impinging on the shock front exerts a ram pressure, in addition to the normal gas pressure, and the shocked material is compressed and heated until its pressure equals the sum of these applied pressures. The condition

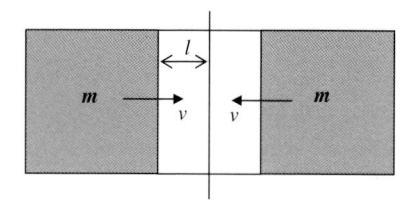

Fig. AD.6   Compression of a gas by two solid masses.

for a final equilibrium state, as seen from Eq. (AD6b), is that the pressure must be uniform throughout the system.

We now consider a simple, and fairly obvious, system where the final SIE is well above that from just a conversion of the specific kinetic energy. This is shown in Fig. AD.6 and is a mirror-symmetric arrangement consisting of two solid slabs of thermally insulating material, each of mass m per unit cross-section, plus gas layers of thickness l moving together with each having speed v. They will continue moving together until the increase in the internal energy of the gas equals the original kinetic energy of the system, at which point the system is momentarily static before it begins to re-expand. This condition gives

$$T_c = T + \frac{(\gamma - 1)(m + l\rho)v^2}{2kl\rho} \qquad (AD10)$$

where T and T_c are the initial and final temperatures of the gas and other quantities are as previously defined.

Values that could relate to astronomical bodies at a planetary scale are $T = 1000\,\text{K}$, $\gamma = \frac{5}{3}$, $m = 10^{10}\,\text{kg m}^{-2}$, $l = 10^6\,\text{m}$, $\rho = 10\,\text{kg m}^{-3}$, $v = 5 \times 10^4\,\text{m s}^{-1}$ and $\mu = 4 \times 10^{-27}\,\text{kg}$ and from these we find $T_c = 2.66 \times 10^6\,\text{K}$. Just converting the specific kinetic energy of the gas into specific internal energy gives $T_c = 2.42 \times 10^5\,\text{K}$. Of course, this is not a purely hydrodynamic system but one in which the solid components act like pistons to compress the gas. Nevertheless, it does illustrate that the maximum value of the specific internal energy is not limited by the specific kinetic energy of the gas component of the colliding bodies. The distribution of density and the types of material present also have a role to play.

Colliding Model Planets

The characteristics of major planets are that the density and temperature both increase from the boundary to the interior with large variations. Here we shall consider the collision of two similar bodies with radius 5×10^7 m and an idealised variation of density and temperature given by

$$\rho(r) = 30\,000 \exp(-4 \times 10^{-15} r^2) \, \text{kg m}^{-3} \quad \text{(AD11a)}$$
$$T(r) = 18\,000 \exp(-2.8 \times 10^{-15} r^2) \, \text{K} \quad \text{(AD11b)}$$

where r is the distance from the centre of the body in metres. The density and temperature at the surface are $1.36 \, \text{kg m}^{-3}$ and $16.4 \, \text{K}$ respectively. Real planetary bodies are spherical, or nearly so, and a spherical planet with the density distribution (AD11a) would have a mass of $119 \, M_\oplus$, some 25% greater than the mass of Saturn. The VNR program was run with 2,000 cells for a head-on collision of two streams of gas with a variation of density and temperature corresponding to that along the diameters of the colliding planets. We take $\gamma = \frac{5}{3}$ and $\mu = 4 \times 10^{-27}$ kg with each stream having a speed of $50 \, \text{km s}^{-1}$. Running the program for 400 s of simulated time without the spreading procedure ($\alpha = 1$) gives the results for density and SIE illustrated in Fig. AD.7, where only the region of interest close to the collision interface is shown.

The SIE shows a very sharp peak at the collision interface and from the VNR results obtained from the collision of uniform gas

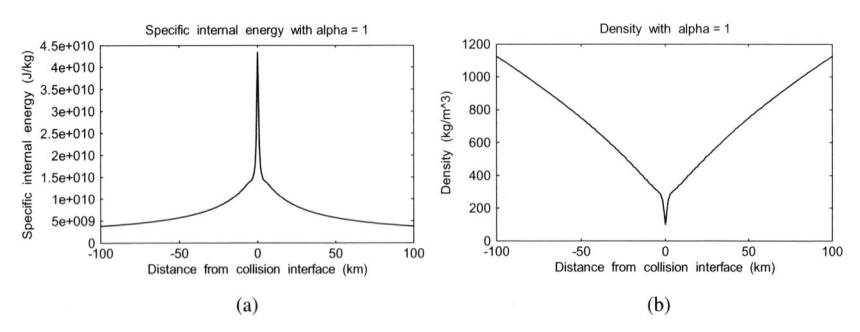

Fig. AD.7 (a) Specific internal energy and (b) density for a one-dimensional model planet without spreading.

streams, the result is bound to be aberrantly high. This is also indicated by the sharp dip in the density. In the case of the uniform streams we had an analytical solution with which to compare the VNR results but here no such comparison can be made. Another characteristic of the planetary simulation collision is that the shocked material is not at a uniform pressure and hence in a state of equilibrium. This can be seen in Fig. AD.8 that shows the pressure distribution within the shocked material; just as for the uniform gas streams, the pressure will be properly given by a VNR calculation without spread (Fig. AD.3(c)). As new material of ever-increasing density and temperature impacts the shocked region, so the pressure distribution within it is further disturbed and modified.

We can get an insight into the distribution of specific internal energy by looking at the results for various values of α, as shown in Fig. AD.9.

As α is reduced from unity, so the single peak at the collision interface separates into two peaks that become more and more distinct. Although the general expectation with supersonic collisions is that the greatest temperature should occur at the collision interface, this is not necessarily true when the incoming material is of ever-increasing density and temperature. Figure AD.10 shows the factor of enhancement of SIE between the beginning and end of a calculation of 400s simulated time with $\alpha = 0.995$. It is quite clear

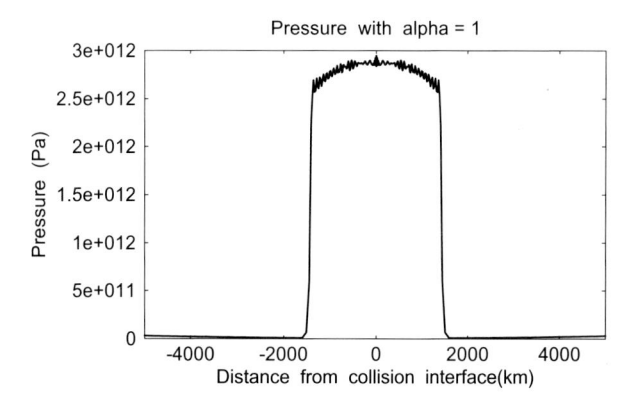

Fig. AD.8 The pressure distribution within the shocked region for non-uniform gas streams.

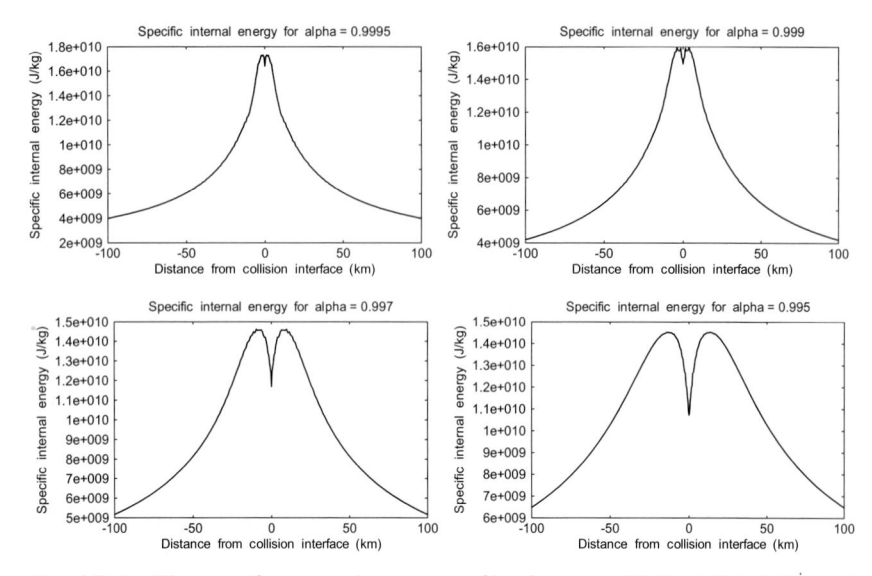

Fig. AD.9   The specific internal energy profiles for $\alpha = 0.9995$, 0.999, 0.907 and 0.995.

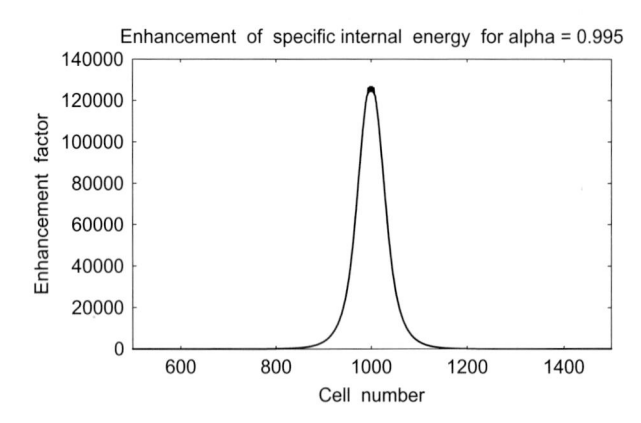

Fig. AD.10   The enhancement of specific internal energy against cell number. Cell 1,000 straddles the collision interface.

that the enhancement is greatest at the collision interface, as one would expect, and falls off with increasing distance. Nevertheless, as seen in the final frame of Fig. AD.9, the peaks of specific internal energy are away from the collision interface.

Maximum specific internal energy as a function of alpha

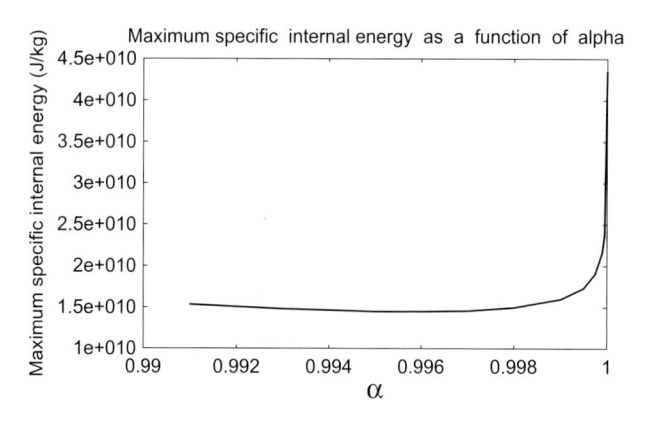

Fig. AD.11 The maximum specific internal energy after 400 s for different values of α.

The SIE shown at the collision interface in Fig. AD.7 is clearly a huge overestimate of what the value would be in a real physical collision. In Fig. AD.11 we show the maximum specific internal energy for 400 s of simulated time against various values of α.

As α is reduced from unity the first effect is to remove the false spike at the collision interface. Once this is removed, two separate and opposite effects set in. Decreasing α further increases the spread of the SIE, which tends to reduce high values and increase low values, but it also brings towards the maximum region higher SIE from outside the shocked material. The net effect is seen in Fig. AD.11 that shows the rapid fall to a level just below 1.5×10^{10} J kg^{-1} followed by a very gradual rise. The minimum SIE, 1.45×10^{10} J kg^{-10} for $\alpha = 0.996$, is a reasonable estimate, and perhaps a lower bound, of the value that should be expected for the collision of the one-dimensional model planets. In terms of temperature, this amounts to 4.20×10^6 K, compared to 2.42×10^5 K from just converting specific kinetic energy into SIE.

Although this analysis does not precisely define the density distribution at all stages of the collision, it does give a clear indication that temperatures capable of triggering D-D reactions are reached when major planets collide.

Appendix AE

Heating by Deuterium-Based Reactions

At a temperature of about 3×10^6 K, and particularly at a high density, D-D reactions can take place in two different ways:

$$_1^2 D + {}_1^2 D \rightarrow {}_1^3 T + {}_1^1 p \tag{AE1a}$$

$$_1^2 D + {}_1^2 D \rightarrow {}_2^3 He + {}_0^1 n. \tag{AE1b}$$

These two reactions take place with equal probability. The first produces a tritium nucleus and a proton and the second the less common stable isotope of helium plus a neutron. At comparatively low temperatures deuterium reacts with tritium according to

$$_1^2 D + {}_1^3 T \rightarrow {}_2^4 He + {}_0^1 n, \tag{AE2a}$$

and with helium-3 by

$$_1^2 D + {}_2^3 He \rightarrow {}_2^4 He + {}_1^1 p, \tag{AE2b}$$

both reactions producing a great deal of energy. The approximate energy output per deuterium atom due this combination of reactions is about $q = 8 \, \text{MeV} = 1.28 \times 10^{-12}$ J.

In a young, differentiated major planet, as the iron + silicate core is approached there will be a region heavily populated by deuterium-enriched hydrogen-containing molecular species. When energy is generated within a mixed mass of material, the principle of equipartition of energy operates whereby the expected energy of each particle of matter is the same, regardless of its mass. If we take a region solely occupied by water (H_2O), ammonia (NH_3), and methane (CH_4), then more than two-thirds of the atoms present must be hydrogen. However, due to ionization of atoms at very high temperatures, on average these molecules will give about 12 particles, including electrons. Taking a mixture of equal quantities of water, ammonia and methane, there will be 9 hydrogen atoms per 36 particles or, if 2% of the hydrogen is deuterium, 0.18 deuterium atoms per 36 particles. The temperature reached in this mixture by D-D reactions will be given by

$$36 \times \frac{3}{2}kT = 0.18 \times 1.28 \times 10^{-12}$$

where k is Boltzmann's constant, $1.38 \times 10^{-23}\,\mathrm{J\,K^{-1}}$.

This gives

$$T = 3.09 \times 10^8\,\mathrm{K}.$$

With some silicate material also present, the proportion of hydrogen will be less but it is clear that temperatures of order 10^8 K can be reached which will then initiate other reactions involving lighter elements contained in silicates. These, in turn, will generate further energy and so a chain of reactions involving heavier and heavier elements will occur.

Appendix AF

The Thermal Evolution of the Moon

The model we have for the formation of the Moon is that it is formed by a steady accretion of satellitesimals and we now assume that at all times the growing core has a spherical form. When it reached a radius x, satellitesimals falling on it did so with the escape speed from the core, v_x, the minimum possible speed of impact. For a core of uniform density ρ, the kinetic energy of the impacting material, assumed to be falling just at the escape speed, is given by

$$\frac{1}{2}v_x^2 = \frac{4}{3}\pi G\rho x^2. \tag{AF1}$$

If the specific heat capacity of the material is C and if there is no change of state, then, with all kinetic energy transformed into heat, the temperature at distance x from the centre of the Moon would be

$$T_x = \frac{4\pi G\rho x^2}{3C} + T_0 \tag{AF2}$$

where T_0 is the initial temperature of the impacting material. From this equation the minimum radius of a satellite that would initially have a molten region can be estimated. With T_m as the melting point and L the latent heat of fusion of the accreting material, this gives the critical radius as

$$x = \left\{\frac{3[C(T_m - T_0) + L]}{4\pi G\rho}\right\}^{1/2}. \tag{AF3}$$

There are various minerals in the material that formed the Moon but we take values $T_m = 2000\,\text{K}$, $T_0 = 100\,\text{K}$, $C = 700\,\text{J}\,\text{kg}^{-1}\,\text{K}^{-1}$,

$L = 10^4 \, \mathrm{J \, kg^{-1}}$ and $\rho = 3000 \, \mathrm{kg \, m^{-3}}$. This gives, from (AF3), $x = 1264 \, \mathrm{km}$. This can only be an approximate figure since the physical properties depend on the type of material being accreted but it is clear that the Moon, with a radius of 1,738 km, would certainly have been molten in its outer regions.

The actual temperature profile in the accreted Moon would have differed from that given by (AF2). Some of the energy of impacting material will be transformed into shock waves that will be transported towards the interior of the growing Moon, where they will dissipate and heat up inner material. There will also be energy continuously radiated out of the forming Moon from the recently created heated surface. These factors were taken into account by Toksöz and Solomon[14] who, in 1973, calculated what the initial temperature profiles would be for bodies of different masses and different rates of formation.

The basic partial differential equation for the rate of change of temperature, T, by heat conduction at distance r from the centre of a spherically symmetric system is

$$\frac{\partial T}{\partial t} = \frac{1}{\rho C_p} \left\{ \frac{1}{r^2} \frac{\partial}{\partial r} \left(r^2 \kappa \frac{\partial T}{\partial r} \right) + H(r,t) \right\} \qquad \text{(AF4)}$$

in which ρ, C_p and κ are the density, specific heat capacity at constant pressure and thermal conductivity of the material and $H(r,t)$ is the rate of heat generation within the material at time t and distance r. When the material is wholly or partially molten, then convection will increase the rate of heat transfer and this can be simulated by an enhanced conductivity, the form of which is modelled on experimental data. The heating term $H(r,t)$ is due to radioactive heating from uranium-238, thorium-232 and potassium-40 and possibly other shorter-lived radioactive elements that may have been around at an earlier time, *e.g.* aluminium-26. The radioactive elements tend to concentrate in the crust and this has been modelled by having their concentrations exponentially declining with distance from the surface. The present concentrations of radioactive elements can be

---

[14]Toksöz, M.N. and Solomon, S.C., 1973, *Moon*, **7**, 251.

estimated from measured heat flows through the Moon's surface, 16–22 W m^{-2} depending on location, and then, knowing the decay rates, the concentrations earlier in the history of the Solar System can be inferred.

Putting all these factors into a computational model, in 1989 John Dormand and I found the way that the temperature distribution varied with time, as shown in Fig. AF.1 The initial thermal profile was based on a fairly rapid accretion of the Moon. Also shown is the profile at the time of the last volcanic outflows on the Moon — about 3,200 million years BP (before present). The melting point of the silicate rocks that constitute the majority of the Moon is pressure-dependent and is higher for higher pressures. The dashed solidus (melting-point) line shown in the figure was that which other workers in the field have suggested.

It will be seen from Fig. AF.1 that the last volcanism occurred when the molten region extended to about 100 km below the surface — incidentally corresponding to the thickness of crust on the far side. The final profile, corresponding to the present, suggests a considerable molten region at the centre but at a temperature not

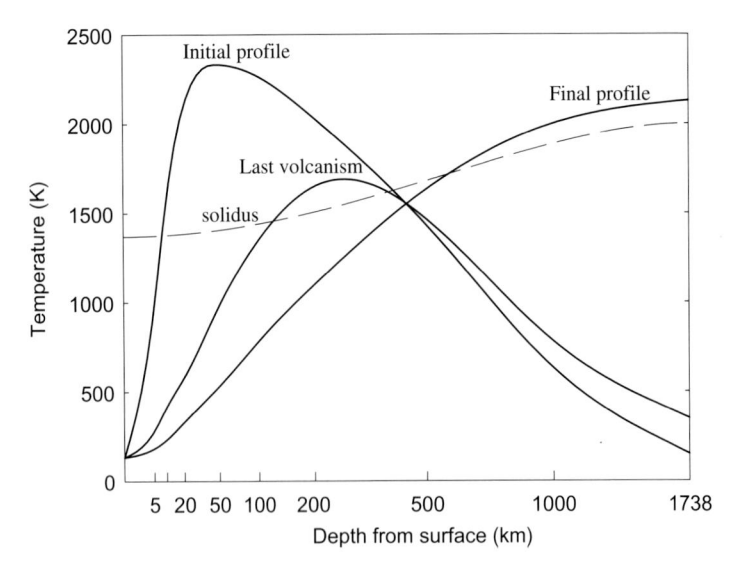

Fig. AF.1 The evolution of temperature profiles in the Moon.

far above the melting point. A more detailed calculation by Andrew Mullis[15] in 1993 suggests that a region of similar extent is occupied by partially molten material.

The Mullis calculation determined another aspect of thermal evolution, the shrinkage of the Moon as it cooled overall. This enabled the volumes of magma occupying the mare basins to be estimated, including the effect due to shrinkage of the Moon. His estimates of the gravitational effects of the mascon regions so formed agreed well with those measured from spacecraft.

---

[15] Mullis, A.M., 1993, *Geophys. J. Int.*, **114**, 196.

Appendix AG

The Abrasion of a Hemisphere of the Moon

The planetary collision, as illustrated in Fig. 11.8, produced an enormous amount of debris. In considering the abrasion of the Moon, we shall just consider the solid component, with total mass $CM_\oplus$[16] coming from the silicate and iron cores. If the Moon, of radius a, were to be at a distance R from the collision and if the debris sprayed out in all directions equally, then the mass striking the Moon would have been

$$M_d = \frac{a^2}{4R^2} CM_\oplus. \tag{AG1}$$

However, the debris is emitted most strongly in the plane of the collision interface so the mass striking the Moon will have been αM_d where the factor α is more likely to be less than unity than more and may be quite small.

If all the energy of the debris were to be transformed into kinetic energy of the Moon's surface material and the acquired speed of all the surface material were exactly v_{esc}, the escape speed from the Moon, then the mass of material abraded from the Moon would have been

$$\alpha M_d \frac{v_d^2}{v_{esc}^2}$$

where v_d ($\gg v_{esc}$) is the root-mean-square of the speed of the debris. In practice, because of other ways that energy can be transformed

---

[16]The symbol $\oplus$ represents Earth so $M_\oplus$ is the mass of Earth.

and because surface material can acquire a greater speed than v_{esc}, it will be less than this by a factor β, which will be less than unity and probably considerably less. Hence the total mass of abraded material will be

$$M_{abrad} = \alpha\beta C \frac{a^2}{4R^2} M_\oplus \frac{v^2}{v_{esc}^2}. \tag{AG2}$$

If the average depth of the material removed from the hemisphere of the Moon facing the collision is $d(\ll a)$, then the total mass of crust abraded is

$$M_{abrad} = 2\pi a^2 d\rho \tag{AG3}$$

where ρ is the average density of the removed material. From (AG2) and (AG3)

$$d = \alpha\beta C \frac{M_\oplus}{8\pi\rho R^2} \frac{v^2}{v_{esc}^2}. \tag{AG4}$$

The quantities v and R on the right-hand side are completely unknown as are the factors α, β and C. However, we can put in some reasonable values to see what they imply. Taking $\rho = 3 \times 10^3 \,\mathrm{kg\,m^{-3}}$, $v = 150\,\mathrm{km\,s^{-1}}$ and $R = 10^6\,\mathrm{km}$, then for $d = 60\,\mathrm{km}$ we find $\alpha\beta C = 0.194$. Various reasonable combinations of the factors can give this value, e.g. $C = 15$, $\alpha = 0.13$ and $\beta = 0.1$. The conclusion to be drawn is that while the precise conditions of the lunar abrasion event cannot be specified, given the proposed planetary collision scenario, such an event is plausible.

Appendix AH

The Rounding-Off of a Highly Eccentric Satellite Orbit

To illustrate the progress of the decay and rounding-off of the Moon's orbit, we take an initial orbit corresponding to an apogee of 10^6 km and a perigee of 10^5 km. This gives

$$a(1+e) = 10^9 \text{ m} \quad \text{(AH1a)}$$

and

$$a(1-e) = 10^8 \text{ m}, \quad \text{(AH1b)}$$

giving $a = 5.5 \times 10^8$ m and $e = 0.818$.

We are now going to find the rate of tidal energy generation in the Moon, as was done for Io in Appendix AA, but in this case, since e is so large, we cannot use Eq. (AA8) but must use a form based on Eq. (AA6), which is

$$W = \frac{9\pi}{32QPY}(GM_m M_\oplus)^2 \frac{r_m}{a^6} \left\{ \frac{1}{(1-e)^6} - \frac{1}{(1+e)^6} \right\}. \quad \text{(AH2)}$$

Q, P and Y are as defined in Appendix AA, $M_\oplus$ is the mass of Earth, 5.97×10^{24} kg, and M_m and r_m are the mass and radius of the Moon, 7.35×10^{22} kg and 1,734 km respectively. As initial values we take $Q = 500$ and $Y = 5 \times 10^{10}$ N m^{-2}. The period is given by

$$P = 2\pi \sqrt{\frac{a^3}{G(M_\oplus + M_m)}}. \quad \text{(AH3)}$$

The situation for this model is that the orbital energy of the system, E, is falling at the rate that energy is being generated in the

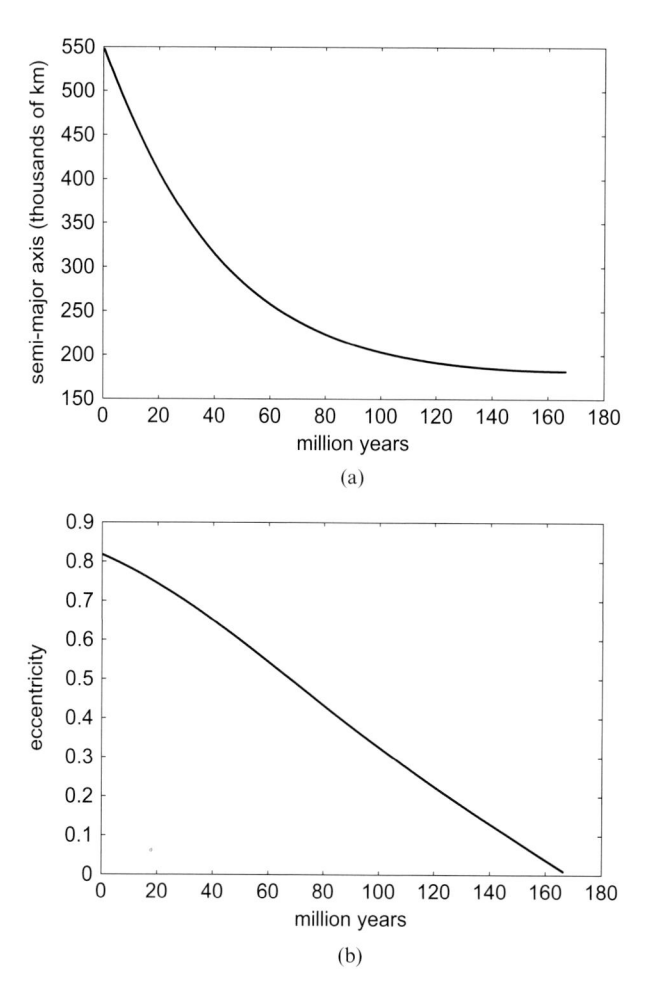

Fig. AH.1 The variation with time of (a) semi-major axis and (b) eccentricity due to the generation of tidal energy in the Moon.

Moon but we assume that the angular momentum remains constant. There will be components of added energy and angular momentum, due to the mechanism displayed in Fig. 12.1, but at first they are comparatively small and so, for now, they may be discounted. The rate of decrease of the semi-major axis can be found from

$$a = -\frac{G(M_\oplus + M_m)}{2E},$$

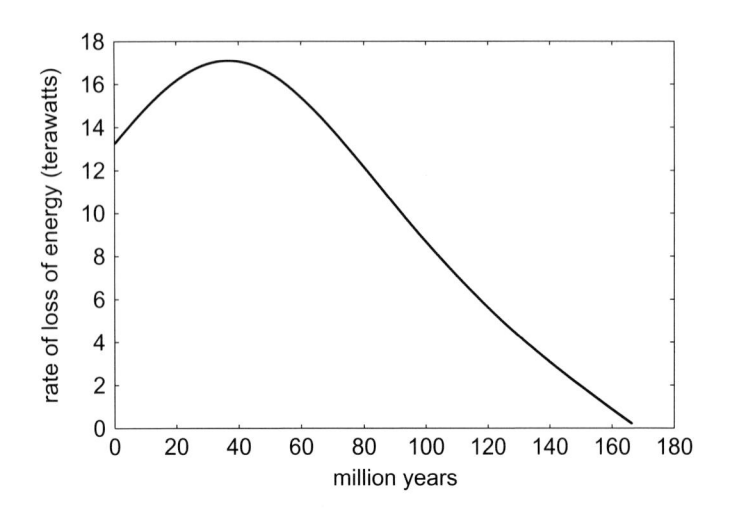

Fig. AH.2 Variation in the rate of tidal energy generation in the Moon.

giving

$$\frac{da}{dt} = \frac{G(M_\oplus + M_m)}{2E^2}\frac{dE}{dt} = -\frac{2a^2}{G(M_\oplus + M_m)}W, \qquad \text{(AH4)}$$

in which W is a positive quantity.

From Eq. (12.1), if the angular momentum is constant, then $a(1 - e^2)$ must be a constant. Differentiating this product with respect to time and equating to zero gives

$$\frac{de}{dt} = \frac{1 - e^2}{2ea}\frac{da}{dt} = \frac{a(1 - e^2)}{eG(M_m + M_\oplus)}W. \qquad \text{(AH5)}$$

Equations (AH4) and (AH5) are a pair of coupled differential equations in a and e and may be solved numerically. Using the initial values for a and e previously given, together with other values as defined, the equations have been solved numerically to give the changes of a and e with time as shown in Fig. AH.1.

The rate of generation of tidal energy in the Moon, which equals the rate of loss of orbital energy, at first increases and then falls as shown in Fig. AH.2. When the rate of orbital energy loss has fallen to a level that equals the rate at which orbital energy (presently about

1.2×10^{11} W or 0.12 terawatts) is gained due to the lunar regression mechanism, described in Section 12.1 in relation to Darwin's theory, then a state of equilibrium is reached. However, it is not stable equilibrium and, since Q and Y both increase as the Moon evolves, the regression mechanism becomes dominant so that, thereafter, the Moon slowly retreats from Earth.

Appendix AI

Continental Drift on Mars

A theorem proposed by Lamy and Burns[17] in 1972 states that the spin axis of a body with internal energy dissipation will move towards the principal axis of the maximum moment of inertia. That this should be so is easily demonstrated. For a body with angular momentum A and moment of inertia I, the total kinetic energy of spin is

$$E = \frac{A^2}{2I}. \tag{AI1}$$

If kinetic energy is being lost by being transformed into heat, so that E falls in value, but angular momentum remains constant as it will do for an isolated spinning system, then I must increase. Thus matter moves so as to increase its distance from the spin axis.

A spinning gaseous body, which would be spherical if not spinning, automatically adopts this configuration of spinning about an axis of maximum moment of inertia. The body flattens along the spin axis, to take up the form of an oblate spheroid, and the moment-of-inertia tensor is of the form

$$I = \begin{pmatrix} I_{xx} & 0 & 0 \\ 0 & I_{yy} & 0 \\ 0 & 0 & I_{zz} \end{pmatrix}. \tag{AI2}$$

If the spin axis is along z, then $I_{zz} > I_{xx} = I_{yy}$.

---

[17] Lamy, P.L. and Burns, J.A., 1972, *Am. J. Phys.*, **40**, 441.

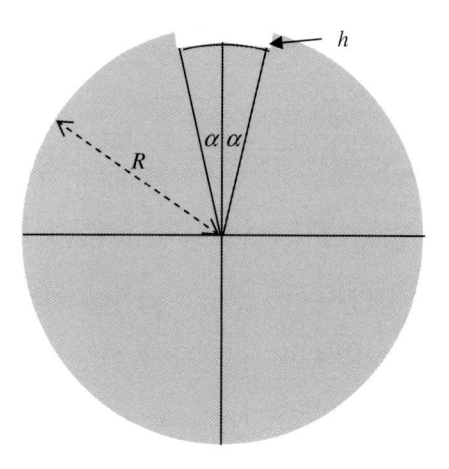

Fig. AI.1 A crater centred on the z-axis of a spherical body.

Figure AI.1 shows a circular crater of depth h centred on the z-axis subtending an angle 2α at the centre of a spherical body of radius R and density ρ. Its moment of inertia components are

$$I_{xx} = I_{yy} = \frac{1}{3}\pi h R^4 \rho(4 - 3\cos\alpha - \cos^3\alpha)$$
$$I_{zz} = \frac{1}{3}\pi h R^4 \rho(4 - 6\cos\alpha + 2\cos^3\alpha)$$
(AI3)

with all other components being zero.

If the crater is centred at the point with latitude λ and longitude ϕ, the components of the tensor are

$$\begin{pmatrix} \sin\lambda\cos\phi & -\sin\phi & \cos\lambda\cos\phi \\ \sin\lambda\sin\phi & \cos\phi & \cos\lambda\sin\phi \\ -\cos\lambda & 0 & \sin\lambda \end{pmatrix} \begin{pmatrix} I_{xx} & 0 & 0 \\ 0 & I_{yy} & 0 \\ 0 & 0 & I_{zz} \end{pmatrix}. \quad \text{(AI4)}$$

A circular basin or crater with uniform depth h will give components with negative h in (AI3) and a circular highland region gives components with positive h. Where there are several features, then the components in (AI4) are found for each individual feature and the components are simply added together. A standard diagonalisation process then finds the principal axes of the resultant tensor.

Table AI.1   Approximate representations of prominent features on Mars.

Feature	Mean height (km)	Semi-angular size (°)	Longitude (°)	Latitude (°)
Northern plains	−4	80	150	58
Tharsis uplift	4	50	95	−16
Tharsis supplement	5	15	107	−3
Elysium plain	5	17	210	25
Olympus Mons	12	5	133	18
Hellas Basin	−4	15	291	−44
Argyre plain	−2	7	42	−52

Features on Mars are not conveniently circular and of uniform height or depth but in 1983 Tony Connell and I approximated prominent features on Mars as indicated in Table AI.1.[18] The Tharsis region is a large plateau in the highlands of Mars containing Olympus Mons and other volcanoes. Because of its shape it was modelled in two parts. The Elysium plain is another raised feature within the southern highlands and the Argyre plain is a large impact basin in the southern highlands.

Using the data in the table, the principal axis of maximum moment of inertia was found to be at 11.9° to the spin axis. Given the crudity in the representation of the structural features the agreement in the alignment is satisfactory. The probability of an alignment within this angle or less, just by chance, is about 0.02. Again, it is possible that the crust may have ceased to become mobile before alignment had been achieved.

---

[18] Connell, A.J. and Woolfson, M.M., 1983, *Mon. Not. R. Astr. Soc.*, **204**, 1221.

Appendix AJ

The Oort Cloud and Perturbing Stars

For a comet with a very extended orbit, such as one coming from the Oort cloud into the planetary region, the eccentricity e is very close to unity. The intrinsic angular momentum (angular momentum per unit mass) of such a comet in an elliptical orbit around the Sun is

$$H = \{GM_\odot a(1-e^2)\}^{1/2} \approx (2GM_\odot q)^{1/2} \quad \text{(AJ1)}$$

in which a is the semi-major axis and $q = a(1-e)$, the perihelion distance. If a perturbation changes a comet's orbit from being unobservable to being observable, then q, and hence the magnitude of H, must have been reduced.

Calculations of stellar perturbation are usually made using the *impulse approximation*. This is based on the assumption that comets being perturbed by stars are near their aphelion and move very little during the passage of the star. Figure AJ.1 shows the passage of a star of mass M_* moving at speed V_* in a straight line past a body, A, with a closest approach distance of D. For a star passing at a distance of order 1 pc from the Sun, with a typical speed in the range of 20–30 km s^{-1}, a straight-line path is a good approximation.

The effect of the star's passage is to impart to the body a velocity with magnitude Δv_A perpendicular to the motion of the star where

$$\Delta v_A = \frac{2GM_*}{DV_*}. \quad \text{(AJ2)}$$

In Fig. AJ.2 the passage of the star is shown relative to the positions of the Sun, S, and a comet, C with closest approaches of the star

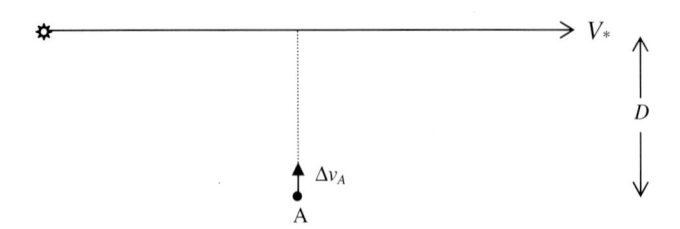

Fig. AJ.1 The passage of a star past a body at closest distance D.

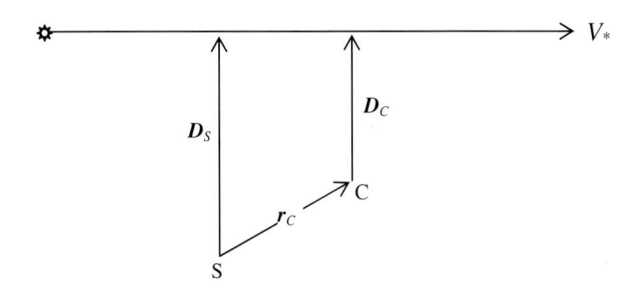

Fig. AJ.2 The Sun and comet relative to the passage of the star. The vectors $\boldsymbol{D}_S$ and $\boldsymbol{D}_C$ are not necessarily coplanar.

to the two bodies indicated by the vectors $\boldsymbol{D}_S$ and $\boldsymbol{D}_C$ respectively. These two vectors will not normally be coplanar. If the star passes at a distance of 1 pc, then appreciable interactions with the bodies will occur for less than 10 pc of the star's path. At a speed of $20\,\mathrm{km\,s^{-1}}$ relative to the Sun, this will take 5×10^5 years. For a comet with a semi-major axis of 40,000 au, the period is 8×10^6 years, so, in the region of the aphelion, it will not have moved very far during the passage of the star.

If the Sun and the comet have changes in velocity $\Delta\boldsymbol{v}_S$ and $\Delta\boldsymbol{v}_C$, then the relative change of the velocity of the comet relative to the Sun is

$$\Delta\boldsymbol{v}_{CS} = \Delta\boldsymbol{v}_C - \Delta\boldsymbol{v}_S. \tag{AJ3}$$

The change in the intrinsic angular momentum of the comet's orbit is

$$\Delta\boldsymbol{H} = \Delta\boldsymbol{v}_{CS} \times \boldsymbol{r}_C. \tag{AJ4}$$

If $\Delta\boldsymbol{H}$ is parallel to $\boldsymbol{H}$, then there is a maximum change of q found by differentiating (AJ1):

$$\Delta q = \sqrt{\frac{2q}{GM_*}}\Delta H. \qquad (AJ5)$$

The direction of $\boldsymbol{r}_C$ relative to the star's motion is very important. For a coplanar system, if $\boldsymbol{r}_C$ is parallel to the star's motion, then $D_S = D_C$ and there is no effect since Δv_{CS} would be zero. However, if $\boldsymbol{r}_C$ is perpendicular to the star's motion, then, although $\Delta\boldsymbol{v}_{CS}$ has the greatest possible value, it is parallel to $\boldsymbol{r}_C$ and hence the magnitude of the vector product in (AJ4) is zero. For maximum change of q, $\boldsymbol{r}_C$ should be at about 45° to the star's motion.

Table AJ.1 shows the minimum possible final perihelia for comets with various values of r_C and initial values of q where the star, of solar mass, has a closest distance of 0.5 pc and a speed of $20\,\mathrm{km\,s^{-1}}$ relative to the Sun.

For a given arrangement of Sun and comet and a particular stellar speed, it can be seen from (AJ2) that the perturbation depends on the ratio M_*/D. For a star of 2 solar masses, Table AJ.1 will apply for a distance of 1 pc. A comet with perihelion less than about 30 au, the orbital radius of Neptune, is likely to have its orbit considerably changed when it wanders into the region of the planets so most new comets, those coming from the Oort cloud, are likely to have had initial perihelia greater than 30 au. From Table AJ.1 it is clear that for an aphelion of, say, 50,000 au, a value of M_*/D much greater than $2\,M_\odot\,\mathrm{pc}^{-1}$, the value used in the table, is required to produce

Table AJ.1 Minimum final perihelia for solar-mass star passing at distance 0.5 pc at $20\,\mathrm{km\,s^{-1}}$.

		q_{init} (au)		
r_C (au)	40	30	20	10
80,000	23.47	15.68	8.31	1.73
60,000	33.25	24.15	15.23	6.62
40,000	37.59	27.92	18.30	8.80
20,000	39.49	29.56	19.64	9.75

a visible comet with perihelion less than about 5 au. A lesser value, but still greater than that used for the table, will be required to give a detectable comet — one that can be seen with a telescope — with perihelion within about 15 au. For very small values of D, the impulse approximation is no longer valid but in 1996 Soenke Eggers and I described a better approximation for closer star passages that can be applied computationally.[19]

---

[19] Eggers, S. and Woolfson, M.M., 1996, *Mon. Not. R. Astr. Soc.*, **282**, 13.

Appendix AK

Planetary Perturbation of New Comets

When a new comet enters the planetary region, it is perturbed by the planets, which can either add to or subtract from its original intrinsic energy. The expectation value for the change of energy is zero, which is to say that the likelihood of gaining energy is the same as that of losing it, no matter how deeply a comet penetrates into the planetary region. However, what can be found by numerical calculations is the average standard deviation of the change of energy as a function of the perihelion distance. This depends on the inclination of the cometary orbit; in Fig. AK.1 the full line shows the average standard deviation of energy change, σ_E, as a function of perihelion distance and the dashed lines the variation of the average change over all possible inclinations.

To consider the implications of the figure we consider a comet with a semi-major axis 40,000 au so that its energy, in inverse distance units, is -2.5×10^{-5} au^{-1} — negative because it is in an elliptical orbit. If the perihelion of its orbit is 10 au, then a possible magnitude of change of intrinsic energy is 10^{-4} au^{-1}. If that change is negative, then the new intrinsic energy is $(-2.5 \times 10^{-5} - 10^{-4})$ au$^{-1} = -1.25 \times 10^{-4}$ au^{-1}. Thus the new orbit is elliptical with a semi-major axis $1/(1.25 \times 10^{-4})$ au $= 8000$ au. Conversely, if the change is positive, then the new intrinsic energy is positive $(-2.5 \times 10^{-5} + 10^{-4})$ au$^{-1} = 7.5 \times 10^{-5}$ au^{-1} and positive energy implies a hyperbolic orbit, which is not closed, so the comet leaves the Solar System, never to return.

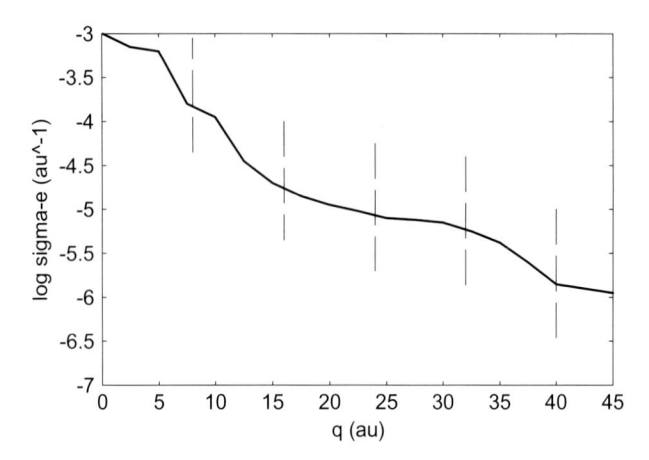

Fig. AK.1 The average value of σ_E for random inclinations as a function of perihelion distance (full line). The vertical lines are the limits for inclinations from 0° (top) to 180° (bottom).

It should be understood that Fig. AK.1 does not give limits to the possible changes of energy. For example, a comet with a perihelion of 30 au that just happened to pass very close to Neptune would have a very large change of energy, well outside the range indicated in the figure.

Appendix AL

Reactions and Decays

Most nuclear reactions are of the form

$$A + b \to C + d \qquad (\text{AL1})$$

and can be represented in shorthand as

$$A(b, d)C \qquad (\text{AL2})$$

where A is the initial nucleus, C the final product nucleus of the reaction and b and d are involved particles of radiation. Some light nuclei may be in both categories and there is a convention that within the brackets a helium nucleus is denoted by α, i.e. an alpha-particle, and a hydrogen nucleus by p, i.e. a proton. Other symbols, other than nuclei, are n for a neutron, e^- and e^+ for negative and positive β particles and γ for a gamma photon. Every reaction that takes place either produces energy or requires energy for it to take place. This energy is described as the Q-value for the reaction and is given in the units MeV (million electron volts) — about 1.6×10^{-13} J. The quantity given in parenthesis at the end of each reaction is the Q-value, positive if energy is produced and negative if energy is required. The following 34 reactions do not fall neatly into a general category and so are described in full.

H(e^-, ν)n	(−0.78)	H(p, $e^+ + \nu$)D	(1.19)
H(p + e^-, ν)D	(0.00)	D(p, n)2H	(−2.23)
D(D, n)$^3_2$He	(3.27)	D(D, p)T	(4.03)
T(D, n)$^4_2$He	(17.59)	T(T, 2n)$^4_2$He	(11.33)

$^{3}_{2}$He(e^{-},ν)T   (−0.02) $^{3}_{2}$He(p, e^{+}+ν)$^{4}_{2}$He   (10.16)

$^{3}_{2}$He(D, p)$^{4}_{2}$He   (18.35) $^{3}_{2}$He(T,D)$^{4}_{2}$He   (14.32)

$^{3}_{2}$He(T,n+p)$^{4}_{2}$He   (12.10) $^{3}_{2}$He($^{3}_{2}$He,2p)$^{4}_{2}$He   (12.86)

$^{4}_{2}$He(2n,γ)$^{6}_{2}$He (0.97) $^{4}_{2}$He(n+p,γ)$^{6}_{3}$Li   (3.70)

$^{4}_{2}$He(D,γ)$^{6}_{3}$Li   (1.47) $^{4}_{2}$He(T,γ)$^{7}_{3}$Li   (2.47)

$^{4}_{2}$He(T,n)$^{6}_{3}$Li   (−4.78) $^{4}_{2}$He($^{3}_{2}$He,γ)$^{7}_{4}$Be   (1.59)

$^{4}_{2}$He(α+n,γ)$^{9}_{4}$Be (1.57) $^{4}_{2}$He(2α,γ)$^{12}_{6}$C   (7.28)

$^{6}_{3}$Li(p,$^{3}_{2}$He)$^{4}_{2}$He (4.02) $^{7}_{3}$Li(D,n)2$^{4}_{2}$He   (15.12)

$^{7}_{3}$Li(T,2n)2$^{4}_{2}$He   (8.87) $^{7}_{3}$Li($^{3}_{2}$He,n+p)2$^{4}_{2}$He (9.63)

$^{7}_{4}$Be(e^{-},γ+ν)$^{7}_{3}$Li (0.05) $^{7}_{4}$Be(D, p)2$^{4}_{2}$He   (16.77)

$^{7}_{4}$Be(T,n+p)2$^{4}_{2}$He   (10.51) $^{7}_{4}$Be($^{3}_{2}$He,2p)2$^{4}_{2}$He   (11.27)

$^{9}_{4}$Be(p,D)2$^{4}_{2}$He   (0.65) $^{12}_{6}$C + $^{12}_{6}$C → $^{24}_{12}$Mg   (13.93)

$^{16}_{8}$O + $^{16}_{8}$O → $^{32}_{16}$S   (16.54) $^{12}_{6}$C + $^{16}_{8}$O → $^{28}_{14}$Si (16.75)

A(n, γ)B reactions with A =

$^{19}_{10}$Ne	(16.87)	$^{20}_{10}$Ne	(6.76)	$^{21}_{10}$Ne	(10.37)	$^{22}_{10}$Ne	(5.20)
$^{21}_{11}$Na	(11.07)	$^{22}_{11}$Na	(12.42)	$^{23}_{11}$Na	(6.96)	$^{24}_{11}$Na	(9.01)
$^{23}_{12}$Mg	(6.53)	$^{24}_{12}$Mg	(7.33)	$^{25}_{12}$Mg	(11.09)	$^{26}_{12}$Mg	(6.44)
$^{27}_{12}$Mg	(1.83)	$^{25}_{13}$Al	(1.92)	$^{26}_{13}$Al$_g$[20]	(13.06)	$^{26}_{13}$Al$_m$[20]	(13.06)
$^{27}_{13}$Al	(7.73)	$^{28}_{13}$Al	(9.44)	$^{29}_{13}$Al	(17.18)	$^{27}_{14}$Si	(17.17)
$^{28}_{14}$Si	(8.47)	$^{29}_{14}$Si	(10.61)	$^{30}_{14}$Si	(6.59)	$^{31}_{14}$Si	(9.22)
$^{29}_{15}$P	(11.33)	$^{30}_{15}$P	(12.31)	$^{31}_{15}$P	(7.94)	$^{32}_{15}$P	(10.11)
$^{33}_{15}$P	(6.56)	$^{31}_{16}$S	(10.37)	$^{32}_{16}$S	(5.20)	$^{33}_{16}$S	(11.07)
$^{34}_{16}$S	(16.87)	$^{35}_{16}$S	(6.76)	$^{36}_{16}$S	(10.37)		

---

[20] $^{26}_{13}$Al$_g$ and $^{26}_{13}$Al$_m$ are excited states of the aluminium isotopes.

A(n, p)B reactions with A =

$^{21}_{11}$Na (4.33) $^{22}_{11}$Na (3.63) $^{25}_{13}$Al (5.06) $^{27}_{14}$Si (5.59)
$^{29}_{15}$P (5.73) $^{30}_{15}$P (5.01) $^{32}_{15}$P (0.57) $^{31}_{16}$S (6.22)
$^{33}_{16}$S (0.53)

A(n, α)B reactions with A =

$^{19}_{10}$Ne (12.14) $^{21}_{10}$Ne (0.70) $^{21}_{11}$Na (2.59) $^{22}_{11}$Na (1.95)
$^{23}_{12}$Mg (7.21) $^{25}_{13}$Al (1.92) $^{27}_{14}$Si (7.19) $^{29}_{15}$P (0.91)
$^{31}_{16}$S (8.14) $^{32}_{16}$S (1.52) $^{33}_{16}$S (3.49) $^{35}_{16}$S (0.89)

A(p, γ)B reactions with A =

D (5.49) T (19.82) $^{6}_{3}$Li (5.61) $^{7}_{4}$Be (0.14)
$^{9}_{4}$Be (6.58) $^{10}_{5}$B (8.69) $^{11}_{5}$B (15.96) $^{11}_{6}$C (0.59)
$^{12}_{6}$C (1.94) $^{13}_{6}$C (7.55) $^{14}_{6}$C (10.21) $^{13}_{7}$N (4.63)
$^{14}_{7}$N (7.30) $^{15}_{7}$N (12.13) $^{16}_{8}$O (0.60) $^{17}_{8}$O (5.61)
$^{18}_{8}$O (7.99) $^{19}_{9}$F (12.85) $^{20}_{10}$Ne (2.43) $^{21}_{10}$Ne (6.74)
$^{22}_{10}$Ne (8.79) $^{23}_{10}$Ne (10.56) $^{24}_{10}$Ne (10.70) $^{23}_{11}$Na (11.69)
$^{24}_{11}$Na (12.06) $^{25}_{11}$Na (14.15) $^{24}_{12}$Mg (2.27) $^{25}_{12}$Mg[21] (6.31)
$^{25}_{12}$Mg[21] (6.09) $^{26}_{12}$Mg (8.27) $^{27}_{12}$Mg (9.55) $^{28}_{12}$Mg (10.48)
$^{29}_{12}$Mg (10.62) $^{26}_{13}$Al (7.46) $^{27}_{13}$Al (11.59) $^{28}_{13}$Al (12.33)
$^{29}_{13}$Al (13.51) $^{30}_{13}$Al (14.35) $^{28}_{14}$Si (2.75) $^{29}_{14}$Si (5.60)
$^{30}_{14}$Si (7.30) $^{31}_{14}$Si (8.65) $^{32}_{14}$Si (9.54) $^{30}_{15}$P (6.13)
$^{31}_{15}$P (8.86) $^{32}_{15}$P (9.57) $^{33}_{15}$P (10.88)

---

[21] These pairs of reactions give two different excited states of $^{26}_{13}$Al.

A(p, n)B reactions with A =

T	(−0.76)	$^{7}_{3}$Li	(−1.64)	$^{9}_{4}$Be	(−1.85)	$^{11}_{5}$B	(−2.77)
$^{13}_{6}$C	(−3.00)	$^{14}_{6}$C	(−0.63)	$^{14}_{7}$N	(−5.93)	$^{15}_{7}$N	(−3.54)
$^{19}_{9}$F	(−4.02)	$^{23}_{10}$Ne	(3.60)	$^{24}_{10}$Ne	(1.69)	$^{23}_{11}$Na	(−4.84)
$^{24}_{11}$Na	(4.73)	$^{25}_{11}$Na	(3.05)	$^{26}_{12}$Mg^{21}	(−4.79)	$^{26}_{12}$Mg^{21}	(−5.01)
$^{27}_{12}$Mg	(1.83)	$^{28}_{12}$Mg	(1.05)	$^{29}_{12}$Mg	(4.87)	$^{28}_{13}$Al	(3.86)
$^{29}_{13}$Al	(2.90)	$^{30}_{13}$Al	(7.76)	$^{31}_{14}$Si	(0.71)	$^{32}_{15}$P	(0.93)

A(p, α)B reactions with A =

$^{7}_{3}$Li	(17.35)	$^{9}_{4}$Be	(2.13)	$^{10}_{5}$B	(1.15)	$^{11}_{5}$B	(8.68)
$^{14}_{7}$N	(−2.92)	$^{15}_{7}$N	(4.97)	$^{16}_{8}$O	(−5.22)	$^{17}_{8}$O	(1.19)
$^{18}_{8}$O	(3.98)	$^{19}_{9}$F	(8.12)	$^{20}_{10}$Ne	(−4.13)	$^{23}_{11}$Na	(2.38)
$^{24}_{11}$Na	(2.18)	$^{25}_{11}$Na	(3.53)	$^{24}_{12}$Mg	(−6.88)	$^{27}_{13}$Al	(1.60)
$^{30}_{13}$Al	(3.56)	$^{31}_{15}$P	(1.92)	$^{32}_{15}$P	(2.45)	$^{33}_{15}$P	(2.96)
$^{35}_{16}$S	(0.32)	$^{36}_{16}$S	(0.54)				

A(α, γ)B reactions with A =

$^{6}_{3}$Li	(4.46)	$^{7}_{3}$Li	(8.67)	$^{7}_{4}$Be	(7.54)	$^{12}_{6}$C	(7.16)
$^{14}_{7}$N	(4.42)	$^{15}_{7}$N	(4.01)	$^{16}_{8}$O	(4.73)	$^{17}_{8}$O	(7.35)
$^{18}_{8}$O	(9.67)	$^{17}_{9}$F	(6.56)	$^{18}_{9}$F	(8.48)	$^{19}_{9}$F	(10.47)
$^{20}_{10}$Ne	(9.31)	$^{21}_{10}$Ne	(9.88)	$^{22}_{10}$Ne	(10.61)	$^{23}_{10}$Ne	(11.86)
$^{24}_{10}$Ne	(11.49)	$^{21}_{11}$Na	(9.15)	$^{22}_{11}$Na	(9.45)	$^{23}_{11}$Na	(10.09)
$^{24}_{11}$Na	(10.86)	$^{25}_{11}$Na	(11.28)	$^{24}_{12}$Mg	(9.98)	$^{25}_{12}$Mg	(11.13)
$^{26}_{12}$Mg	(10.64)	$^{27}_{12}$Mg	(10.79)	$^{28}_{12}$Mg	(11.50)	$^{25}_{13}$Al	(10.46)
$^{26}_{13}$Al	(10.42)	$^{27}_{13}$Al	(9.67)	$^{28}_{13}$Al	(9.88)	$^{29}_{13}$Al	(10.55)
$^{30}_{13}$Al	(11.37)	$^{28}_{14}$Si	(6.95)	$^{29}_{14}$Si	(7.12)	$^{30}_{14}$Si	(7.92)

$^{31}_{14}$Si (8.32) $^{32}_{14}$Si (9.00) $^{29}_{15}$P (6.48) $^{30}_{15}$P (6.66)

$^{31}_{15}$P (7.00) $^{32}_{15}$P (7.64) $^{33}_{15}$P (7.85) $^{32}_{16}$S (6.64)

$^{33}_{16}$S (6.79) $^{34}_{16}$S (7.21) $^{35}_{16}$S (6.82) $^{36}_{16}$S (6.80)

A(α, n)B reactions with A =

$^{7}_{3}$Li (−2.79) $^{9}_{4}$Be (5.70) $^{10}_{5}$B (1.06) $^{12}_{6}$C (−8.50)

$^{13}_{6}$C (2.21) $^{14}_{7}$N (−4.74) $^{15}_{7}$N (−6.42) $^{17}_{8}$O (0.59)

$^{18}_{8}$O (−0.70) $^{21}_{10}$Ne (2.55) $^{22}_{10}$Ne (−0.48) $^{23}_{10}$Ne (5.42)

$^{24}_{10}$Ne (2.99) $^{23}_{11}$Na^{21} (−2.97) $^{23}_{11}$Na^{21} (−3.20) $^{24}_{11}$Na (3.13)

$^{25}_{11}$Na (1.85) $^{25}_{12}$Mg (2.65) $^{26}_{12}$Mg (0.03) $^{27}_{12}$Mg (4.20)

$^{28}_{12}$Mg (2.29) $^{27}_{13}$Al (−2.64) $^{28}_{13}$Al (1.95) $^{29}_{13}$Al (0.45)

$^{30}_{13}$Al (4.80) $^{31}_{14}$Si (1.33) $^{35}_{16}$S (0.22)

A(α, p)B reactions with A =

$^{18}_{9}$F (1.74) $^{19}_{9}$F (1.68) $^{22}_{11}$Na (3.15) $^{23}_{11}$Na (1.82)

$^{24}_{11}$Na (1.30) $^{25}_{11}$Na (0.80) $^{25}_{12}$Mg (−1.21) $^{26}_{12}$Mg (−2.87)

$^{25}_{13}$Al (7.72) $^{26}_{13}$Al (4.82) $^{27}_{13}$Al (2.37) $^{28}_{13}$Al (1.24)

$^{29}_{13}$Al (1.01) $^{29}_{15}$P (4.20) $^{30}_{15}$P (1.52) $^{31}_{15}$P (0.63)

Decays

T $^{6}_{2}$He $^{7}_{4}$Be $^{8}_{4}$Be $^{8}_{5}$B $^{9}_{5}$B $^{11}_{6}$C $^{14}_{6}$C $^{12}_{7}$N

$^{13}_{7}$N $^{14}_{8}$O $^{15}_{8}$O $^{17}_{9}$F $^{18}_{9}$F $^{19}_{10}$Ne $^{23}_{10}$Ne $^{24}_{10}$Ne $^{21}_{11}$Na

$^{22}_{11}$Na $^{24}_{11}$Na $^{25}_{11}$Na $^{23}_{12}$Mg $^{27}_{12}$Mg $^{28}_{12}$Mg $^{29}_{12}$Mg $^{25}_{13}$Al $^{26}_{13}$Al$_m$

$^{26}_{13}$Al$_g$ $^{28}_{13}$Al $^{29}_{13}$Al $^{30}_{13}$Al $^{27}_{14}$Si $^{31}_{14}$Si $^{32}_{14}$Si $^{29}_{15}$P $^{30}_{15}$P

$^{32}_{15}$P $^{33}_{15}$P $^{31}_{16}$S $^{35}_{16}$S

Appendix AM

Cooling and Grain Formation

Cooling Rates

The calculations given here are very speculative and involve guesswork for most of the involved parameters. Fortunately, the conclusions drawn are not very sensitive to the chosen parameters so we may have reasonable confidence in the general outcome.

The inwardly moving planetary material initially enveloped the nuclear reactions. When the reactions built up sufficient pressure, the inward motion was reversed and all the surrounding material, in whatever state it happened to be, was pushed outwards. If the maximum average temperature in the reaction region was T_m, its volume was V_o and the mean particle density was n_o, then the total thermal energy contained in the exploding region would have been

$$E_o = \frac{3}{2}kT_m n_o V_o. \tag{AM1}$$

If most of this energy were imparted to the combined mass of the planets, M, then the root-mean-square speed given to planetary material would have been

$$v_p = \left(\frac{3kT_m n_o V_o}{M}\right)^{1/2}, \tag{AM2}$$

in which k is Boltzmann's constant.

The modelling we have done suggests approximate values of $T_m = 5 \times 10^8$ K, $n_o = 2 \times 10^{30}$ m^{-3} and $M = 4 \times 10^{27}$ kg. The effective volume of the nuclear reactions is more difficult to estimate but 4×10^{18} m^3, equivalent to a sphere of radius about 1,000 km, should be in the right ballpark. These numbers give $v_p = 6.4\,\mathrm{km\,s^{-1}}$. This is just a root-mean-square value for the overall motion of the whole planet; material from the heart of the reaction region would have been thrown out much faster.

We now consider the expansion of the material in the reaction region. It would have expanded rapidly, and adiabatically, and the change of volume from V_o to V would have given a change of temperature from T_m to T where

$$\frac{T}{T_m} = \left(\frac{V_o}{V}\right)^{\gamma-1} = \left(\frac{r_o}{r}\right)^{3(\gamma-1)} = \left(\frac{r_o}{r}\right)^2 \qquad \text{(AM3)}$$

where γ, taken as $\frac{5}{3}$, is the usual ratio of specific heats and r and r_o are the radii of spherical volumes V and V_o respectively.

The speed of efflux of a gas into a vacuum is given by

$$v_c = \left(\frac{2}{\gamma-1}\right)^{1/2} c = \left[\frac{6kT}{(\gamma-1)\mu}\right]^{1/2} \qquad \text{(AM4)}$$

where c is the speed of sound in the gas and μ is the mean particle mass. If the exploding region, taken as spherical, moves outwards with this speed, then we can write

$$v_c = \frac{dr}{dt} = \left[\frac{6kT}{(\gamma-1)\mu}\right]^{1/2}. \qquad \text{(AM5)}$$

From (AM3) we find

$$\frac{dr}{dt} = -\frac{r_o T_m^{1/2}}{2T^{3/2}}\frac{dT}{dt} \qquad \text{(AM6)}$$

and, combining (AM5) and (AM6),

$$\frac{dT}{dt} = -\frac{2}{r_o T_m^{1/2}}\left[\frac{6k}{(\gamma-1)\mu}\right]^{1/2} T^2. \qquad \text{(AM7)}$$

By integration we can find the time it takes for the temperature to fall from T_m to T_c, the temperature at which the first condensations should take place. This is

$$t_c = \frac{r_o T_m^{1/2}}{2} \left[\frac{(\gamma-1)\mu}{6k}\right]^{1/2} \left(\frac{1}{T_c} - \frac{1}{T_m}\right)$$

$$\approx \frac{r_o T_m^{1/2}}{2} \left[\frac{(\gamma-1)\mu}{6k}\right]^{1/2} \frac{1}{T_c} \qquad \text{(AM8)}$$

since $T_m \gg T_c$. Inserting $r_o = 10^6$ m, $\mu = 4 \times 10^{-27}$ kg and $T_c = 2500$ K, with other values as previously given, we find $t_c = 25400$ s or about 7 hours. While this may seem a short time we should recall that the minerals in chondrules are unequilibrated, indicating that the cooling process from when the chondrules formed was extremely rapid — estimated to be of the order of hours (Section 16.2.1). It is very likely that the cooling stage before chondrules formed was also very rapid.

Grain Formation

By the time that the temperature has fallen to 2,500 K, stable entities, such as SiO_2 and MgO, would have formed and would be coming together to form grains. If a forming grain has a radius a and density ρ and is being bombarded by particles with an average density in the surrounding region σ, with individual particles moving at average speed v, then it can be shown that the rate of increase of the radius of the grain is constant and equal to

$$\frac{da}{dt} = \frac{\sigma v}{4\rho}. \qquad \text{(AM9)}$$

At the time the temperature falls to 2,500 K, which we shall take as the beginning of grain formation, the original radius of the reaction region has increased from 10^6 m to about 4.5×10^8 m, indicating that the density has fallen to 1.1×10^{-8} of its original value. The density of the unexpanded reaction region was taken as 10^4 kg m^{-3} and Table 17.1 shows that about one-third of the material in it will go on to form silicates. From this we deduce that the density of silicate-forming material when grains start forming is $\sigma = 3.7 \times 10^{-5}$ kg m^{-3}. If an average silicate-forming particle is taken as SiO_2, with a mass

$\mu = 1.06 \times 10^{-25}$ kg, then its average (actually root-mean-square) speed is given by

$$v = \sqrt{\frac{3kT}{\mu}} = 990\,\mathrm{m\,s^{-1}}. \qquad \text{(AM10)}$$

For a grain of density $3{,}000\,\mathrm{kg\,m^{-3}}$ this gives the rate of growth of the grain as $3.05 \times 10^{-6}\,\mathrm{m\,s^{-1}}$. Thus a grain of radius 1 mm would have taken just over 5 minutes to grow.

The analysis has several uncertainties associated with it and the estimated times for growth seem suspiciously low. However, even if the rate of grain growth were underestimated by four orders of magnitude, we can still be reasonably sure that grains will form and have cooled on a timescale much less than the half-life of sodium-22.

Index

47 Uma   72
51 Pegasus   72

abrasion by bombardment   215
absorption of radiation   336, 364
accretion   387
accretion by bombardment   215
accretion cross section   358
accretion radius   358
achondrites   258, 308
Adams, John Couch   172
Airy, George Biddell   172
albedo   165
Alfvén, H.   132
aluminium unstable isotopes   285
aluminium-26   274, 276, 417
Amalthea   144, 158
Ames Research Center   213
amino acids   260
ammonia   155, 273, 414
Ananke   161
Anders, E.   290
angular momentum   xxii, xxiv, 132, 303, 313, 348
angular momentum conservation   314
anthropic principle   xxxii, 196
apastron   108, 331
apogee   208
Apollo asteroids   178
Apollo space missions   212
areal density   96, 136
Argyre plain   428
artificial viscosity   342, 405
asteroid cooling rates   263
asteroids   78, 177, 241, 245, 308

ataxites   262
Aten group   178
atomic cooling   23
atomic energy levels   335

Bailey, M. E.   251, 252
Bate, M.   32, 59
Bell Burnell, Jocelyn   65
Bellona   197, 206, 221, 305
Bellona satellites   207
Benz, W.   206
benzene   260
beryllium   xxxiii
Big Bang   4, 20, 191
binary orbit period   350
binary star frequencies   60
binary star orbits   330
binary stars   299
binary system formation   28, 33
binary system frequency   300
binary systems   3, 338, 348
black hole   52, 299
blue shift   334
Bode, Johann Elert   174
Bode's law   174, 175
bombardment in Solar System   247
Bonnell, I.   32, 59
boron   xxxiii
Bromm, V.   32
brown dwarfs   89
Brown, Robert   137
Brownian motion   137, 383
Bruno, Giordano   xix
Brush, Stephen G.   295
Burns, J.A.   227, 426
Butler, P.   72

calcium-aluminium-rich inclusions (CAI)   261, 279, 281
Callisto   145, 146, 161, 219
Caloris Basin   229, 231, 232
Calypso   164
Cameron, A.W.G.   xxxv, 206
Cameron, Andrew Collier   116
Capture Theory   85, 90, 301
carbon   309
carbon-14   287
carbonaceous chondrites   247, 258, 260
Cassini division   167
Castets, A.   194
Caughlan, G.R.   272
Caux, E.   194
Ceccarelli, C.   194
centre of figure   217, 227
centre of mass   217, 227, 330, 340
Cepheid variables   8
Ceres   174, 241, 307
Chamberlin, T.C.   xxv
Chandrasekhar limit   65, 67
Chandresekhar, S.   52, 58
Chaotic Terrain   229, 231, 232
Charon   176
chassignites   261
Chiron   178
chondrites   258, 308
chondrule cooling time   260
chondrules   258, 259, 308
Clube, S.V.M.   251
CODAG   xxxix, 139
cold dense cloud L134N   194
colliding planets   401, 410
collision cross section   60
collision interface   402
comet coma   181
comet dust tail   181
comet Mrkos   181
comet plasma tail   181
comets   78, 181, 241, 248, 307
commensurable orbits   114, 164
compensation level   210
condensation sequences   xxxiii

condensations in a filament   87
Connell, A.J.   227, 428
continental drift on Mars   227, 426
cooling after planetary collision   440
Copernicus crater   211
Copernicus, Nicolaus   xix, 297
cosmic rays   10, 21
counterglow (gegenschein)   78
Crab Nebula   51
Cremin, A.W.   62, 381

D/H in planetary atmospheres   195
Dalton, John   269
Dantona, F.   109, 366
dark cool cloud   55
Darwin, Charles   v
Darwin, George   203
Davis, D.R.   206
De Revolutionibus Orbium Coelestium   xix
degenerate state   48
Deimos   171
dense cool cloud (DCC)   20
dense cool cloud formation   25
Descartes, René   xxii
deuterium reactions   198, 272, 414
deuterium   44, 91, 192, 269
Dialogue Concerning the Two Chief World Systems   xxi
dimensional analysis   318
Dione   164
Discordia   176
disks around young stars   xxxviii
disruption of planetary system   126, 376
Dodd, R.J.   245
Donn, B.   xxxix, 137, 384
Doppler effect   3, 27, 69, 299, 333, 339, 360
Doppler shift velocity profile   71
Doppler, Christian   333
Dormand, J.R.   196, 210, 310, 418
dust carpet formation   xxvii, 135, 140, 327
dust carpet instability   xlii

dust cooling   21
dust in ISM   19
dust settling   137
dust-grain absorption   324
dust-grain scattering   324
dwarf planets   172, 173, 219, 241, 243
dwarf planet orbits   244
Dyson, Frank   120

Earth and Venus formation   196
Earth axial tilt   157
Earth formation   145, 304
Earth orbit   332
Earth spin   199
Earth   131, 152
Earth, formation by SNT   xli
eccentricity   69, 302, 331
eclipsing binary   339
ecliptic   399
Eddington accretion mechanism   358
Eddington, Arthur   120, 358
Eggers, S.   432
Einstein, Albert   120, 296
Elara   161
electromagnetic radiation   9
electronic transitions   336
ellipse   68, 331
elliptical galaxy   328
Elysium plain   428
embedded cluster   55, 57, 85, 300
endothermic reactions   50
energy generation in stars   42
enstatite chondrites   258
enstatite   259
Enyo   197, 206, 305
equilibration   260
equipotential surfaces   xxviii, 316
Eris   173, 176, 242
escape speed   214, 420
Euler rotation matrix   376
Europa tidal heating   161
Europa   158, 219
evolution of lunar orbit   204, 215
exoplanet characteristics table   74

exoplanet formation   372
exoplanet imaging   77
exoplanet mass   360
exoplanet orbit   360
exothermic reactions   50
exponential distribution   369
Exposition du Système du Monde   xxiv

falls   256
filament instability   xxvii, xxix
finds   256
First Point of Aries   399
fluorine unstable isotopes   277
foci of ellipse   69, 331
Fomalhaut   79, 81
Fowler, William   272
Fraunhofer lines   4, 336
free-fall collapse   26, 345
free-fall time   26, 346
free-floating planets   89, 93
frequency   333
frequency of planetary systems   119
FU Orionis   41

galactic (open) cluster   2
galactic cluster evaporation   58
galactic halo   3, 339
galactic nucleus   3
galactic nucleus   328
galactic plane   3, 328
Galilean satellites   132, 140, 143, 158
Galileo spacecraft   158
Galileo, Galilei   xx, 132, 297
Galle, Johann   172
Ganymede   161, 203, 219, 221
Gaussian curve   96, 102
Gaussian distribution   369
gegenschein (counterglow)   78
geometrical cross section   60
geometrical moment of inertia   353
Giant Molecular Clouds   251
Gingold, R.A.   341
Gleise 581   120
globular cluster evaporation   58

globular cluster   2
Golanski, Yann   25
Gold, Thomas   66
Goldreich, P.   xl, 142
Goodricke, John   7
grain formation   442
grain surface chemistry   194
grains in molecular clouds   324
gravitational cross section   380
gravitational instability   141, 318
gravitational potential   316
Great Dark Spot   155
Great Red Spot   155
Greaves, J.S.   79, 301
greenhouse effect   226

Haisch, K.E.   99
Halley, Edmund   182
Halley's comet   182
hard binaries   60
hardened orbits   128
HARPS   120
Hartmann, W.R.   206
Haumea   176, 242
Hayachi, Chushiro   21, 39
Hayashi plot   40
HD172555   305
HD209458   75
HD80606   75
heat transfer   417
heavy carbon   286
heavy nitrogen   287
Helene   164
helium flash   48
helium   154, 192, 269
helium-3   414
Hellas Basin   223, 428
hemispherical asymmetry of Moon   212, 420
hemispherical asymmetry   306
Herbig, G.H.   41
Herbst, E.   194
Herschel crater   165
Herschel, William   172
Hertzsprung-Russell diagram   39

Hewish, Anthony   65
hexahedrites   262
Hidalgo   178
high iron chondrites   258
high-density regions   90
Hipparcos   6
HL Tau   80, 301
Holden, P.   272, 311
Hoyle, Fred   xxxvi
HR8799   82
Hydra   175
hydrogen   192, 269
hydrogen-shell burning   46
Hygeia   177
Hyperion   164
hysteresis   158, 391

Iapetus   165, 219
Iben, I.   62
Icarus   178
impact model of Moon formation   206
impulse approximation   429
inclination   69, 185
interactions between planets   185
intersecting orbits   187
interstellar medium (ISM)   19
intrinsic angular momentum   133
intrinsic energy   433
Io tidal heating   391, 394
Io volcanoes   158
Io   146, 219, 220
ionic cooling   23
iron meteorites   180, 256, 262, 309
iron   50
ISM (intersetellar medium)   19
isostacy   210
isotope   44
isotopic anomalies   269, 309

Jacobi ellipsoid   29
Jeans critical mass   xxxi, xxxiv, 20, 55
Jeans tidal theory   xxvi

Jeans, James   xxvii, xliii, 29, 132, 299, 301
Jeffreys, Harold   xxxi
Juno   177
Jupiter rings   162
Jupiter satellites   143, 158
Jupiter spin   155
Jupiter   75, 131, 154
Jupiter, formation by SNT   xli

Kalas, Paul   81
kamacite   262, 309
Kant, Immanuel   xxii
Kelvin-Helmholtz contraction   41, 61, 353, 366
Kepler, Johannes   xxii
Kepler's equation   376
Keplerian orbit   134
Keplerian speed   97, 108, 367
kernal   341
Kimura, H.   xxxix
Kirkwood, David   179
Klrkwood gaps   179
Kroupa, P.   61
Kuiper belt   172, 176, 241, 248, 252
Kuiper, Gerard   176

Lada, C.J.   99
Lada, E.A.   99
Lagrangian method   404
Lamy, P.L.   227, 426
Laplace, Pierre-Simon   xxii
latent heat of fusion   416
Le Verrrier, Urbain   172
Leavitt, Henrietta   8
Lee, T.   279
Lefloch, B.   194
libration   208
light nitrogen   287
light year   2
lithium   xxxiii, 192, 269
Little Green Men   65
Loinard, L.   194
long-period comets   182, 249

low iron and low metal chondrites   258
low iron chondrites   258
Lowell, Percival   173
Lucas, P.   88
Lucy, L.B.   341
luminosity   8
Luna 3 spacecraft   211
lunar crust   213, 418
lunar highlands   209
lunar surface heat flow   418
lunar volcanism   418
Lynden-Bell and Pringle mechanism   322
Lynden-Bell, D.   xxxv

Maclaurin spheroid   29
magnesium isotopic anomaly   279
magnesium stable isotopes   279
magnesium   309
magnesium-25 anomalies   281
magnetic field line   34
magnetic transfer of angular momentum   xxxvi
magnetite   259
main sequence stars   2, 14, 39
main-sequence lifetime   356
major planet spin periods   155
major planets   154
Makemake   176, 242
Mann, I.   xxxix
Marcy, G.   72
Mare Moscoviense   211, 223
Mars   152, 219, 242
Mars atmosphere   223, 261
Mars core   228
Mars crust   222
Mars density   219
Mars Global Surveyor   222
Mars hemispherical asymmetry   221
Mars orbit as satellite   233
Mars orbital eccentricity   154, 232
Mars origin   305
Mars past climate   225
Mars polar caps   224

Mars scarp   222
Mars spin axis   226
Mars surface   224
mascons   210, 419
maser emission   26, 299
mass frequency function   13
mass-dependent fractionation   270
massive star formation   58, 300
Mathilde   180
Mayor, M.   72
Mazzitelli, I.   109, 366
McCaughrean, M.J.   61
McCord, T.B.   238
Meakin, P.   xxxix, 137, 384
Melita, M.D.   114, 149
Mercury core   228
Mercury density   228
Mercury orbital eccentricity   154, 232
Mercury origin   305
Mercury spin period   231
Mercury   75, 131, 152, 219, 221, 242
mesosiderites   263, 265, 309
metallic hydrogen   154
metallicity   5
meteorites   255, 308
methane   155, 273, 415
Metis   158
microlensing   120
migration   xli
Milky Way   1, 328
Millar, T.J.   194
Mimas   164
Miranda surface   168
Miranda   167
molecular cooling   23
moment-of-inertia factor   35
moment of inertia   426
moment-of-inertia tensor   426
Monaghan, J.J.   341
Moon capture by Earth   205
Moon origin   305
Moon shape   217
Moon   171, 203, 219, 422
moonquakes   213

Moulton, Forest   xxv
M-type asteroid   248, 265
Mukhopadhyay, L.   194
Mullis, A.M.   211, 419
Muxlow, T.W.B.   301

nakhlites   261
Napier, W. McD.   245, 251
neon in meteorites   282
neon in silicon carbide   288
neon stable isotopes   282
neon   309
neon-22 (neon-E)   282
Neptune formation time   145
Neptune satellites   169, 235
Neptune   155, 173, 307
Neptune formation by SNT   xli
Neptune-Pluto relationship   235
Neptune-Triton tidal interaction   238
Neptune-Triton-Pluto relationship   236
Nereid orbital eccentricity   236
Nereid   169, 170, 238, 307
neutron star   51, 65, 299
Newton, Isaac   xxii, 296
nitrogen unstable isotopes   287
nitrogen   309
Nix   175
non-equilibrated chondrules   260
Northern plains   428
nuclear reactions   435
nucleotides   261

Oberon   168, 219
Occam's razor   296
Olympus Mons   225, 448
Oort cloud survival   251
Oort cloud   183, 249, 252, 429
Oort, Jan   182
open (galactic) cluster   2
optical spectrometer   70
orbital characteristics   68
orbital decay and round-off   105, 442
orbital precession   399
orbits of high eccentricity   107, 110

Index

ordinary chondrites   259
Origin of Species   v
Orion Nebula   56, 61, 89, 300
Oxley, S.   86, 123
oxygen isotopes   269
oxygen isotopic anomalies   276
oxygen unstable isotopes   287
oxygen   309

Pagana, L.   194
Pallas,   177
pallasites   263, 265, 309
Papanastassiou, D.A.   279
parallax method   6
Parise, B.   194
parsec   7
Peale, S.J.   158
pear-shaped configuration   29
Pele   158
periastron   107
perigee   208
perihelion   331
period of precession   187
permille   271
Pert, G.J.   372, 397
perturbation of Oort cloud   429
perturbation of planets   127
Petr, M.G.   61
Phobos   171
Phoebe   165
photon   335
photon energy   361
photon momentum   361
photosphere   4
Piazzi, Giuseppe   174
Pinwheel Galaxy   328
planet formation time   145
planetary perturbation of comets   433
planetary collision   196, 304
planetary cores   152
planetary crust   152
planetary densities   152
planetary disk   135
planetary mantles   152

planetary nebula   49
planetary perturbation   250
planetary spin axis tilts   155, 303
planetary system frequency   302
planetary transits   75
planetesimal accumulation   xli
planetesimals   xxxvii
planets   151
planetary microlensing   121
Pluto orbital eccentricity   235
Pluto orbital inclination   235
Pluto   172, 173, 175, 243, 307
Poincaré, Henri   320, 354
Polydeuces   164
Popper, Karl   296, 298
Population I   5, 55, 85
Population II   5
Population III   5
potassium-40   417
potential energy   320
Poynting-Robertson effect   78, 79, 304, 361
precession of orbits   185, 186
Pringle, J.E.   xxxv
prograde orbits and spin   155
proportion of stars with planets   121
Proteus   169, 236
protoplanet disk   89
protostellar disk mass   186
protostar   56, 86
protostar 16293E   194
protostar formation   27, 298
protostar fragmentation   348
protostar spin   185
PSRB257+12   67
Ptolemy   297
pulsar   51, 66

quality factor   392
Queloz, D.   72

radian   313
radiation pressure   109, 361
radioactive decay   435
Rankine-Hugoniot equations   402

red giant 12, 47
red shift 334
regular satellites 157
relativistic Doppler effect 334
resisting medium 302
retrograde orbits and spin 156
reverse nuclear reactions 273
Rhea 219
Richards, A.M.S. 301
rills 211
Rios, W.K.M. 301
Roberts, H. 194
Roche model 30
Roche, Edouard xxv
Roche, P. 88
Runge-Kutta integration 375
Russell, Henry Norris xxxii

Safronov, Victor xl, 142, 144, 145, 387
Saha, M.N. 274
satellite formation 303
satellites 387
satellitesimal growth 389
satellitesimals 141, 383, 387
Saturn 154
Saturn ring system 155, 166
Saturn satellites 162
Saturn spin 155
scattering of radiation 364
Schmidt, Otto xl
Schofield, N. 148
Schrödinger wave equation 335
Schrödinger, Erwin 335
Seaton, M.J. 23
segregation within planets 246
semi-latus rectum 331
semi-major axis 69, 331
semi-minor axis 69, 331
shergottites 261
short-period comets 182, 249
silicon 309
silicon carbide 284
silicon stable isotopes 284
silicon three-isotope plot 284

Slattery, W.L. 206
smoothed-particle hydrodynamics 25, 32, 86, 341
smoothing length 341
SMOW 270, 278
SNC meteorites 261
sodium unstable isotopes 285
sodium-22 274, 282
soft binaries 60
softened orbits 128
Solar Nebula Theory xxxiii, 89
solar prominences xxv
solar wind 34
Solomon, S.C. 417
spallation 264
specific internal energy 403, 408
spectral classes 11
spectral lines 11, 336
spectroscopic binaries 3
SPH with radiation transfer 86
spin slowdown 351
spiral galaxy 328
spiral nebulae xxv
spiral waves xlii, 105
Spitzer, Lyman xxxii
spread factor 406
stable equilibrium 397
standard deviation 96
starlight heating 21
Stefan's constant 13
stellar brightness 8
stellar clusters 2
stellar disk decay 103, 369
stellar disk masses 98
stellar disk model 101
stellar disk structure 96, 369
stellar disks 77
stellar distances 5
stellar luminosity 356
stellar mass loss 46
stellar masses 338
stellar material 4
stellar radiation 365
stellar radii 12
stellar spin axes 33, 114, 303

Index

stellar spin loss   36
stellar spin rates   15, 299
stellar temperature   9
stellar wind   34, 96, 109, 351, 366
Stock, J.D.R.   206
stones   180
stony meteorites   256, 258
stony-iron meteorites   180, 256, 263
S-type asteroid   248, 265
supernova   51, 68
surface chemistry   271

taenite   262, 309
Talbot, R.J.   62
Tang, M.   290
Taurus Auriga region   61
Telesto   164
temperature of ISM   20
terrestrial planet formation   190
terrestrial planets   152
Tethys   164
Tharsis uplift   428
Thebe   158
thermal evolution of Moon   209, 416
thorium-232   417
three-isotope plot   271
tidal bulges   204
tidal disruption   140, 373, 386
tidal energy   424
tidal interactions   188
Tielens, A.G.G.M.   194
tilts of spin axes   186
Titan atmosphere   162
Titan   164, 219
Titania   219
Toksös, S.M.   417
Tombaugh, Clyde   173
transformation method   372
translational kinetic energy   320, 353
Trapezium Cluster   56, 57, 300
triple-α process   48
tritium   44, 199, 414
Triton polar cap   236
Triton volcanoes   236

Triton   169, 170, 203, 219, 235, 307
troilite   209, 259
Trojan asteroids   179, 395
T-Tauri stars   34
turbulence   27, 55, 299
types I and II migration   95, 100

Ulysses spacecraft   xxxix
Undina   177
unstable aluminium isotopes   280
unstable sodium isotopes   280
uranium-238   417
Uranus   155, 173
Uranus axial tilt   157, 189
Uranus rings   169
Uranus satellites   167
Uranus spin axis   167

Valhalla   161
Valles Marineris   224
variable stars   7
vector   314
Vega   79
Venus   xx, 152
Venus atmosphere   153
Venus axial tilt   156
Venus D/H ratio   193, 272
Venus formation   304
Venus phases   297
Venus spin   199
Venus, uncompressed density   220
Vesta   177
Virial Theorem   320, 353
viscosity   99
visual binaries   3
Von Neumann-Richtmeyer method   404

Wallace, Alfred Russell   vi
Ward, W.B.   xl, 142
WASP-17   116
Wasserberg, G.T.   279
water vapour   155
water   273, 414
wave mechanics   335

wavelength 333
Weidenschilling, S.J. xxxix, 137, 139, 384
white dwarf 49, 299
Widmanstätten figure 262, 265, 309
Wien's law 10
William of Occam 296
Williams, I.P. 62, 381
Wolszczan, Aleksander 67
Woolfson, M.M. 86, 105, 114, 123, 129, 148, 206, 210, 228, 237, 372, 376, 377, 381, 397, 428, 432

XL Tau 80
XZ Tau 301

young stellar object 42, 299
Young's modulus 394

Zimmerman, B.A. 272
Zinnecker, H. 59
Zinner, E. 290
zodiacal light 77

β-Pictoris 79, 82
δ-Cephei 8
ε-Eridani 79
γ-rays 21